ENCYCLOPEDIA OF ORGANIC CHEMISTRY

ENCYCLOPEDIA OF ORGANIC CHEMISTRY

Vol. 5

By

Sananda Chatterjee

DAE, DCA, CMCNet

Scientific Consultant

Institute for Natural Sciences

Kolkata

(West Bengal)

DISCOVERY PUBLISHING HOUSE PVT. LTD.

NEW DELHI-110 002

Published by:
Tilak Wasan
DISCOVERY PUBLISHING HOUSE PVT. LTD.
4383/4B, Ansari Road, Darya Ganj
New Delhi-110 002 (India)
Phone : +91-11-23279245, 43596064-65
Fax : +91-11-23253475
E-mail : parul.wasan@gmail.com
discoverypublishinghouse@gmail.com
web : www.discoverypublishinggroup.com

***First Edition:* 2012**

ISBN: 978-81-8356-875-3 (Set)

Encyclopedia of Organic Chemistry

Printed at:
Shree Balaji Art Press
Delhi

Preface

Organic Chemistry is that branch of chemistry, which deals with the reactions related to the life form, or one of the source of the life form, carbon (we can say the end of it).

This carbon mainly conjugates with different elements such as nitrogen, hydrogen and oxygen sometimes metals and other non-metals also). Organic chemistry studies mainly the extraction, structure, properties, of these compounds.

In this book chapters deals with the different methods of purification, extraction of compounds, some processes were used and devised by Alchemists.

The extraction preservation of natural dyes are examples of alchemical work, these been discussed in the chapter "dyes indicators and pigments".

Dyes is the mal-pronunciation of the damathi word dyum meaning colour, it was, the ancient Egyptians who use different natural pigments to extract different colours.

In this book the organic compounds has been mainly classified into two broad groups, the aromatic and the aliphatic compounds. The extraction of alcohol is definitely an alchemical process'.

The destructive distillation of wood is one of the finest processes devised by the alchemists for the preparation of methanol is one of the important topics of this book.

Looking ahead in to the modem world, the world of future, the subject comes in vision is the biochemistry, the base of which is organic chemistry, the chapter containing carbohydrates, and the chapter, chemical compounds of biological and biochemical interest is the best element of focus for the students stepping foot in the advanced staircase of biochemistry.

The last chapter, chapter 41, discusses the different organic reactions and their suitably mechanisms, this will help those students who are very much afraid of organic formulas and reactions. This chapter also deals with the chirality one of the interest of Dr. Palash Gangopadhyay (author's friend).

Organic Chemistry is not a mere subject, it is the rhythm of life tuned in the essence of carbon, which looks black but carries the dazzleness of diamond in its heart.

SANANDA CHATTERJEE

Contents

CHAPTER

27

Aromatic Amines and their Derivatives

Introduction

Like aliphatic amines aromatic compounds, benzene, toluene and its derivatives. When ammonia reacts with benzene derivatives nitro amines are produced.

The amino derivatives are of two types

(*a*) Aryl amines and aromatic amines in which the $-NH_2$ group (or substituted $-NH_2$ group) is attached directly to a carbon of the benzene ring

(*b*) Aryl – alkyl or arlykyl amines in which the $-NH_2$ group is attached to a carbon of the side chain.

Both with respect to the methods of preparation and reactions of - NH_2 group arlykyl amines are similar to the aliphatic amines. The aromatic amines are generally prepared by reduction of the aromatic nitro compounds, which are readily obtained by direct nitration of aromatic hydrocarbons. They differ considerably in a number of respects from aliphatic amines. In this chapter we will discuss aromatic amines in detail.

Like the aliphatic amines the aromatic amines can also divided into primary secondary and tertiary amines according to the one, two or three hydrocarbon groups attached to the amino N-atom. Thus

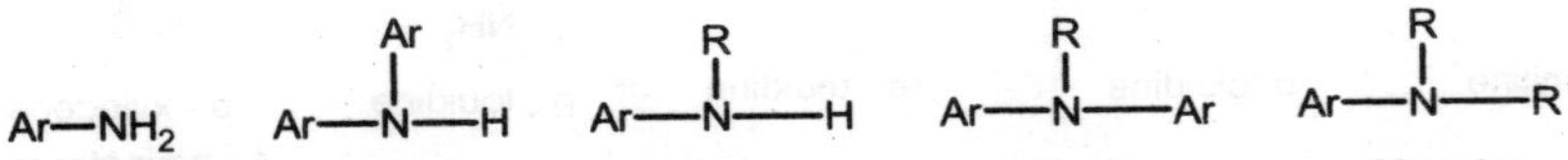

$$\underset{1^\circ \text{ amine}}{Ar{-}NH_2} \qquad \underset{2^\circ \text{ amine}}{Ar{-}\overset{\displaystyle Ar}{\overset{|}{N}}{-}H} \qquad \underset{2^\circ \text{ amine}}{Ar{-}\overset{\displaystyle R}{\overset{|}{N}}{-}H} \qquad \underset{3^\circ \text{ amine}}{Ar{-}\overset{\displaystyle R}{\overset{|}{N}}{-}Ar} \qquad \underset{3^\circ \text{ amine}}{Ar{-}\overset{\displaystyle R}{\overset{|}{N}}{-}R}$$

The primary amines are further designated as mono-amines, diamine and triamine, depending upon the number of NH_2 groups attached to the aromatic ring.

Nomenclature

(1) The aromatic amines may be named by naming the aryl group bonded to nitrogen and adding the suffix – amine. If the group appears more than once, the prefix, di, tri, etc used

NH_2 H   N   N

pheylamine   diphenyl amine   tri phenyl amine

When two or three NH_2 group are present in the benzene ring the suffix used is diamine or triamine. The positions of NH_2 groups are indicated by prefixing o – m or p or by using numbers

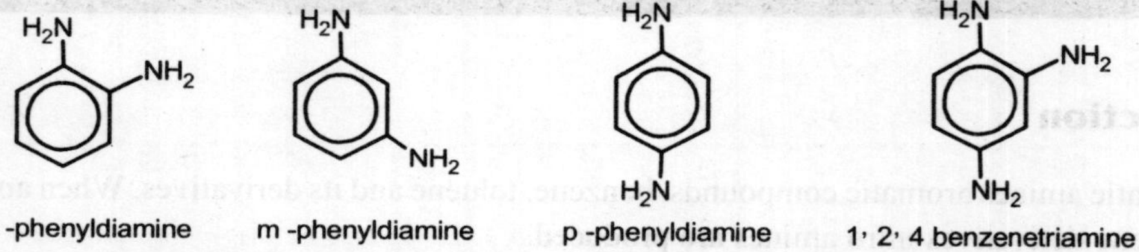

(2) The aromatic amines may be named as amino derivatives of the parent hydrocarbon

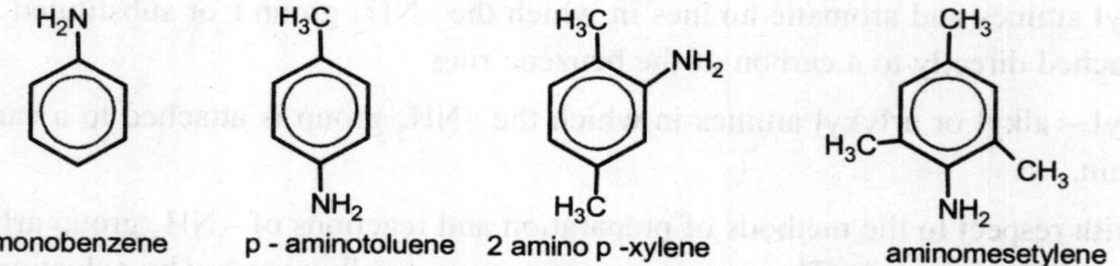

(3) The common names are frequently used for similar and technical aromatic amines.

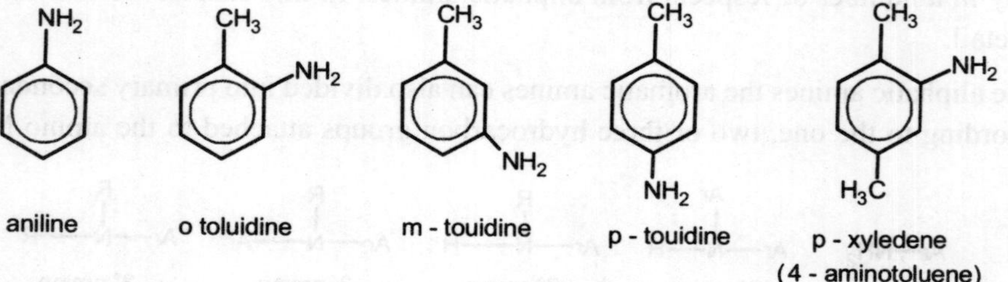

(4) Arlykyl 2° and 3° amines are names as alkyl derivatives of the parent compound. Aniline the prefix N being used to indicate the alkyl group is specifically located on nitrogen. If more than one nitrogen is present they are differential as N, N', N" etc.

MONOAMINES

Introduction

Among the monoamines the compounds corresponding one single amine attached to it for e.g. aniline diphenyl amine etc. we are going to discuss all these monoamines under this subhead.

ANILINE

Introduction and History

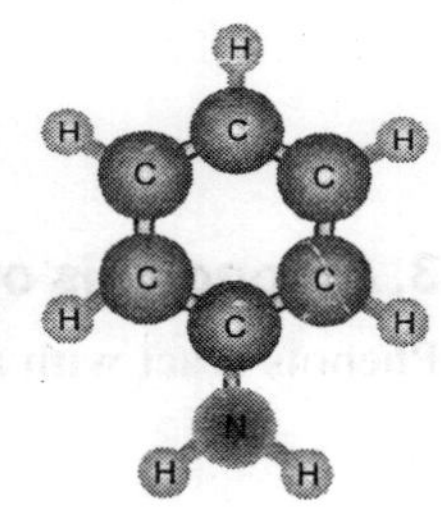

Aniline was first isolated from the destructive distillation of indigo in 1826 by Otto Unverdorben (Pogg. Ann., 1826, 8, p. 397), who named it crystalline. In 1834, F. Runge (Pogg. Ann., 1834, 31, p. 65; 32, p. 331) isolated from coal tar a substance which produced a beautiful blue colour on treatment with chloride of lime; this he named kyanol or cyanol. In 1841, C. J. Fritzsche showed that by treating indigo with caustic potash it yielded oil, which he named aniline, from the specific name of one of the indigo-yielding plants, Indigofera anil, anil being derived from the Sanskrit nîlâ, dark-blue, and nîlâ, the indigo plant. About the same time N. N. Zinin found that on reducing nitrobenzene, a base was formed which he named Benzdam. August Wilhelm von Hofmann investigated these variously prepared substances, and proved them to be identical (1855), and thenceforth they took their place as one body, under the name aniline or phenylamine.

Preparation

1. Reduction of Nitro Compounds

Aniline is prepared by the reduction of nitrobenzene in the presence of tin and hydrochloric acid.

$$C_6H_5NO_2 + 6(H) \xrightarrow{Sn + HCl} C_6H_5NH_2 + H_2O$$

nitrobenzene → aniline

The reduction of a nitro compound can also be effected with hydrogen in the presence of platinum or Raney Nickel at room temperature.

2. Amonolysis of Aryl hydrides

Aryl halides react with high pressure to form the corresponding amino compounds.

$C_6H_5Cl + NH_3 \xrightarrow[200\,°C]{Cu_2O} C_6H_5NH_2 + HCl$

chlorobenzene aniline

If the ring also contains nitrogen group in ortho and para positions to halogen atom Amonolysis takes conveniently under ordinarily temperature conditions.

2-chloro-1,3,5-trinitrobenzene $\xrightarrow{NH_3}$ 2,4,6-trinitroaniline

3. Amonolysis of Phenols

Phenols react with ammonia in the presence of zinc chloride at about 300ºC to forms the aniline.

$C_6H_5OH + NH_3 \xrightarrow[300\,°C]{ZnCl_2} C_6H_5NH_2 + H_2O$

4. Degeneration of Amides

It is also called Hoffmann's Rearrangement, like primary amines aliphatic amines can be obtained by the degeneration of aryl with bromine or chlorine in alkaline solution.

benzamide $\xrightarrow[aq\ NaOH]{Br_2}$ aniline

5. Reduction of Azo-Compounds

The reduction of azo-compounds by the way of hydrazo compounds yields primary aromatic amines. Thus catalytic hydrogenation of azobenzene gives aniline.

$$C_6H_5-N{=}N-C_6H_5 \xrightarrow[40\ °C\ 15\ atm]{H_2\ Ni} C_6H_5-NH-NH-C_6H_5 \xrightarrow[40\ °C\ 15\ atm]{H_2\ Ni} C_6H_5NH_2$$

diphenyldiazene → 1,2-diphenylhydrazine → aniline

Since Azo- and hydrazo compounds are prepared from nitro compounds, this method is limited to specific cases.

6. Action of Hydroxylamine with hydrocarbon

Aromatic hydrocarbons react directly with hydroxylamine in the presence of a catalyst ($FeCl_3$, $AlCl_3$ to give aniline.

$$C_6H_6 + H_2N-OH \xrightarrow[AlCl_3]{FeCl_3} C_6H_5NH_2$$

benzene + hydroxylamine → aniline

Laboratory Preparation

Theory: In the Lab aniline is prepared by the action of tin and hydrochloric acid on nitro benzene.

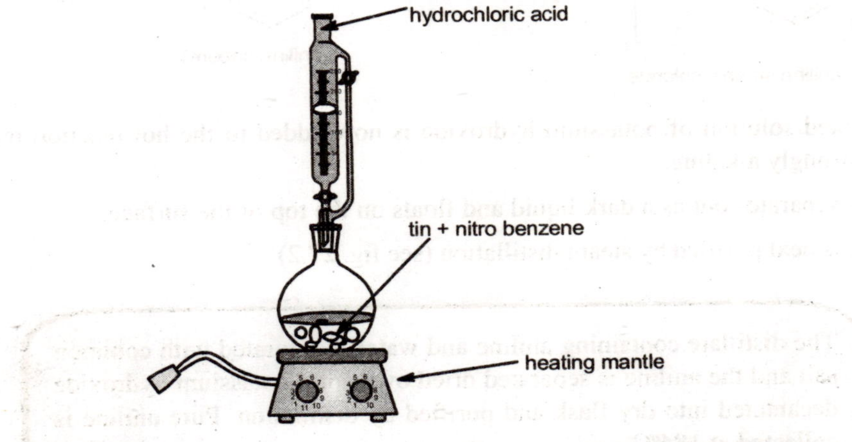

Fig. 27.1. Showing the Laboratory Preparation of Aniline from Nitrobenzene

Procedure: Finely granulated tin (30 gms) and nitrobenzene (15 gms) taken in a 500 ml flask, about 70 ml hydrochloric acid is dropped with continuous stirring and cooling (if necessary). After addition is complete, the mixture is heated over a heating mantle to complete the reaction for 45 mins. At this stage some solid aniline stannic chloride is also produced.

Reaction Equation

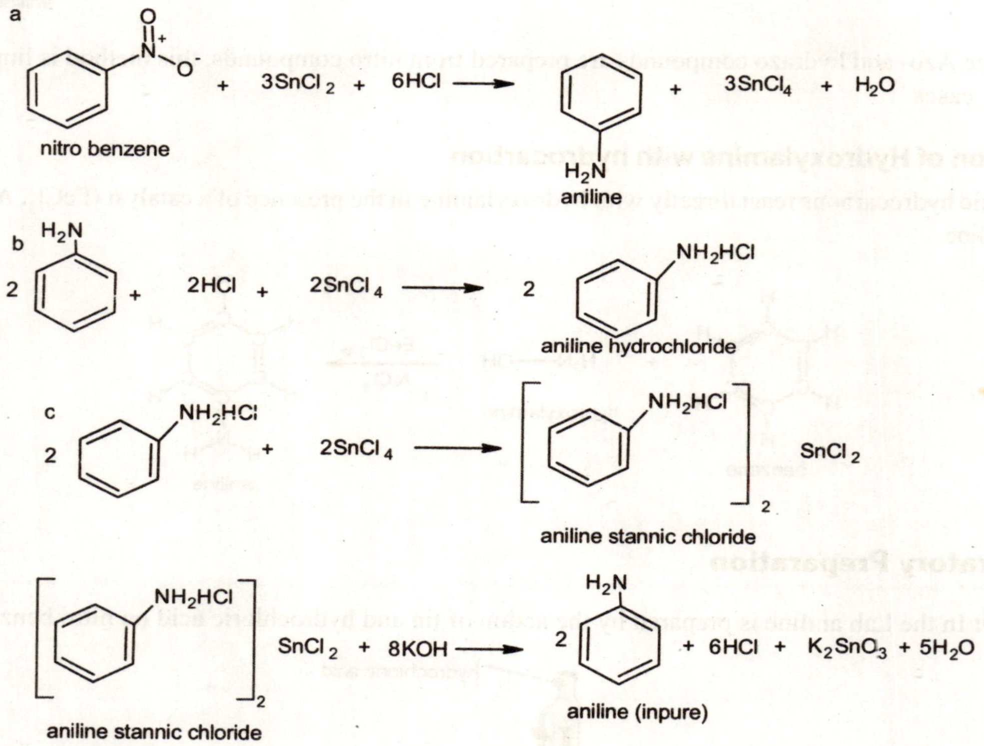

a

nitro benzene + $3SnCl_2$ + $6HCl$ ⟶ aniline (H_2N–C_6H_5) + $3SnCl_4$ + H_2O

b

2 aniline (H_2N–C_6H_5) + $2HCl$ + $2SnCl_4$ ⟶ 2 aniline hydrochloride ($C_6H_5NH_2HCl$)

c

2 $C_6H_5NH_2HCl$ + $2SnCl_4$ ⟶ $[C_6H_5NH_2HCl]_2$ $SnCl_2$

aniline stannic chloride

$[C_6H_5NH_2HCl]_2$ $SnCl_2$ + $8KOH$ ⟶ 2 aniline (H_2N–C_6H_5) + $6HCl$ + K_2SnO_3 + $5H_2O$

aniline stannic chloride ⟶ aniline (inpure)

A concentrated solution of potassium hydroxide is now added to the hot reaction mixture until the product is strongly alkaline.

Aniline separates out as a dark liquid and floats on the top of the surface.

Aniline is next purified by steam distillation.(see fig. 27.2)

> The distillate containing aniline and water is saturated with common salt and the aniline is separated dried over solid potassium hydroxide decantated into dry flask and purified by distillation. Pure aniline is collected at 184°C.

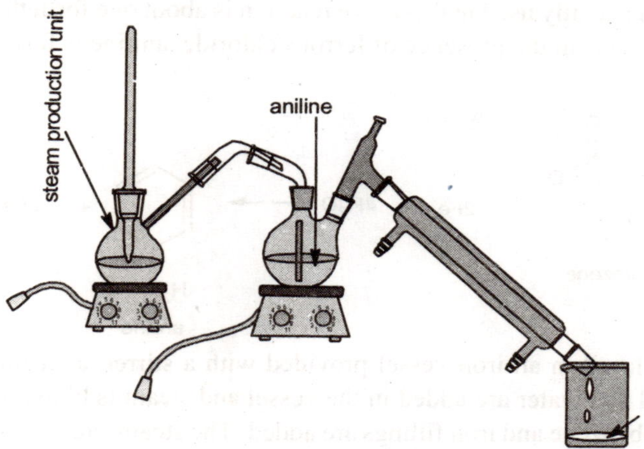

Fig. 27.1a. Showing the Steam Distillation of Aniline purification.

Manufacture of Aniline

1. By reduction of nitro benzene: Aniline is prepared same as the lab process but on large scale by the reduction of nitro benzene with iron instead of zinc and HCl.(See fig. 27.2).

$$\text{nitro benzene} + 3Fe + 3HCl \longrightarrow \text{aniline} + 3FeCl_2 + H_2O$$

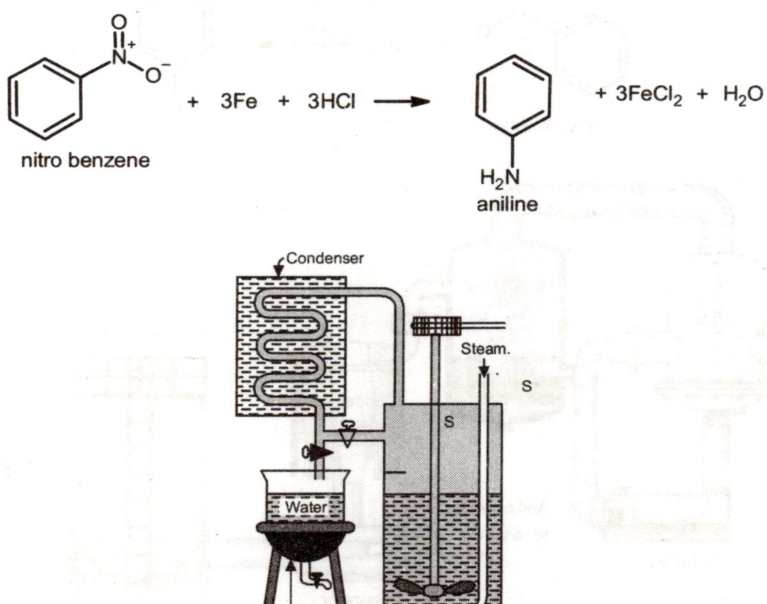

Fig. 27.2. Showing the Manufacturing Process of Aniline.

The quantity of HCl actually used in the above reaction is about one fortieth of that calculated from the equation. This is because in the presence of ferrous chloride, aniline is formed by other reaction as well

$$C_6H_5NO_2 \text{ (nitrobenzene)} + 2Fe + 4H_2O \longrightarrow C_6H_5NH_2 \text{ (aniline)} + 2Fe(OH)_3$$

Nitro benzene is placed in an iron vessel provided with a stirrer, a steam pipe and a condenser. Some iron fillings, HCl and water are added in the vessel and steam is blown in. During the course of the reaction, more nitrobenzene and iron fillings are added. The steam causes some if the liquid to distill and the distillate is returned to the iron vessel as long as any nitrobenzene is left. When only aniline and water distill, the whole of the liquid is distilled over with steam. Aniline collects as a layer at the bottom of the receiver. The bottom layer is separated dried over solid KOH and purified by distillation.

2. By Vapour Phase reduction of Nitro benzene: Aniline can also prepared by the reduction of nitro benzene with hydrogen and copper (copper impregnated on silica) in the gas phase.

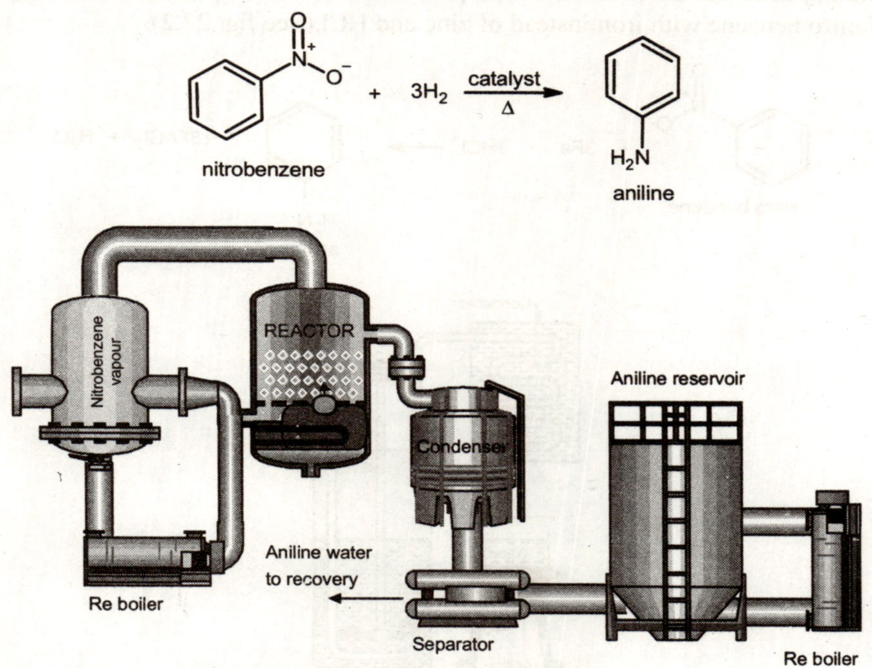

Fig. 27.3. Showing the Industrial Process of Aniline Preparation by Vapour Phase Reduction of Nitrobenzene

This reaction is carried out at 270ºC and 294 atmos pressure, using excess of hydrogen.

Nitrobenzene and hydrogen are introduced in the first chamber (see fig. 27.3), where nitrobenzene vapours is made to go to the reactor where the catalyst was introduced. Aniline with hydrogen comes out cooled in the condenser and hydrogen gas is separated.

which returns back to the first chamber. From the separator the crude aniline is send to the purification tower called aniline still where from the pure aniline is obtained.

3. By Amonolysis of Chlorobenzene: Aniline can also be manufactured by Amonolysis of Chlorobenzene using copper catalyst (Cu_2O or $CuCl_2$) at 150 - 250ºC

$$\underset{\text{chlorobenzene}}{C_6H_5Cl} + 2NH_3 \xrightarrow[150\text{-}250°C]{\text{catalyst}} \underset{\text{aniline}}{C_6H_5NH_2} + NH_4Cl$$

> This process of aniline preparation is uneconomical unless Chlorobenzene is available as the bi-product from the phenol producing plant.

Structure of Aniline

The molecular structure of aniline is described as below. Here the nitrogen atom may be (see fig. 27.4) orbitals are forming σ bonds with a carbon of the ring and the two H atom. The pure P orbital of the nitrogen containing the resulting molecular orbital extends over the six carbon atoms of the ring as also the nitrogen atom of NH_4 group. In the above formulation of the structure of aniline, hybridization at the nitrogen atom is sp^2, so that the benzene ring, the nitrogen atom as also the H lies in the same plane. But actually the benzene ring, the nitrogen atom on the same plane, while the two

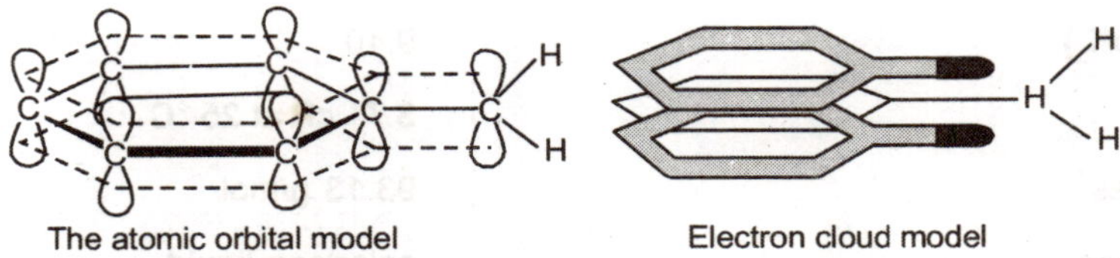

Fig. 27.4. Structure of Aniline.

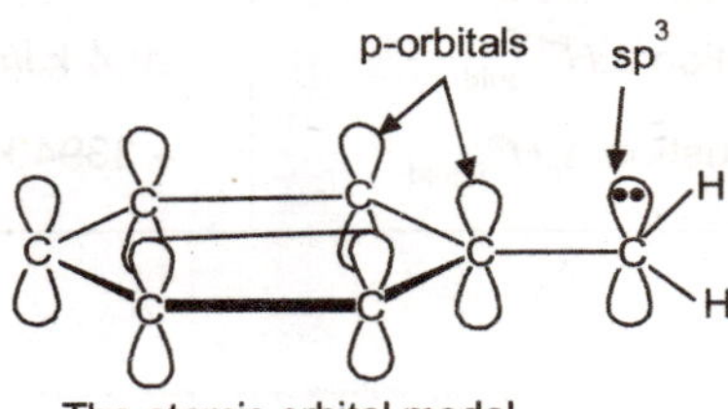

Fig. 27.5. Aniline has a pyramidal structure at the nitrogen atom.

Hence it stand to reason that the hybridization at the N-atom could be approximately sp^3 and the three of the sp^3 orbitals form δ bond with carbon of the ring and the two H atoms the fourth one containing the unshared electron pair interacts with the π system of the ring. Therefore, the nitrogen remains pyramidal.

The electronic structure of aniline can be represented as a resonance hybrid of the following four canonical forms

(i) (ii) (iii) (iv) (v)

(2 kekule form) Resonance hybrid

Physcial Propeties

Properties	
Density and phase	1.0217 g/ml, liquid
Solubility in water	3.6 g/100 mL at 20°C
Solubility in ethanol, acetone	Miscible
Melting point	– 6.3 °C
Boiling point	184.13 °C
Basicity (pK_b)	9.40
Viscosity	3.71 cP at 25 °C
Molar mass	93.13 g/mol
Appearance	colorless liquid
Thermodynamic data	
Standard enthalpy of formation $\Delta_f H^{\Phi}_{liquid}$	20.4 kJ/mol
Standard enthalpy of combustion $\Delta_c H^{\Phi}_{liquid}$	– 3394 kJ/mol

Chemcial Properties

1. Basic Character: Although aniline is neutral to indicators but still it produces some basic character by formation of salts while reacting with acids

(a) Aniline (H_2N) + HCl (Hydrogen chloride) → Aniline hydrochloride (H_2NHCl)

(b) aniline (H_2N) + H_2SO_4 (Sulphuric acid) → Aniline sulphate ($H_2N\,H_2SO_4$)

(c) Aniline (H_2N) + H_2PtCl_6 (Chloroplatinic acid) → Aniline chloroplatinate ($H_2N\ H_2PtCl_6$)

Mechanism for Basicity of Aniline

While discussing the basic character of aniline basicity is defined as the accepting of the proton that is H^+ for water. Thus aniline is protonated to give anilinium ion.

Aniline (NH_2) + H—O—H ⇌ Anilinium ion ($\overset{+}{N}H_3$) + OH

The value of equilibrium constant K_p faster which determines the concentration of OH ions or the basicity of aniline.

$$K_p = \frac{[C_6H_5 - NH_3][OH]}{[C_6H_5 - NH_2]}$$

Compared to ammonia (NH_3) the basicity of aniline is significantly reduced

NH_3: $K_b = 1.8 \times 10^{-5}$ Aniline ($C_6H_5NH_2$): $K_b = 4.2 \times 10^{-10}$

The explanation behind this is the non-bonded electrons on the nitrogen atom delocalized into the π-system of the benzene ring (see fig. 27.5 orbital model). Therefore these electrons which entangle the basic carbons of the ring as also the nitrogen atoms are less available for inter reaction with protons from water. Thus the value of K_b for aniline is decreased making it less basic

Non bonded electrons delocalized

(reduced electron density or N atom makes non bonded electrons less available

hence we can also see the resonance characteristic of the aniline of the five resonance forms

I II III IV V

Thus aniline structure is greatly stabilized. On the other hand the anilinium ion formed by protonation of aniline is less resonance stabilized as it exists only in two Kekule forms and no more.

Resonance form of anilinium ion

Hence, the aniline structure which exists in five resonance forms is more resonance stabilized than the anilinium ion which can exist in two forms only. That is

$$:NH_2\text{-}C_6H_5 + H_2O \rightleftharpoons C_6H_5\text{-}\overset{+}{N}H_3 + OH$$

More resonance stabilized than anilinium ion

Anilinium ion less resonance than aniline

This protonation of aniline requires energy to over come the excess resonance energy (Δ^H res $PhNH_3^+ > \Delta$ res Ph NH_2) of aniline. But the resonance stabilization of either ammonia or ammonium ion is not possible as aniline or anilinium ion respectively, the protonation of ammonia requires less energy relative to the aniline is a weaker base than ammonia, and also aliphatic amines.

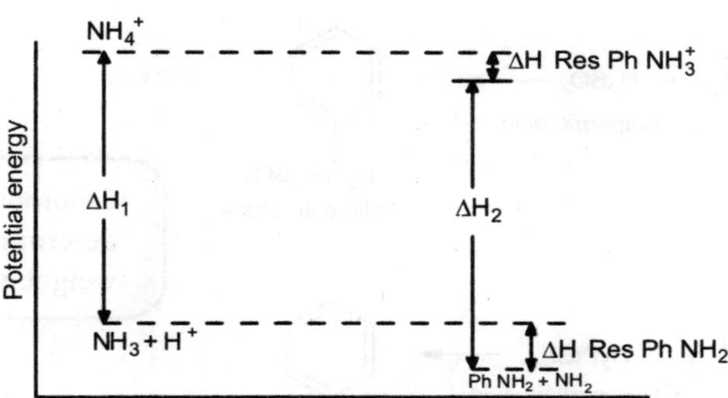

Fig. 27.6. Showing the Resonance stabilized comparison between ammonia and aniline.

How is the basicity of aniline affected by substituents on the ring?

An electron releasing group ($-NH_2$, $-OCH_3$, $-CH_3$) pushes electrons towards the N-atoms and makes the non-bonded electron pair more available for protonation. Hence such a substituent increases the basicity. Thus R_b value for p-methoxyaniline is 2.0×10^{-9} as compaired to aniline 4.2×10^{-10}.

How is basicity of aniline affected by the presence of substituents in N-atom?

When an alkyl group (electron releasing) is present on N-atom of aniline, the basicity of aniline is increased. Thus the basicity of p-methylaniline is 2.0×10^{-9} as compaired to that aniline (4.2×10^{10})

When more than one benzene ring are bonded electrons suffer greater delocalization and much less available for protonation. The presence of large phenyl group in the molecule also screens the non-bonded electrons on N-atoms and thus reduces the basicity further. As a result the diphenylamine is only a very weak base but triphenylamine is completely away form basic character.

Salt Formation

Although aromatic amines are weather bases than ammonia or aliphatic amines they form well defined crystalline salt on reaction with strong mineral acids as HCl or H_2SO_4 for example :

$$C_6H_5NH_2 + HCl \longrightarrow [C_6H_5NH_3]^+ Cl^- + C_6H_5NH_3Cl$$

Aniline — Phenyl ammonium chloride anilinium chloride

(*a*)

$$C_6H_5NH_2 + HCl \longrightarrow C_6H_5NH_2 \cdot HCl$$

Aniline + Hydrogen chloride → Aniline hydrochloride

(b) aniline (H_2N–C_6H_5) + H_2SO_4 (Sulphuric acid) → $H_2N\,H_2SO_4$ Aniline sulphate

Chloroplatinate is used for the determination of molecular weight of aniline.

(c) Aniline (H_2N–C_6H_5) + H_2PtCl_6 (Chloroplatinic acid) → $H_2N\,H_2PtCl_6$ Aniline chloroplatinate

These salts of aniline are soluble in water and undergo hydrolysis to a great extent, giving acidic reaction.

$$H_3C{-}NH_3{-}Cl \longrightarrow C_6H_5NH_2 + H^+ + Cl^-$$

Like salts of aliphatic amines, these salts liberate the original aromatic amine, when treated with an alkaline solution

2. *Acylation*: The primary and secondary aromatic amines react with aryl chloride or anhydrides, when the hydrogen atom attached to N – atom is replaced by aryl group.

$$C_6H_5{-}NH{-}H + Cl{-}C(=O){-}CH_3 \longrightarrow C_6H_5{-}NH{-}C(=O){-}CH_3 + HCl$$

Aniline → *N*-phenylacetamide acetanilide

3. *Acetylation*: It reacts with acetyl chloride, chloride acetic anhydride or glacial acetic acid in the presence of $ZnCl_2$, to give acetanilide.

$$C_6H_5{-}NH{-}H + Cl{-}C(=O){-}CH_3 \longrightarrow C_6H_5{-}NH{-}C(=O){-}CH_3 + HCl$$

Aniline → *N*-phenylacetamide acetanilide

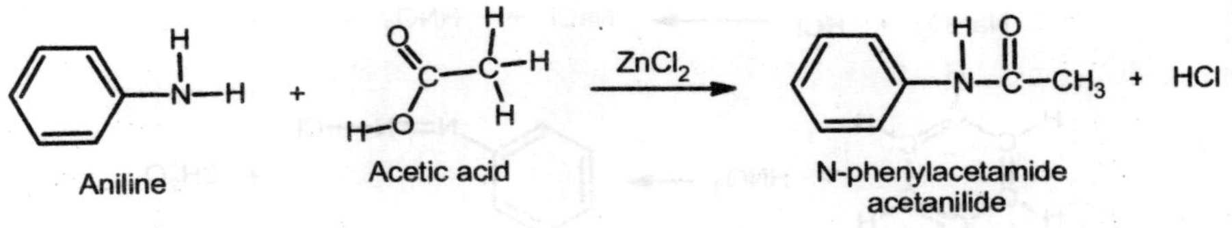

Aniline + Acetic acid —$ZnCl_2$→ N-phenylacetamide (acetanilide) + HCl

With benzoyl chloride, aniline gives benzanilide

Process is called benzolation

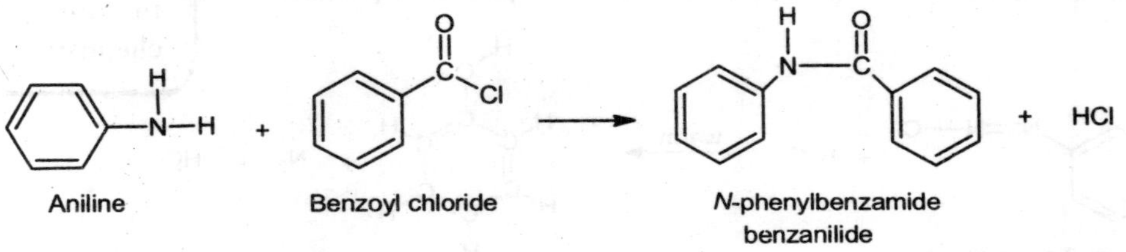

Aniline + Benzoyl chloride → *N*-phenylbenzamide (benzanilide) + HCl

Carbalamine Reaction

If aniline is boiled with chloroform and alcoholic potassium hydroxide, the repulsive odour of phenyl isocyanide is developed (the test of aniline)

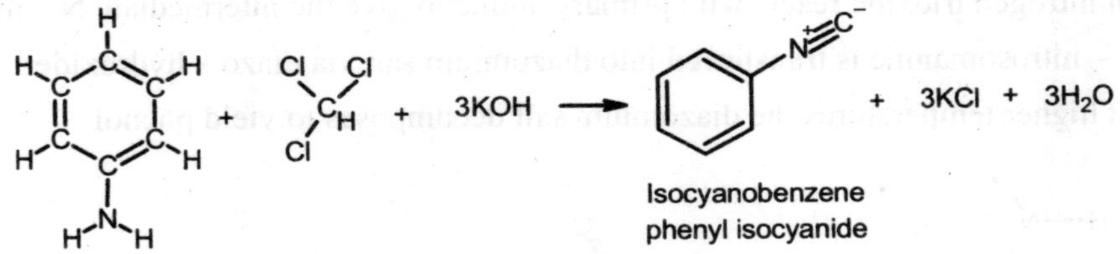

Aniline + $CHCl_3$ + 3KOH → Isocyanobenzene (phenyl isocyanide) + 3KCl + $3H_2O$

Mustard Oil Reaction

If aniline is warmed with alcoholic CS_2 and excess of mercuric chloride is then added and the mixture again heated the characteristic smell of phenyl thiocyonate (phenyl mustard oil) is produced.

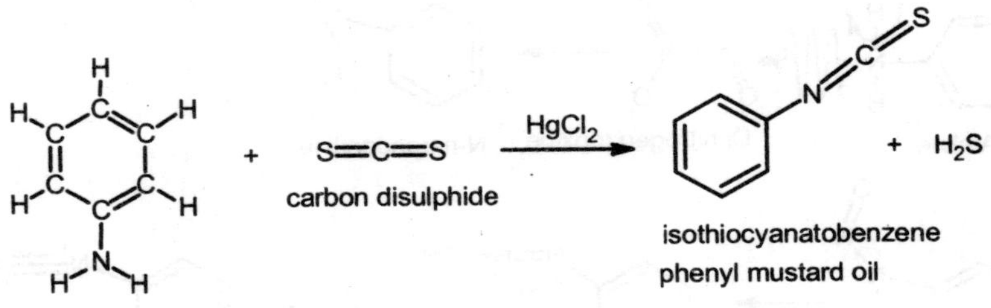

Aniline + S=C=S (carbon disulphide) —$HgCl_2$→ isothiocyanatobenzene (phenyl mustard oil) + H_2S

Diazo Reaction

With nitrous acid in the cold aniline gives diazomium compounds (see next chapter)

$NaNO_2 + HCl \longrightarrow NaCl + HNO_2$

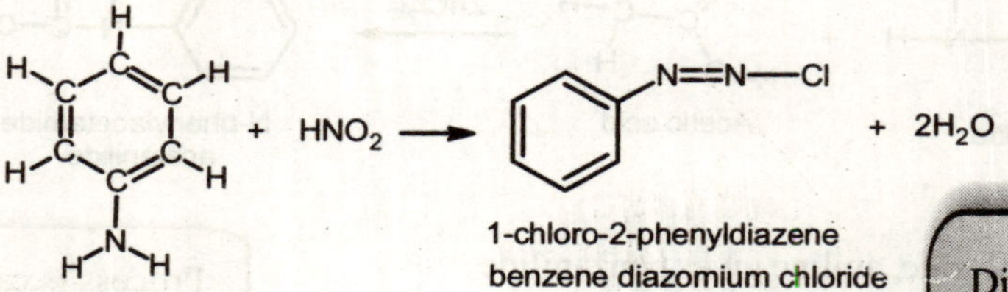

Aniline + HNO_2 ⟶ (benzene ring)–N=N–Cl + $2H_2O$

1-chloro-2-phenyldiazene
benzene diazomium chloride

Diazomium salts are of great importance in the synthetic chemistry

On warming diazomium compounds with water phenol is produced.

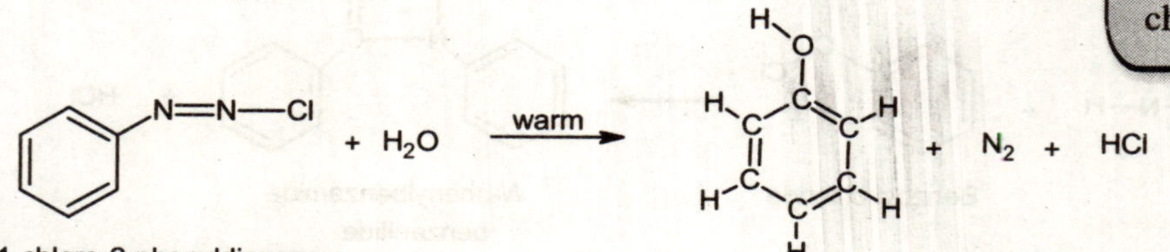

(benzene ring)–N=N–Cl + H_2O —warm→ Phenol + N_2 + HCl

1-chloro-2-phenyldiazene

Phenol

Mechanism

(*i*) Nitrous acid produces di nitrogen trioxide

(*ii*) Di nitrogen trioxide reacts with primary amine to give the intermediate N – nitrosoamine.

(*iii*) N – nitrosomanine is transferred into diazomium salt via diazo – hydroxide;

(*iv*) At higher temperatures the diazomium salt decomposes to yield phenol

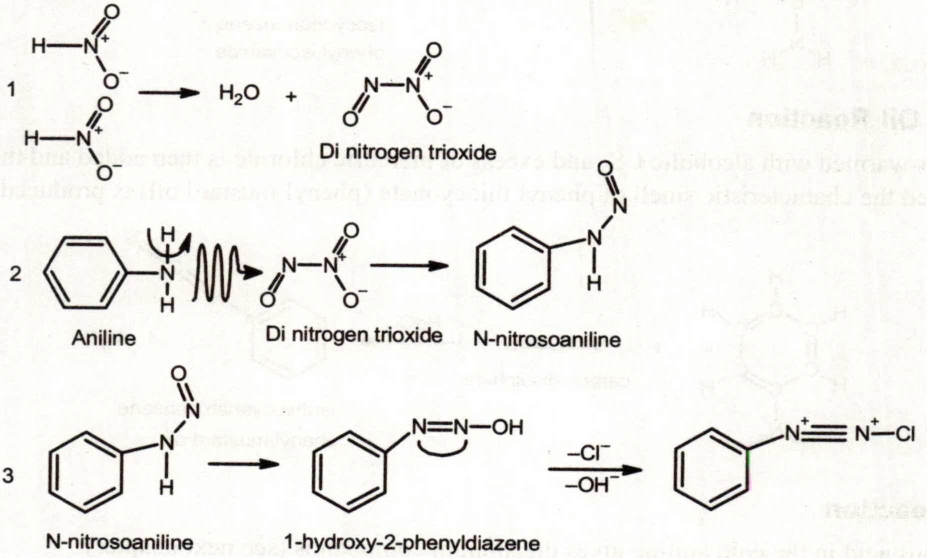

1 H–N⁺(=O)–O⁻ + H–N⁺(=O)–O⁻ ⟶ H_2O + O=N–N⁺(=O)–O⁻

Di nitrogen trioxide

2 Aniline + Di nitrogen trioxide ⟶ N-nitrosoaniline

Aniline Di nitrogen trioxide N-nitrosoaniline

3 N-nitrosoaniline ⟶ (benzene ring)–N=N–OH —$-Cl^-$ / $-OH^-$→ (benzene ring)–$N^+\equiv N^+$–Cl

N-nitrosoaniline 1-hydroxy-2-phenyldiazene

4 Dazomium cation → N_3 + C_6H_5 $\xrightarrow{+OH^-}$ C_6H_5OH Phenol

Aliphatic primary amines do not form stable diazomium salts under similar conditions. They react with similar nitrous acid to yield alcohols and nitrogen. This reaction is therefore, used as test between aliphatic and aromatic primary amines.

4. With Benzaldehyde: Aniline undergoes condensation forming Schiff's Base

Aniline + Benzaldehyde → N-[(1Z)-phenylmethylene]aniline schiff's base + H_2O

5. With Hinsberg's Reagents: Aniline gives substituted sulphonamide

Aniline + Benzenesulfonyl chloride → *N*-[(phenylsulfinyl)oxy]aniline phenyl benzene sulphonamide

Reaction of Nulceus

9. Halogenation: Like aromatic compounds in general aniline can also be halogenation under suitable condition.

H_2N + $3Cl_2$ → 4-chloroaniline + 3HCl

H_2N + $3Br_2$ → 2,6-dibromoaniline + 3HBr

It is to be noted that the presence of –NH_2 group in the nucleus makes the remaining five hydrogen atoms more readily replaceable by other substituents

2-iodoaniline

10. *Sulphonation*: With fuming H_2SO_4 at 180 - 200°C it gives sulphamic acid

Aniline —heat, 180 -200 °C→ 4-aminobenzenesulfonic acid + H_2

11. *Nitration*: Conc. HNO_3 attacks aniline violently giving di and tri nitroaniline and ultimately decomposes it.

aniline → 2,4,6-trinitroaniline + H_2O

Mono-nitroaniline can be prepared only if the $-NH_2$ group in the nucleus makes the remaining five hydrogen atoms more readily replaceable by other substituents.

12. *Sulphonation*: Primary aromatic amines can be sulphonated without prior of $-NH_2$ group, presumably because sulphuric acid is weaker oxidizing agent than nitric acid. Thus aniline can be

Aniline + H_2SO_4 → aniline hydrogen sulphate —Δ, $-H_2O$→ phenylsulphamic acid —rearangement→ 4-aminobenzenesulphonic acid sulphonated to give p – amonobenzenesulphonic acid or sulphanilic acid. Aniline also forms aniline hydrogen sulphate. It is assumed that heating at 180 - 200ºC forms phenylsulphamic acid. This undergoes rearrangements to yield p – aminobenzenesulphonic acid.

Hoffmann Martius Rearrangent

When N – Alkyl or N, N – dialkyl anilines are heated in strong acid media at 300ºC , intermolecular migration of alkyl groups occur. Thus when N – N – dimethylaniline hydrochloride is strongly heated, one methyl group migrates preferentially to the para position of the ring. The N – methylaniline hydrochloride so produced than undergoes migration of the remaining methyl group to the ortho position since the para position has been blocked. This reaction is known as ***Hoffman – Martius rearrangement*** and may used for the preparation of homologous of aniline

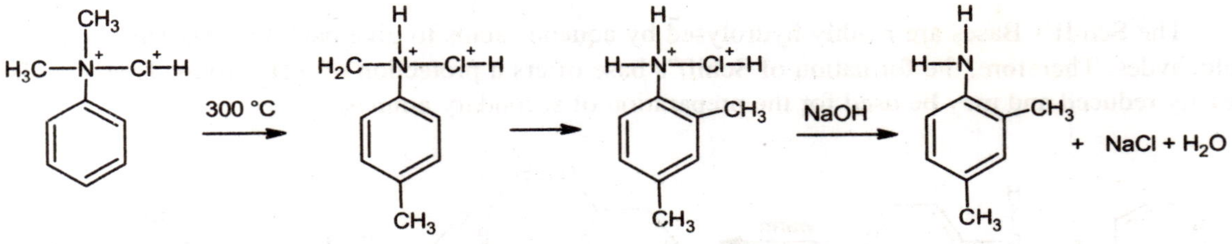

300 °C; NaOH; + NaCl + H_2O

N,N dimethyl hydrochloride — [methyl(4-methylphenyl) ammonio]chloronium — [(2,4-dimethylphenyl) ammonio]chloronium — 2,4-dimethylaniline

13. Oxidation: Aniline is readily oxidized; in fact it is so sensitive in this respect that it undergoes slow oxidation on storage and thus change its colour from colourless to dark yellow on exposure to air. The vigorous oxidation of aniline with potassium dichromate and sulphuric acid results in the formation of quinones

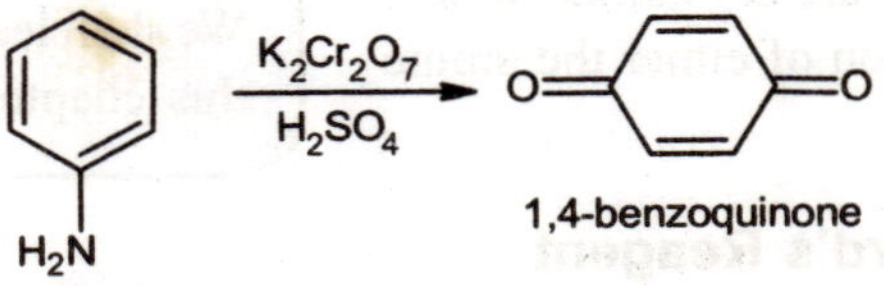

Aniline $\xrightarrow[H_2SO_4]{K_2Cr_2O_7}$ 1,4-benzoquinone

Caro's acid (permonosulphuric acid) oxidizes the primary amino group to the group. These reactions are used for preparing aromatic nitro compounds, which are not formed by direct nitration. Thus, p – nitrobenzene may be made as indicated below :

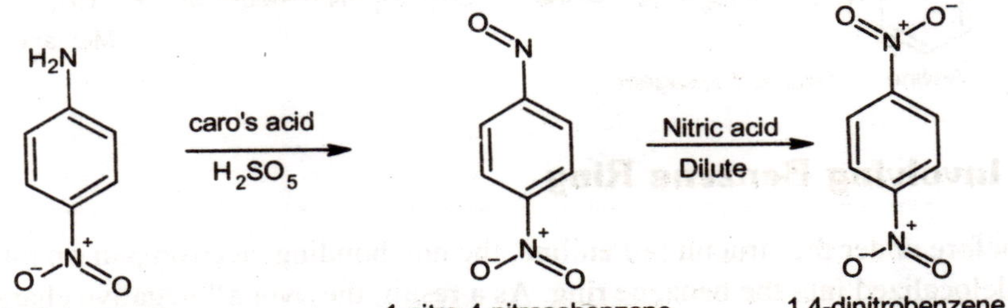

4-nitroaniline $\xrightarrow[H_2SO_5]{\text{caro's acid}}$ 1-nitro-4-nitrosobenzene $\xrightarrow[\text{Dilute}]{\text{Nitric acid}}$ 1,4-dinitrobenzene

14. Reaction with Aldehydes: **Primary aromatic amines, aniline react with aldehydes in a manner similar to that of the primary aliphatic amines, giving condensation products. Thus when aniline reacts with an aldehydes on warming forms anils or Schiff's base. These products are most stable when aldehyde is an aromatic one.**

Aniline + Benzaldehyde —Warm→ N-[(1 Z)-phenylmethylene]aniline schiff's base

The Schiff's Bases are readily hydrolysed by aqueous acids to give back the original amine and aldehydes. Therefore, the formation of Schiff's base offers a protection of NH_2 group. They are also easily reduced and may be used for the preparation of secondary amines.

Aniline + Benzaldehyde —warm→ N-[(1 Z)-phenylmethylene]aniline schiff's base —H_2/Ni→ N-methylethanamine secondary amine

The Schiff's base thus obtained from aromatic amines and aromatic aldehydes are crystalline solids, often useful for the identification of either the amine or the aldehydes.

We shall learn more about anils latter in this chapter.

15. Reaction with Grignard's Reagent

Primary and secondary aromatic amines react with Grignard's Reagent to form hydro carbon Thus

$$\text{Aniline} + H_3C{-}Mg{-}Br \longrightarrow H_3C{-}CH_2{-}\overset{H}{N}{-}Mg{-}Br + CH_4$$

Aniline, Grignard's reagent, Methane

Reactions Involving Benzene Ring

As discussed before under the structure of aniline, the non bonding electron pair on nitrogen atom of –NH_2 group is delocalized into the benzene ring. As a result, the over all negative character of benzene

ring is greatly activated to the attack of electrophiles (E^+) and there would be less possibly of attack by a aniline molecule stated earlier, places a negative charge on the para positions of the – NH_2 group. An electrophile will, therefore, attack the carbon atoms of the ring at these positions preferentially

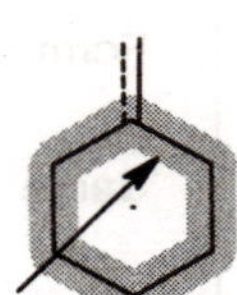

Activated electrophillic attack

Delocalization of non-bonding electron pair of N - into benzene ring makes it negative and activates it by electrophilic attack

The resonance hybrid bears negative charge on the ortho and para position, with which are the points of attack by electrophilic

Thus aryl amines undergo electrophilic substitution in the benzene ring readily and the –NH_2 group directs the new entrant to the ortho and para positions. Thus,

Aniline resonance hybrid — Intermediate — Ortho substitution product

Coupling Reaction: Aniline couples with benzene diazomium chloride in the para position forming p-aminoazobenzene, or Topfer's reagent. Used for the determination of free HCl in the Gastric juice in the intestine.

We shall read about azo dyes in the dye chapter

benzene diazomium chloride + aniline → 4-[(Z)-phenyldiazenyl]aniline p - amino azobenzene + HCl

The azo-compounds are important in the Dye chemistry.

Comparison between Aniline and Aliphatic Primary Amines

Action of	Aniline : $C_6H_5.NH_2$	Aliphatic Amines (Primary) : RNH_2
Similarities:		
1. Acids— HCL, H_2SO_4 or H_2PlCl_6	Forms Salts	Form Salts
2. Alkyl halides	Gives Secondary and Tertiary Amines and also quaternary salts	Same
3. With Acetyl chloride	Gives substituted amides	Same
4. Chloroform and KOH. (lcoholic)	Carbylamine reaction (Unpleasent smell)	Same
5. CS_2	Mustard oil smell	Same
6. Hinsberg's reagent	Substituted benzene sulphonamide	Same
Differences :		
1. Litmus	Weak base (neutral to litmus)	Stronger base, turns red litmus blue
2. Air & light	Turns brown	Remain uneffected
3. HONO	It gives diazonium salts in the cold and phenol on warming	They form alcohols and nitrogen gas is evolved
4. Nitration, Sulphonation, Halogenation	Can be easily nitrated, sulphonated and halogenated	Such reactions are not shown
5. NaClO	Purple violet colour	No colour
6. H_2SO_4 and $K_2Cr_2O_7$	Deep blue or black colour	No colour
7. Benzene diazonium chloride	Coupling lakes place	No coupling takes place
8. C_6H_5CHO Benzaldehyde	Schiff's base is formed the presence of an aromatic ring in aniline causes it to differ from aliphatic Amines	No action

Derivatives

SALTS OF ANILINE

Aniline, being a weak base, combines with acids to form salts. Several of these are of some technical importance for various uses.

Aniline acetate, $C_6H_5NH_2.CH_3COOH$, formula weight 153.17, is a colourless liquid that is miscible with water and alcohol; density, 1.071. On exposure to air or light it gradually darkens. On standing it is slowly converted to acetanilide. This reaction occurs much more rapidly on heating and is the basis for the commercial manufacture of acetanilide.

Aniline hydrochloride (also known as aniline salt, aniline chloride, phenyl ammonium chloride), $C_6H_5NH_2.HCl$, formula weight 129.59, is a white crystalline compound; m.p., 198°C.; b.p.,245°C. It is soluble in one part of water, and also in alcohol. On exposure to light or air it gradually darkens. It is prepared by treating aniline with hydrochloric acid in lead, or preferably, enamel. Aniline salt is used for the manufacture of various intermediates and dyes, and also for the dyeing of cotton and wool black by means of after oxidation to form aniline black on the fiber. Among the dyes in whose manufacture it appears as an intermediate are Rhodamine G *(CL* 750), Phenylene Blue *(CL* 81.8), Rosinduline G *(CL* 831), Indamine Blue B *(C.I.* 859), Induline *(CL* 860), and Nigrosine *(C.I.* 864).

Aniline sulphate $(C_6H_5NH_2)_2.H_2SO_4$, formula weight 284.33, is a white crystalline powder that is soluble in 15 parts of water, somewhat soluble in alcohol, and practically insoluble in ether. It darkens on exposure to light and air. On heating it forms sulphanilic acid (p-aminobenzenesulphonic acid). Formerly it was used as an analgesic but it is not used medicinally any more.

ACYLATED ANILINES

The primary use of the anilides (N-acyl derivatives of aniline) is as analgesics and antipyretics *(q.v.)*. However, they find some use as intermediates in various syntheses, particularly of dyes and pharmaceuticals.

Formanilide (phenylformamide), C_6H_5NHCOH, formula weight 121.13, is a white crystalline material; m.p., 48-50°C.; b160, 271 °C.; b12o, 216°C. It is soluble in water and alcohol and is prepared by boiling aniline with strong formic acid. It is used medicinally as a local anesthetic, analgesic, and antipyretic.

ACETANILIDE

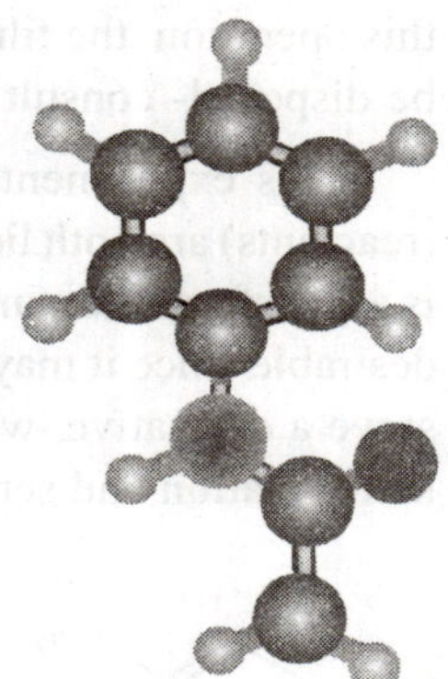

Introduction

Acetanilide is an odourless solid chemical of leaf or flake-like appearance. It is also known as N-phenylacetamide, acetanil, or acetanilide, and was formerly known by the trade name antifebrin.

Preparation

Recrystallization is a widely-used technique to purify a solid mixture. The desired product is isolated from its impurities by differences in solubility. Insoluble impurities and coloured impurities can be removed from hot solvent through the use of activated carbon and filtration. Soluble impurities remain in the cold solvent after recrystallization. The desired product should be as soluble as possible in hot solvent and as insoluble as possible in cold solvent. The selection of solvent is, therefore, critical to the successful recrystallization.

Recrystallization is a purification procedure, which requires solubility of the impure solid in a heated solution and crystallization of the solid upon cooling. Clearly, this operation depends upon solute-solvent in traction involving a number of parameters including concentration, polarity of solute and solvent (like dissolves like), etc.

Choice of a solvent or solvent pair for recrystallization experiments generally involves preliminary tests using a small sample and various solvent systems. To determine the proper solvent or solvent system, the following steps are commonly performed.

(*i*) The crude crystals should have low solubility in the chosen solvent at room temperature.

(*ii*) The crude crystals should have high solubility in the chosen solvent when heated to boiling.

(*iii*) The crude crystals should not react with the solvent

(*iv*) The solvent should boil at temperature below the solid melting point.

(*v*) The solvent should moderately be volatile so crystals dried readily.

(*vi*) The solvent should be non-toxic, non-flammable, and inexpensive

The procedure illustrated in this experiment involve recrystallization, **gravity filtration, suction filtration**, **melting** and **mixture melting points**, as well as **calculations of theoretical and percentage yields.**

In **suction filtration**, a Büchner funnel is employed to collect the desired crystals resulting from a reaction or recrystallization attempt. Be sure to "wet the filter paper" with the solvent/solid mixture to be filtered. When performing a suction filtration, it is usually advisable to install a trap between the aspirator and the suction flask. In any case always break the vacuum before turning the water off. In this operation, the filtrate or "mother liquor" may be concentrated to obtain a second crop, etc. (or may be disposed- consult with you instructor).

This experiment involves four functional groups common in organic chemistry. The substrate (reactants) are both liquids and one of the products is solid. The reaction of aniline with acetic anhydride is a transformation in which products, acetanilide and acetic acid, are obtained. A solid product is often desirable since it may be recrystallized and a melting point determined. Solids prepared in this manner serve a derivative, whose melting point may be correlated with known values and thus is a means of identification and serves as a test for homogeneity or purity.

aniline, C_6H_7N + acetic anhydride ⟶ acetanilide, C_8H_9NO + acetic acid

Experimental Procedures

Using a medicine dropper, place 0.15 to 0.20 g of aniline (about 10 drops) (d = 1.02 g/ml) in a large tared test tube and determine the weight to the nearest mg. Add 5 ml of distilled water to the test tube and then add 20 drops of acetic anhydride again using a medicine dropper (Fig. 27.6). stir, the mixture using stirring rod for 5 minutes until solid forms.

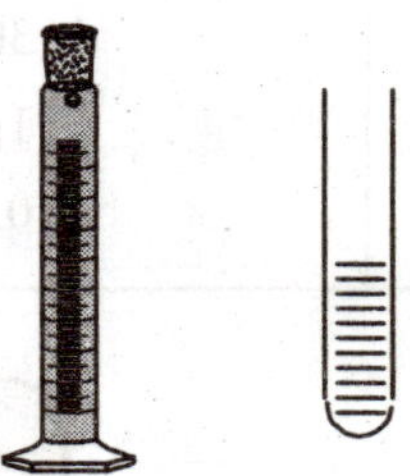

Fig. 27.6. showing mL graduate cylinder and test tube

The product crystallized in the same test tube. Add 5 ml of water and heat the test tube in a hot water bath (250 mL beaker) (Fig. 27.7) with occasional stirring until the entire solid dissolved. Set the test tube aside to cool for 3-5 minutes and then chill it in an ice bath. When crystallization is complete, collect the product by vacuum filtration using a small Büchner funnel (Fig. 27.8). Allow the sample to dry completely. Weigh the dry product, calculate the percentage yield and determine its melting point. Collect to product in a paper and write your name and submit it to your instructor. The aqueous filtrate may be flushed down the drain.

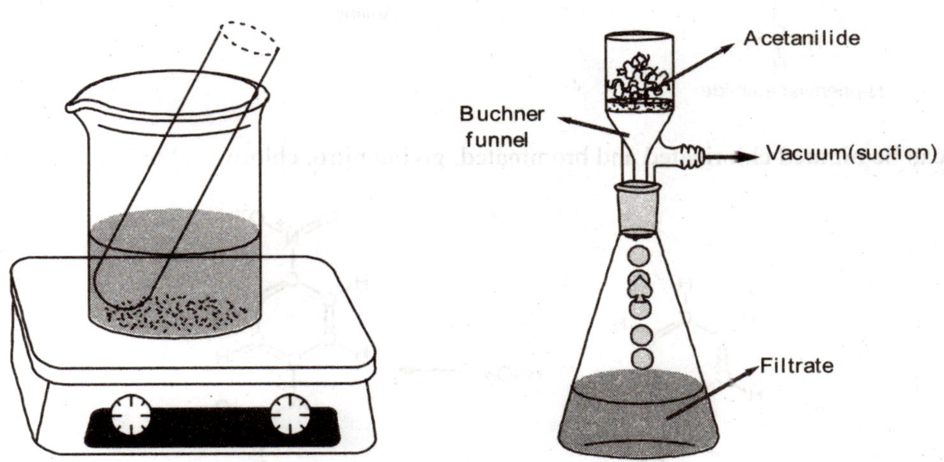

Fig. 27.7. Hot water bath. **Fig. 27.8. Büchner funnel and suction flask.**

Physical Properties

Molecular mass	135.17 g/mol
Density	1.219 g/cm³
Melting point	113- 115ºC
Boiling point	304 °C
Solubility	1g/185mL
Water solubility	0.1g/100mL at 22 °C

Chemical Properties

Acetanilide gets its name from this acetic acid or anhydride + aniline

1. When heated with acids or alklies acetamide gets hydrolysed to aniline and acetic acid

N-phenylacetamide + H_2O —KOH→ Aniline + Acetic acid

2. It can be nitrated chlorinated and brominated, giving nitro, chloro, and bromoacetanilide

N-phenylacetamide + HNO_3 → *N*-(4-nitrophenyl)acetamide nitroacetanilide

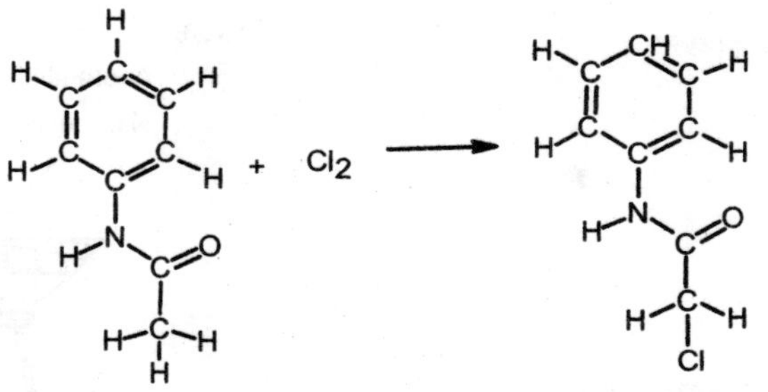

$$C_6H_5NHCOCH_3 + Cl_2 \longrightarrow C_6H_5NHCOCH_2Cl$$

2-chloro-*N*-phenylacetamide

chloroacetanilide

DERIVATIVES OF ACETANILIDE PARANITROACETANILIDE

For a spectacular chemical demonstration the decomposition of paranitroacetanilide to yield a voluminous ash is hard to beat. Although this compound can be purchased, preparing it yourself is an interesting exercise in organic synthesis.

Preparation

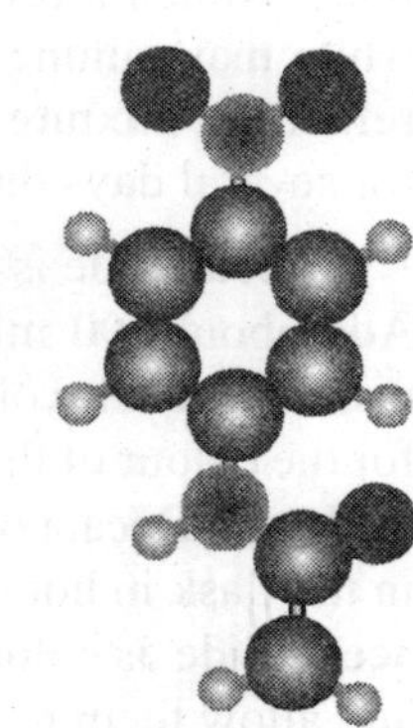

The synthesis is accomplished in two stages, starting with aniline (aminobenzene, $C_6H_5NH_2$): First we acetylate aniline with glacial acetic acid to yield acetanilide:

(1) $C_6H_5NH_2 + CH_3COOH \rightarrow CH_3COONHC_6H_5$

This is then nitrated with nitric acid:

(2) 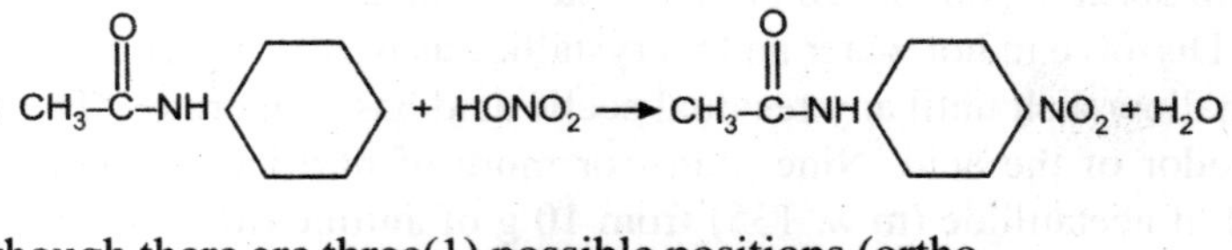 $CH_3{-}\overset{O}{\overset{\|}{C}}{-}NH{-}C_6H_5 + HONO_2 \longrightarrow CH_3{-}\overset{O}{\overset{\|}{C}}{-}NH{-}C_6H_4{-}NO_2 + H_2O$

Although there are three(1) possible positions (ortho- meta- and para-) for the second substituent (Figure 1), in this case the nitro group, to attach to the benzene ring, formation of the para-isomer is favoured. In general the point of entry of a secondary substituent on the benzene ring is influenced by the nature of the primary group, here acetylamino-. It's a common stratagem in organic synthesis to select a group that yields the desired isomer. With this established, the substituents can then be modified en route to the final product.

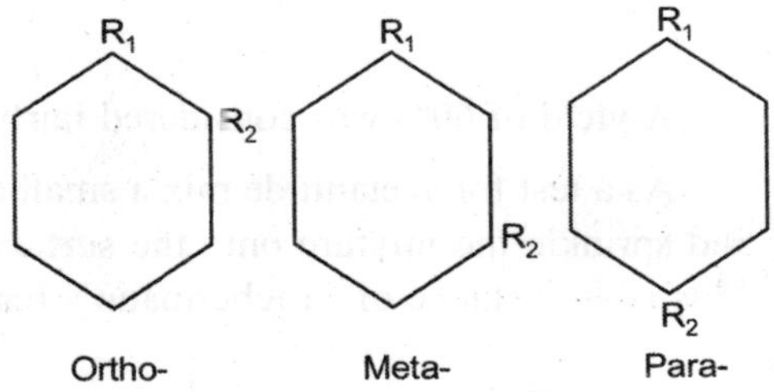

Ortho-   Meta-   Para-

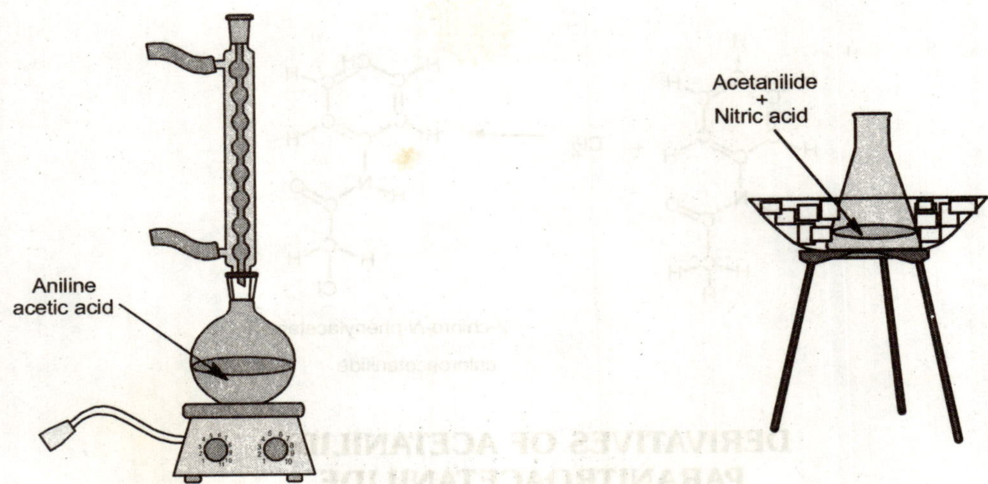

Fig. 27.9. Preparation of Acetanilide **Fig. 27.10. Preparation of paranitroacetanilide**

Place 10 g (9.7 mL) of aniline and 15 g (14.2 mL) of glacial acetic acid (2) in a 125 mL boiling flask. Attach a reflux condenser(2) to the flask and heat to a boil with a Bunsen burner or hot plate while maintaining a flow of cold water through the jacket of the condenser (Figure 2). Continue to reflux the mixture for several hours. The product will be a rose coloured homogeneous liquid. If left for several days the liquid may crystallize into a solid mass.

Acetanilide is quite soluble in hot water, but only sparingly soluble (0.5 g/100 mL) in cold water. Add about 100 mL of distilled or deionized water to the flask and heat until the crude acetanilide dissolves. Dark-coloured droplets of immiscible liquid may be seen in the solution, apparently accounting for the colour of the crude material. Allow the mixture to cool slightly; this unidentified by-product will settle out. Decant off the solution of acetanilide. To improve the yield, dissolve any acetanilide remaining in the flask in hot water and again decant. Cool the combined solutions in an ice bath to precipitate the acetanilide as colorless leaflets. Dissolve in hot water and recrystallize as before. Filter off the crystals and allow them to air-dry in a shallow dish until any residual acetic acid has evaporated. The resulting product should be free of any odor of the acid. Nine grams or more of purified product should be recovered. The theoretical yield of acetanilide (m.w. 135) from 10 g of aniline (m.w. 93) is

x/10 135/93

x 14.5 g

A yield of 60% was considered fairly respectable for one very inexperienced organiker.

As a test for acetanilide mix a small amount of acetanilide with an equal amount of sodium nitrite and sprinkle the mixture onto the surface of concentrated sulfuric acid. Note the red colour formed. This is an instance of "Liebermann's reaction", used as a test for secondary amines:

$$(C_6H_5)(CH_3CO)NH + HONO \longrightarrow (C_6H_5)(CH_3CO)N\text{-}NO + H_2O \qquad (3)$$

A nitrosoamine is formed; this reacts with phenol (C_6H_5OH) to produce a coloured indicator which turns red in acid, blue-green in base. The phenyl group already present in acetanilide reacts similarly to phenol.

To nitrate the acetanilide, place 15 mL of conc. sulfuric acid in a 250 mL Erlenmeyer flask and cool by placing the flask in an ice water bath. To keep the flask from overturning, weight it down with a lead "doughnut". These are available from laboratory suppliers, but you can make your own by melting scrap lead and pouring into a suitable form. Another alternative is to buy a stick of bar solder and bend it into a circle. Next gradually add 7 grams of acetanilide while stirring with a thermometer until all, or nearly all, has dissolved. 7.5 mL of conc. nitric acid are then added dropwise with continuous stirring. At the outset much heat is evolved and care must be taken to keep the temperature below 20 C by adding the acid slowly and keeping ice in the bath. Allow the flask to stand in ice water for about 15 minutes after addition of the acid. Then pour the contents of the flask, a slightly syrupy reddish liquid, into about 500 mL of ice water while stirring vigorously; p-nitroacetanilide will precipitate as a granular yellowish-white solid.

Filter off the precipitate on a double layer of finely-woven cotton cloth in a large glass or plastic funnel, and wash on the filter with ice water. Take up the cloth and squeeze it gently to remove excess liquid. Take care not to lose yield by forcing the precipitate through the cloth. Finally, place the cloth with the precipitate in a shallow dish and allow to air dry for two or three weeks. When the product is sufficiently dry, break up the cake and pulverize. Bottle the p-nitroacetanilide without further purification.

Paranitroacetanilide can be used to produce a "Pharaoh's Serpent" that far outstrips the tiny ones sold as fireworks; these use the combustion of mercuric thiocyanate to produce a voluminous ash. For a showy demonstration place a gram or two of p-nitroacetanilide in a small evaporating dish and add enough conc. sulfuric acid to make a thick paste. Avoid too thin a mixture as this will simply fizzle instead of deflagrate. Place the dish on a gauze mat on a tripod and heat with a small Bunsen flame or alcohol lamp. At first the mixture will darken and bubble slightly. Then, quite suddenly and without warning, it will deflagrate with a hissing sound and a large and dense cloud of white smoke. Only when the smoke starts to dissipate will the "serpent" come into view, a column of black ash the diameter of the dish and extending up two feet or more. It stays in place attached to the dish, but is so light than the dish doesn't overturn. Despite its light weight the ash is remarkably rugged and can be broken loose and handled gently without collapsing.

This demonstration should only be done in a well-ventilated area, as the white fumes of sulfuric acid are extremely irritating.

Notes:

(1) Given the conventional, Kekulé, representation of the structure of benzene as a ring of six carbon atoms alternately linked by single and double bonds, one might ask why there should not be four instead of three distinct isomeric forms for the disubstituted ring. The answer is a phenomenon called resonance wherein the bonds between adjacent carbon atoms may be thought of as existing in a intermediate state between single and double.

(2) The Allihn condenser shown in Figure 2 is a pretty piece of glassware, but there's no specific requirement that one be used . An ordinary straight-tube (Liebig) condenser or any of the water-jacketed columns available at bargain prices from the SAS Store will do nicely.

(3) It is important to keep glacial acetic acid from contact with skin as it attacks the epidermis, loosening it to form extensive blisters. If contact does occur, immediately wash the affected area with soap and water. As in any lab. work, wear protective goggles and clothing. The same caution applies to other concentrated acids (e.g., H_2SO_4, HNO_3).

Similar precautions apply also to aniline, which is toxic and can be absorbed through the skin.

Uses and Application

Acetanilide is used as an inhibitor in hydrogen peroxide and is used to stabilize cellulose ester varnishes. It has also found uses in the intermediation in rubber accelerator synthesis, dyes and dye intermediate synthesis, and camphor synthesis. Acetanilide was used as a precursor in penicillin synthesis and other pharmaceuticals and its intermediates.

Acetanilide has analgesic and fever-reducing properties; it is in the same class of drugs as acetaminophen (paracetamol). Under the name acetanilid it formerly figured in the formula of a number of patent medicines and over the counter drugs. In 1948, Julius Axelrod and Bernard Brodie discovered that acetanilide is much more toxic in these applications than other drugs, causing methemoglobinemia and ultimately doing damage to the liver and kidneys. As such, acetanilide has largely been replaced by less toxic drugs, in particular acetaminophen, which is a metabolite of acetanilide and whose use Axelrod and Brodie suggested in the same study.

In the 19th century it was one of a large number of compounds used as experimental photographic developers.

Acetoacetanilide (acetyl acetanilide), $C_6H_5NHCOCH_2COCH_3$, formula weight 177.20, is a white crystalline material; m.p.,85°C. It is slightly soluble in water, but is soluble in most organic solvents and caustic solutions. It is prepared by boiling acetoacetic esters with aniline, and is used as an intermediate in the preparation of pyrazolones and pyrimidines. It is used in the preparation of such dyes as Dianil Yellow 3GN and of Hansa yellows.

BENZYLAMINE

Preparation of Benzylamine

28% Aqueous Ammonia, 810g
Benzyl chloride, 84.3g
49% Aqueous NaOH, 52.3g
Diethyl Ether, 200g

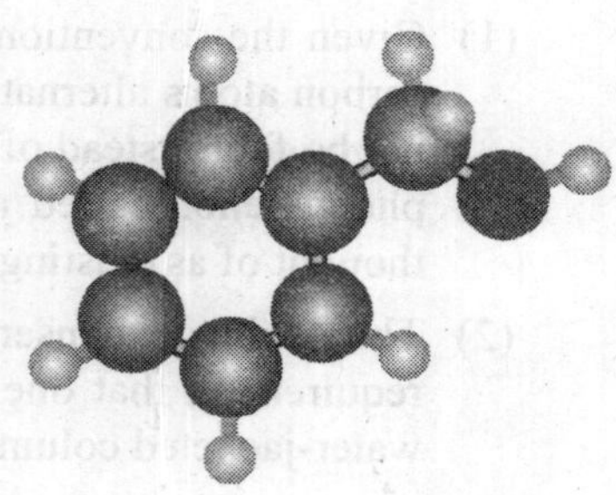

A three-necked flask was equipped with a reflux condenser dropping funnel and an agitator. All of the aqueous ammonium hydroxide solution was poured into the flask, then the benzyl chloride was introduced drop by drop; over a period of two hours with constant stirring of the mixture. The exothermic heat of reaction kept the temperature between 30-34°C during this period. When it is desired to introduce the benzyl chloride at a faster rate, conventional cooling means can be employed to control the temperature. A large excess of ammonia was employed as the molar ratio of reactants was 20:1. An additional two hours was allowed to insure completion of the following reaction:

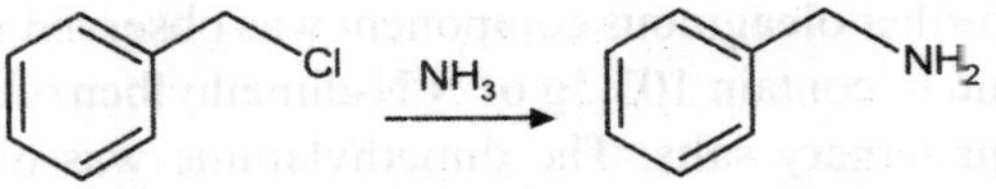

Cl $\xrightarrow{NH_3}$ NH_2

Then the equimolecular quantity of caustic soda solution was added. When quiescent, the mixture split into an aqueous layer and an oily layer. After separating these two layers by use of a separatory funnel, the aqueous phase was made available for further repeated use by merely adding sufficient ammonium hydroxide to bring the total quantity of ammonia up to the original figure. The oily layer was steam distilled until no further oily constituent was visible in the condensed distillate as it dripped down the condenser tube. This distillate was saturated with sodium chlcride and successively extracted with an initial 80 and three 40 gram batches of ethyl ether. Upon evaporation of the ether from the extract, 52g of crude benzylamine remain which was distilled at atmospheric pressure. 41g grams of substantially pure benzylamine distilled over in the boiling range 185-192°C, while the residue was found to contain an additional 2.3g of benzylamine and 8.7g of other matter which was believed to consist entirely of dibenzyl amine. Thus the total yield of benzylamine was 43.3g or 60.7% based on the weight of benzyl chloride.

Example 2, Preparation of N-Methylbenzylamine

25% Aqueous Methylamine, 6 moles

Benzyl Chloride, 1 mole

50% Aqueous NaOH, 1 mole

Diethyl Ether, 200 grams

The procedure of Example I was repeated using the above materials. The reaction here took the course:

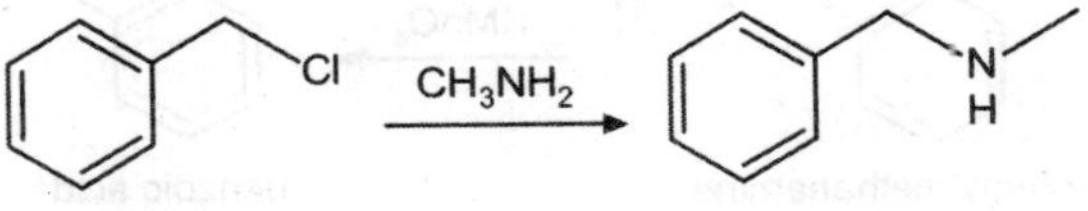

Cl $\xrightarrow{CH_3NH_2}$ N H

N-Methylbenzylamine was obtained in 75% yield and with a boiling point of 82°C/27 mmHg.

Example 3, Preparation of N,N-dimethylbenzylamine

25% Aqueous Dimethylamine, 1088 grams

Benzyl Chloride, 126.6 grams

In the apparatus of Example 1, the benzyl chloride was added dropwise over a two-hour period to the amine (molar ratio 1 to 6) at a rate sufficient to maintain the temperature below 40°C. Stirring was continued at room temperature for an additional hour to insure completion of the reaction denoted by the equation below.

Thereafter the reaction mixture was cooled in a separatory funnel while standing in a refrigerator maintained at 5°C. and separated into two layers. The upper oily layer, weighing 111.5g, was removed and steam distilled until no further oleaginous component was observed in the distillate as it came over. The crude distillate was found to contain 103.5g of N,N-dimethylbenzylamine (76.1% of theory), 3.3g of dimethylamine and no quaternary salts. The dimethylamine was distilled off below 29°C under atmospheric pressure from the N,N-dimethylbenzylamine (bp 82°C/18mmHg).

Physical Properties

Boiling point:	185°C
Melting point:	10°C
Relative density (water = 1):	0.98
Solubility in water:	miscible
Vapour pressure, Pa at 25°C:	87
Flash point:	60°C
Octanol/water partition coefficient as log Pow: 1.09	

Chemical Properties

It closely similar to the primary aliphatic amines in chemical properties.

1. When heated with potassium permanganate it is converted to benzoic acid.

CH_2-NH_2 $\xrightarrow{KMnO_4}$ $C(=O)OH$

1-phenylmethanamine benzylamine → benzoic acid

2. Benzylamine when catalysed with Rany's Nickel its amine bonding easily broken

CH_2-NH_2 $\xrightarrow[+ H_2]{\text{Rany's Nickel}}$ CH_3 + NH_3

1-phenylmethanamine benzylamine → Toluene + Ammonia

3. Benzylamine when allowed to react with nitrous acid (sulphuric acid and sodium nitrite) benzyl alcohol is produced, and nitrogen gas is liberated, like aliphatic amines.

$$C_6H_5CH_2-NH_2 + HNO_2 \longrightarrow C_6H_5CH_2OH + N_2 + H_2O$$

1-phenylmethanamine
benzylamine

phenylmethanol
benzyl alcohol

Uses and Application

Benzylamine and its derivatives are used as chemical intermediate for the manufacture of dyestuffs, pigments, optical brighteners, textile auxiliaries, agrochemicals, amino acids and other organic compounds.

Halogen Derivatives

The halogen derivatives of aniline find only slight technical use, chiefly as intermediates in the preparation of azo and azoic dyes. They are generally produced by reduction of the corresponding nitro compounds by the Bechamp reduction (iron and very dilute hydrochloric acid), but the p or o derivatives may also be made by ammonolysis (*q.v.*) of bromochlorobenzenes.

o-Chloroaniline, o-$ClC_6H_4NH_2$, formula weight 127.57, is a colourless liquid; m.p., - 2°C.; b.p., 207°C.; density, 1.213; refractive index, 1.5895. It is practically insoluble in water, but is soluble in acids and most organic solvents. Its acetyl derivative melts at 87-88°C. and its benzoyl derivative at 99°C. It is also known as Fast Yellow G Base.

m-Chloroaniline, m-$ClC_6H_4NH_2$, is a colourless liquid; m.p., -10°C.; b.p., 230-231°C.; density 1.223; refractive index, 1.5931. It is practically insoluble in water, but is soluble in acids and most organic solvents. Its acetyl derivative melts at 72-73°C. and its benzoyl derivative at 118°C. It is also known as Fast Orange G Base.

p-Chloroaniline, p-$ClC_6H_4NH_2$, is a colourless crystalline material; m.p., 70-71 °C.; b.p., 232°C. It is soluble in hot water and most organic solvents. Its acetyl derivative melts at 169-170°C. and its benzoyl derivative at 192-193°. It is used in the manufacture of Napthol AS-E, which is.the p-chloroanilide of hydroxynaphthoic acid.

2,5-Dichloroaniline, 2,5-$Cl_2C_6H_3NH_2$, formula weight 162.02, is a solid; m.p., 51°C.; b.p., 251°C. It is also known as Fast Scarlet GG Base, and it is prepared by nitrating p-dichlorobenzene to 2,5-dichloronitrobenzene and reducing the nitro compound with iron and acid. It is used for the preparation of Direct Black 2G (*C.1.* 588), Direct Green 2GB (*C.I.* 589), Chloramine Black HW (*C.I.* 591), Rapid Orange GNR, Rapid Fast Brown GGH, and Rapid Fast Scarlet GH. In 1945, 296,000 lb. of the base and its hydrochloride was produced in the U.S.

p-Bromoacetanilide, p~$BrC_6H_4NHCOCH_3$. See "Derivatives" under *Acetanilide*.

Nitroanilines

The nitro derivatives of aniline (frequently called nitranilines) are used to a rather large extent, mainly as intermediates for the preparation of the various types of azo and azoic dyes.

o-Nitroaniline, O-$NO_2C_6H_4NH_2$, formula weight 138.12, is an orange-yellow crystalline compound; m.p., 71-72°C. It is soluble in hot water, acids, alcohol, and chloroform. It is also known as Fast Orange GR Base. and it is prepared by treating o-nitrochlorobenzene with ammonia under pressure or by desulphonation of 2-nitroaniline-4-sulphonic acid with strong sulfuric acid. Its acetyl derivative melts at 9394°C. It is used to a very slight extent technically.

m-Nitroaniline, m-$NO_2C_6H_4NH_2$, formula weight 138.12, is a yellow crystalline product; m.p., 114 °C.; b.p., 286°C. It is soluble in 880 parts of water, 16 parts of alcohol, 16 parts of ether, and is very soluble in mineral acids. It is also known as Fast Orange R Base, and it is prepared by partial reduction of m-dinitrobenzene with sulfides (see Amination by reduction). It finds considerable use as an organic intermediate, particularly in the preparation of azo and other dyes, such as Chrome Printing Yellow G (C.I. 36), Diamine Fast Bordeaux (C.I. 543), and Naphthol AS-BS. In 1945, 265,000 lb. of this product was produced in the U. S. The acetyl derivative melts at 151-153°C.

p-Nitroaniline; p-$NO_2C_6H_4NH_2$, formula weight 138.12, is a bright-yellow crystalline material; m.p.,146-148°C. It is soluble in 1250 parts of cold water, 45 parts of boiling water, 20 parts of alcohol, and 22 parts of ether. It is also soluble in benzene, methanol, and mineral acids. It is also known, as Fast Red GG Base, and it is best prepared by treating p-nitrochlorobenzene with excess ammonia under pressure (see *Ammonolysis*). It may also be prepared by nitrating acetanilide to form p-nitroacetanilide (m.p. 214-216°C.), and hydrolyzing this with caustic or mineral acid. *p*Nitroaniline is used to a considerable extent an intermediate in the preparation of azo and azoic dyes, such as Diamine Black HW (*C.I.* 592), Direct Green B (*C.I.* 593), Thiazol Yellow R (*C.I.* 43), Chrome Printing Orange R (*C.I.* 40), and Rapid Fast Red BB, and also for the production of p-phenylenediamine, a rather important intermediate and dye. In 1945, 127,000 lb. of this intermediate was produced in the U.S.

2,4-Dinitroaniline, 2,4-$(NO_2)_2C_6H_3NH_2$, formula weight 183.12, is a yellow solid; m.p., 188°C. It is very difficultly soluble even in boiling alcohol. It is prepared by treating 2,4-dinitrochlorobenzene with ammonia. Due to its weakly basic character, it can be diazotized only in concentrated sulfuric acid. 2,4-Dinitroaniline finds use ill the preparation of various azo dyes.

2,4,6-Trinitroaniline (picramide), 2,4,6-$(NO_2)C_6H_2NH_2$, formula weight 228.12, is a yellow product; m.p.,188°C. It may be prepared by treating 2,4,6-trinitrochlorobenzene with ammonia or 1,3,5-trinitrobenzene with hydroxylamine.

Sulphonic Acid Derivatives

The anilinesulphonic acids find very extensive use as organic intermediates, especially in the preparation of azo dyes.

2-aminobenzenesulfonic acid
othoanilic acid

3-aminobenzenesulfonic acid
metanilic acid

4-aminobenzenesulfonic acid
sulpahnilic acid

Orthanilic acid (o-aminobenzenesulphonic acid, aniline-2-sulphonic acid) (I), formula weight 173.19, is best prepared by treatment of o-nitrochlorobenzene with sodium polysulphide, oxidation of the resulting o,o'-dinitrodiphellyl disulfide to o-nitrobenze sulphonic acid, and reduction of this to orthanilic acid. This product finds the most limited use of the three possible isomers. It is used for the preparation of a few azo-dyes.

Metanilic acid (m-aminobenzenesulphonic acid, aniline-3-sulphonic acid) (II), formula weight 173.19, is a white crystalline product, soluble in 68 parts of cold water, but only very slightly soluble in alcohol. It is prepared by sulphonating nitrobenzene with 25% oleum at 60-70°C. and then reducing the resulting m-nitrobenzenesulphonic .

monomethylaniline have been used. "Anhydroformaldehyde-aniline," which is formed from aniline and formaldehyde, may be reduced (see also *Anils*). However, this reduction requires zinc dust or a catalytic method with activated nickel. Another method is the methylation of p-toluenesulfonylaniline and hydrolysis of this product with sulfuric acid to give monomethylaniline.

Dimethylaniline (DMA), $C_6H_5N(CH_2)_2$, formula weight 121.18, is a yellowish to brownish oily liquid; m.p., 2°C.; b.p.; 192-194 °C.; density, 0.956; refractive index, 1.5582; flash point, 61°C. It is practically insoluble in water, but is quite soluble in alcohol, chloroform, ether, and aromatic organic compounds. It was discovered by Hofmann in 1850. At present it is manufactured by heating aniline, methanol, and sulfuric acid at temperatures ranging from about 205°C. to about 230° using a steel autoclave. The bases are then liberated with caustic, the methanol is distilled off, and the dimethylaniline purified by vacuum distillation. On reacting for six hours at 205°C. a yield of about 95% of theory is obtained, and the dimethylaniline has a purity exceeding 99.5%. Temperatures of the order of 250°C. give nuclear alkylation. Some of the methanol used in this process is lost as dimethyl ether. The small amounts of aniline and monomethylaniline left may be acylated, as with acetic anhydride, and the tertiary base distilled from the less volatile acylamines. Dimethylaniline is a very valuable intermediate, and it is used for the manufacture of vanillin, nitrosodimethylaniline, Miehler's ketone, Methyl Red (*C.I.* 211), Malachite Green (*C.I.* 657), Methyl Violet (*C.I.* 680), Crystal Violet (*C.I.* 681), Methylene Blue (*C.I.* 922). It is also a good solvent and used in methylation. It is used as an intermediate in the preparation of rubber accelerators, explosives, and medicinal products. The alkylating agent phenyltrimethylammonium chloride, $C_6H_5N(CH_2)2Cl$, is prepared from dimethyl aniline and is used for alkylating morphine to codeine, antipyrine, and theobromine. By nitration, the explosive tetryl is formed.

Monoethylaniline (EA), $C_6H_5NHC_6H_5$, formula weight 121.18, is a colourless, very refractive liquid that rapidly becomes brown on exposure to light and air; it solidifies below -80°C.; m.p., -53.5°C.; b.p., 204.7°C.; ddensity, 0.958; refractive index 1.5559. It is practically insoluble in water, but is soluble in most organic solvents. Its preparation and purification are very similar to that of monomethylaniline. It may be purified by fractional distillation. Monoethylaniline is a very useful intermediate in the preparation of azo and triphenylmethane dyes. In 1945, 1,062,000 lb. of this product was produced in the U.S.

Diethylaniline (DEA), $C_6H_5N(C_2H_5)_2$, formula weight 149.23, is a colourless to yellow liquid; m.p., -38°C.; b.p., 215-216°C.; density 0.935; refractive index, 1.5421 It is somewhat soluble in most organic solvents. Diethylaniline is usually prepared by heating aniline hydrochloride with excess ethyl alcohol under pressure at 180°C. It may be separated from monoethylaniline by careful fractional distillation or by any of the methods' used for separating mono- from dimethyl aniline. Diethylaniline is used for the preparation of Brilliant Green (C.I. 652), Ethyl Violet (C.I. 682), Acid Violet (C.I. 698), Iris Violet (C.l. 847), and Celestine Blue B (C.I. 900).

Ethylbenzylaniline (EBA) N -ethyl- N -phenylbenzylamine), $C_6H_5N(C_2H_5)CH_2C_6H_5$, formula weight 211.29, is a light yellow, oily liquid; b7l0, 285°C.; b22, 186°C.; liquid at O°C.; density, 1.034. It is practically insoluble in water, soluble in 4.4 parts of alcohol, and soluble in most of the organic solvents. Ethylbenzylaniline is prepared by treatment of ethylaniline with benzy chloride in the presence of soda solution. Forthis purpose, crude ethylaniline mav be used and the result ng ethylbenzylaniline separated from diethylaniline by fractional distillation. Ethylbenzylaniline is an intermediate in the preparation of such dyes as Acid Green GG (*C.I.* 670), Xylene Blue AS (*C.I.* 673), and Neptune Blue B (*C.I.* 714).

Diphenylamine, $C_6H_5NHC_6H_5$, formula weight 169.22, is a white crystalline product with a floral odor, darkening on exposure to light; m.p.,53-54°C.; b.p., 302°C. It is practically insoluble in water, but is quite soluble in most organic solvents. Since diphenylamine is a weak base, it forms salts only with strong acids. It is prepared by heating aniline hydrochloride with a small excess of aniline at 21Q-240°C. under 6 atm. pressure. Residual aniline is extracted with dilute acid and the weakly basic diphenylamine is distilled for purification. It is used as a stabilizer for nitrocellulose and celluloid, and also as an intermediate for such dyes as Metanil Yellow (*C.I.* 138), Aniline Yellow (*C.I.* 143), Curcumine (*C.I.* 144), Alkali Green (*C.I.* 665), Cotton Brown (*C.l.* 938). and Pyrogene Indigo (*C.I.* 961). Diphenylamine is used for the preparation of the insecticide phenothiazine, $C_6H_4NHC_6H_4S$. Also known as thiodiphenylamine, this compound is prepared by the reaction of diphenylamine with sulfur (13).

C-alkyl Derivatives

The C-alkyl derivatives of aniline are very useful intermediates, particularly in the field of dyes, where they find use in almost every known class of colours.

Toluidines. The three isomeric monoaminotoluenes are prepared by iron and dilute acid reduction of the corresponding nitrotoluenes. The nitrotoluenes are prepared by nitration of toluene and separated by fractional distillation.

o-toluidine m-toluidine p-toluidine

o-Toluidine (o-methylaniline) (IV), formula weight 107.15, is a light yellow liquid that darkens readily on exposure to light or air; b.p.,200-202°C.; flash point, 87°C.; density, 1.008; refractive index, 1.5688. It is slightly soluble in water (3% at 25-30°C.), but is soluble in dilute acids and most organic solvents. o-Toluidine is used as an intermediate in the preparation of Chrysoidine (*C.I.* 60), Oil Yellow OB (*C.I.* 61), Brilliant Crocein 9B (*C.I.* 313), Safraninc T (*C.I.* 841), Methylene Blue T (*C.I.* 925). and Methyl Indigo B (*C.I.* 1192), Naphthol AS-D, Naphthol AS-GR, Naphthol AS-LB, Rapid Fast B.rown IBH, Rapid Fast Brown GGH, Rapidogen Orange G, Rapidogen Red G, Rapidogen Green B, and Rapidogen Brown IB.

m-Toluidine (m-methylaniline) (V) is a colorless liquid; it solidifies at about -50°C.; b.p., 203-204°C.; density, 0.990; refractive index, 1.5711. It is slightly soluble in water, but is soluble in dilute acids and most organic solvents. It is used M an intermediate in the preparation of other organic compounds and dyes. such as Union Red (*C.I.* 266).

p-Toluidine (p-methylaniline) (VI) is a solid appearing as white lustrous needles; m.p.,44-45°C.; b.p., 200-201°C. It is soluble in about 135 parts of water and quite soluble in dilute acids and most organic solvents. It may be separated from a mixture containing the ortho isomer by the addition of enough oxalic acid to precipitate the more insoluble para isomer. It is very useful as an intermediate', particularly in dyes, where it is used in the manufacture of such colours as Primuline (*C.I.* 812), Indamine 6R (*C.I.* 868), Alizarine hisol R (*C.I.* 1073), Alizarine A!'trol B (*C.I.* 1075), Cyananthrol R (*C.I.* 1076), Alizarine Cyanine Green G (*C.I.* 1078), Anthraquinone Violet (*C.I.* 1080), Alizarine Viridine FF (*C.I.* 1084), Alizarine Sky Blue B (*C.I.* 1088), and Alizarine Rubinole G (*C.I.* 1091).

Xylidines, or monoaminoxylenes, are used to a considerable extent as intermediates in the production of azo dyes, aviation gasoline additives to raise the octane rating, rubber accelerators, wood preservatives, wetting agents for textiles, antioxidants for paint pigments, and frothing agents in ore dressing.

Mixed xylidines, containing varying amounts of the six isomers, are produced by nitration of commercial xylene and reduction of this mixture. The major portion of mixed xylidines is m-4-xylidine, and this can be separated from the mixture" as the acetate by means of acetic acid. Addition of hydrochloric acid to the remainder separates out the p-2-xylidine. On concentrating the filtrate and subliming, the m-2xylidine hydrochloride is sublimed off. The residue is then made alkaline and steamdistilled to drive over the remaining; xylidines. These are converted to the formyl derivatives. The o-3-xylidine crystallizes out whereas the o-4-xylidine remains liquid. Both may be hydrolyzed to the free base by means of alcoholic caustic soda (9). Mixed xylidines are used in the preparation of Sudan II (*C.I.* 73), Wool Scarlet R (*C.I.* 76), and Resorcin Brown (*C.I.* 234).

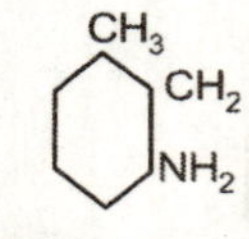

(VII) *o*-3-Xylidine 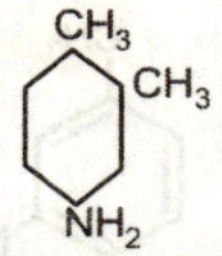 (VIII) *o*-4-Xylidine 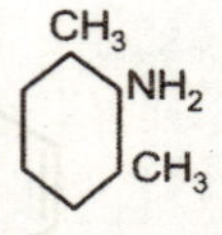 (IX) *m*-2-Xylidine

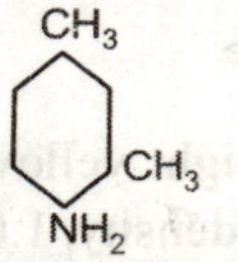

(X) *m*-4-Xylidine 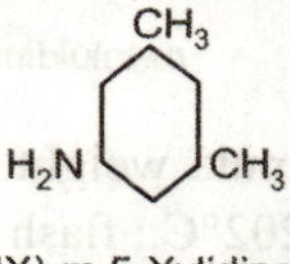 (IX) *m*-5-Xylidine 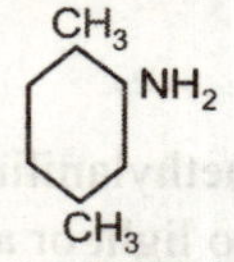 (XII) *p*-2-Xylidine

o-3-Xylidine (2,3-xylidine, 2,3-dimethylaniline, vic-o-xylidine) (VII), formula weight 121.18, is a liquid; b.p. 221-222°e.; d~; 0.991. Its acetyl derivative melts at 135°C.

o-4-Xylidine (3,4-xylidine, 3,4-dimethylaniline, as-o-xylidine) (VIII) is a solid; m.p., 51°e.; b.p., 226°C It may be prepared by nitration of pure o-xylene and subsequent reduction, or by bromination of pure o-xylene and replacement of the bromine by means of ammonia under pressure. It is used as a starting material in the manufacture of riboflavin (vitamin B2).

m-2-Xylidine (2,6-xylidine, 2,6-dimethylaniline, vic-m-xylidine) (IX) is a liquid; b739,214°e. Its acetyl derivative melts at 177°C. and its benzoyl derivative at 168-168.5°C. On oxidation it gives rise to m-xyloquinone.

m-4-Xylidine (2,4-xylidine, 2,4-dimethylaniline, as-m'-xylidine) (X) is a liquid; .bp, 216°C.; density 9.6, 0.9783; refractive index, 1.5607. Its acetyl derivative melts at 129-130ºC. and its benzoyl derivative at 192°e. m-4-Xylidine may be separated from those xylidines having a free position para to the amino group, such .as. p-xylidine, by treatment with formaldehyde; It does not react and may be steam-distilled off.

m-5-Xylidinc (3,5-xylidine, 3,5-dimethylaniline, sym-m-xylidine) (XI) is an oil; n.p., 220-221 °C.; d~. 0.9935. It is prepared by deamination of 5-nitro-4-amino-l,3-dimethylbenzene and subsequent reduction. Its acetyl derivative melts at 144°C.

p-2-Xylidine (2,5-xylidine, 2,5-dimethylaniline, p-xylidine) (XII) is a solid; m.p., 15.5°C.., b.p., 213.5°C.; density 0.9790; refractive index, 1.5593. It may be prepared by nitration of pure p-xylene and subsequent reduction. It is moderately soluble in water. On oxidation with chromic acid it yields p-xyloquinone. Its acetyl derivative melts at 139°C. and its benzoyl derivative at 140°C. It is used for the preparation of Diamine Beta Black B (C.I. 549) and Naphthogene Blue 4R (C.I. 534).

In 1945 there was produced in the U.S. 487,000 lb. of xylidines, of which 92,000 lb. was m-4-xylidine and 17,000 lb. was p-2-xylidine.

Cumidines find very little technical use. *o-Cumidinc* (o-isopropylaniline) (XIII), formula weight 135.20, is a liquid; bp,214°C.; bl3, 95°C.; density, 0.9760. Its acetyl derivative melts at noc. *p-Cumidine* (p-isopropylaniline) (XIV) is a liquid; b.p., 225°C.; density, 0.953. Its acetyl derivative melts at 102.5 °C. and its benzoyl derivative at 152°C. Both the ortho and para isomers are obtained by nitrating isopropylbenzene and reducing the resulting mixture of 0- and p-nitro derivatives. Oxalic acid precipitates out the more insoluble para isomer. *Pseudocumidinc* (2,4,5-trimethylaniline) (XV) is obtained by nitration of pseudocumene and subsequent reduction; m.p., 58°C.; b. p., 234-235°C. It is soluble in 0.8 part of water at 20°C. Oxidation with chromic acid yields p-xyloquinone. Acetylpseudocumidine is used as an intermediate in the manufacture of Cumidine Scarlet (C.I. 80).

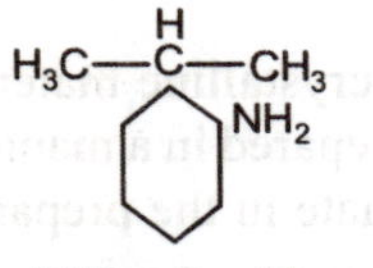

(XIII) *o*-Cumidine

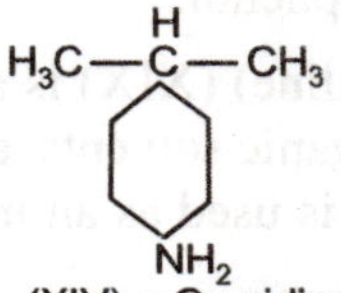

(XIV) *p*-Cumidine

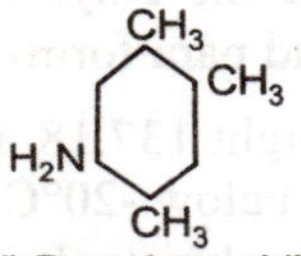

(XV) Pseudocumidine

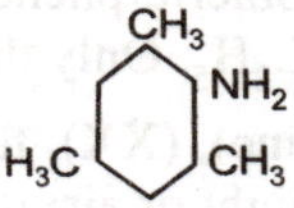

(XVI) Mesidine

Mesidine (2,4,6-trimethylaniline, arninomesitylene) (XVI), formula weight 135.20, is an oil; b.p., 232-233 °C.; density, 0.9633. It acetyl derivative melts at 216217°C. and its benzoyl derivative at 204 °C. It is obtained by nitration of mesitylene and reduction to mesidine. Mesidine is used in the manufacture of Acid Rosarnine A.

C-alkoxy Derivatives

C-Alkoxy derivatives of aniline are in general useful intermediates, especially in the dye, pharmaceutical, and perfume fields.

Anisidines. The three isomeric anisidines may be regarded as monomethoxy derivatives of aniline or as monoamino derivatives of anisole,

o-Anisidine (o-methoxyaniline) (XVII) formula weight 123.15, is a yellowish liquid becoming brownish on exposure to air; m.p., 5°C.; b.p., 225°C.; density, 1.098. It is practically insoluble in water, but is soluble in dilute acids and most organic solvents. It can be prepared by nitration of anisole and separation of the resulting isomers, by methylation of o-nitrophenol, or by action of methanol and caustic on Onitrochlorobenzene, in each case with subsequent reduction by iron and dilute acid.

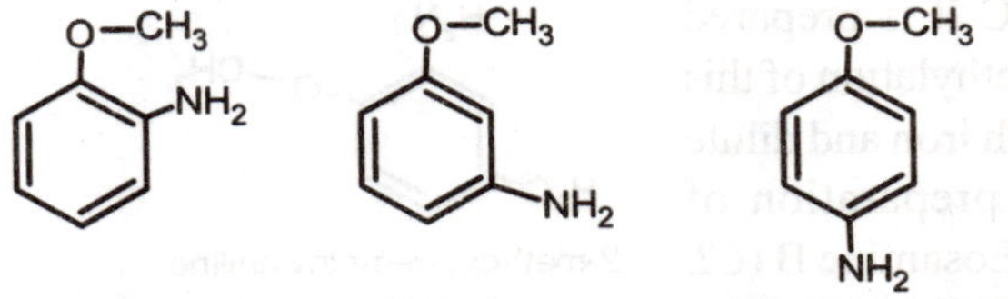

2-methoxyaniline 3-methoxyaniline 4-methoxyaniline

Anisidines

It is also known as Fast Red BB Base, and it is used as an intermediate in the preparation of such dyes as Chrome Fast Yellow 2G (*C.I.* 112), Sudan R (*C.I.* 113), Croceine Scarlet 10B (*C.I.* 291), Diamine Fast Yellow 4G (*C.I.* 349), Naphthol AS-OL, and Rapidogen Red R.

m-Anisidine (m-methoxyaniline) (XVIII) is an oily liquid remaining fluid at -lOoC.; b.p., 251°C. It is not very soluble in water, but is quite soluble in dil\lh~ acids and most organic solvents. It is prepared by the methylation of m-aminophenol.

p-Anisidine (p-methoxyaniline) (XIX) is a white crystalline material; m.p., 57°C.; b.p., 246°C. It is quite soluble in water and organic solvents, and is prepared in a manner similar to o-anisidine, except that the para isomer is used. It is used as an intermediate in the preparation of such dyes as Naphthol ASSG and Rapid Fast Pink LB.

Phenetidines. The three isomeric phenetidines are the ethyl analogs of the anisidines and are derived from phenetole, $C_6H_5OC_2H_5$. Only the ortho and para forms are of any importance.

o-phenetidine (o-ethoxyaniline), (XX), formula weight 137.18, is a colourless liquid that rapidly becomes brown on exposure to light or air; it solidifies below -20°C.; b.p., 228-230°C. It is insoluble in water, but is soluble in dilute acids and most organic solvents. It is prepared in a manner similar to o-anisidine, and is used as a dye intermediate in the preparation of Rapid Fast Orange RH, Rapid Fast Scarlet RH, and Rapidogen Scarlet R.

O—CH₂—CH₃ NH₂ O—CH₂—CH₃ NH₂

2-ethoxyaniline

4-ethoxyaniline

Phenetidines

p-Phenetidines (p-ethoxyaniline) (XXI) is a colourless liquid becoming red to brown on exposure to light and air; m.p., about 3°C.; b.p.,253-255°C. It is insoluble in water, but is soluble in dilute acids and most organic solvents. It is prepared in a. manner similar to o-anisidine except that the para isomer is used and ethyl is substituted for methyl. It is a very useful intermediate, being employed in the manufacture of acetophenetidin (m.p., 134-135°C.) (see *Analgesics*), phenocoll (aminoacetophenetidin), phenacaine, dulcin, and various dyes, among them Alizarine Yellow 5G (*C.I.* 122) and Fast Acid Blue R (C.I. 760).

Cresidine (5-methyl-o-anisidine, l-amino-2-methoxy-5-methylbenzene) (XXII), formula weight 137.18, is a white crystalline product; m.p., 51.5°C.; b.p., 235°C It is prepared by nitration of p-cresol, methylation of this product, and reduction with iron and dilute acid. It is used for the preparation of various azo dyes, such as Eosamine B (*CL* 119), Coccinine B (*CJ.* 120), and Diamino Fast Violet BBN (*C.I.* 325).

H₂N O—CH₃ H₃C

2-methoxy-5-methylaniline

cresidine

O—CH₂—CH₂—CH₃ NH₂ N⁺ O⁻ O

5-nitro-2-propoxyaniline

2-Amino-4-nitro-1-n-propoxybenzene

5-Nitro-2-n-propoxyaniline (2-amino-4-nitro-1-n-propoxybenzene) (XXIII), formula weight 196.20, which has been prepared by Verkade and his co-workers (11), is the sweetest substance on record (see *Sweetening agents*), and leaves no bitter aftertaste. It is an orange material (m.p., 47.5-48.5°C.) and is 4100 times as sweet as sucrose. In order to prepare this ether, 2,4-dinitrochlorobenzene is treated with alkali propoxide to form 2,4-dinitropropoxybenzene. This is then reduced with sodium disulfide at 70-90°C. in propyl alcohol to form 2-amino-4-nitropropoxybenzene (1). Analogous ethers are the methyl (220 times as sweet as sucrose), ethyl (950), n-butyl (1000), isopropyl (600), and allyl (2000).

Anils

Anils are Schiff bases (called also azomethine bases) having the general formula RN=CHR', in which R is an aryl group (Ar) and R' is aryl or alkyl. The anils are derived from aniline (*q.v.*) or substituted anilines.

The anils are coloured crystalline compounds soluble in alcohol or ether and insoluble in water, they are weak bases, forming salts (such as hydrochlorides) in non-aqueous mediums, and they are hydrolysed in alkaline solution to the original amine and carbonyl compounds used in their formation. On catalytic hydrogenation they yield N - substituted anilines, some of which can be prepared by this methods.

N=CH + H_2 → NH—CH_2

N-[(1*Z*)-phenylmethylene]aniline *N*-benzylaniline

The anils are prepared by the condensation of aromatic primary amines

H_2N + H_3C–C(=O)–CH_3 → H_3C–C(=O)–N–CH_3 + H_2O

Aniline Acetone *N*-methyl-*N*-phenylacetamide

An aromatic such as benzaldehyde yields an anil by the condensation of aniline. With aliphatic aldehydes (for e.g. formaldehyde and acetaldehyde) it produces amino resins (see chapter Resins plastics and condensates), ketones like acetones also combine with aniline to yield similar compounds, but at higher temperature and in the presence of catalysts like Zinc chloride or iodine.

Aniline + Acetone → N-methyl-N-phenylacetamide + H_2O

Uses of Aniline

Originally the great commercial value of aniline was due to the readiness with which it yields, directly or indirectly, valuable dyestuffs. The discovery of mauve in 1856 by William Perkin was the first of a series of dyestuffs which are now to be numbered by hundreds. Reference should be made to the articles dyeing, fuchsine, safranine, indulines, for more details on this subject. In addition to its use as a precursor to dyestuffs, it is a starting-product for the manufacture of many drugs such as Acetaminophen/Paracetamol (Tylenol).Currently the largest market for aniline is preparation of methylene diphenyl diisocyanate (MDI), some 85% of aniline serving this market. Other uses include rubber processing chemicals (9%), herbicides (2%), and dyes and pigments (2%).

Schiff's Base

A Schiff base, or Schiff's base, is a functional group or type of chemical compound containing a carbon-nitrogen double bond with the nitrogen atom connected to an aryl group or an alkyl group but not hydrogen. The Schiff base is synonymous with an azomethine. The Schiff base is named after Hugo Schiff and has the following general structure:

$$R_1R_2C=N-R_3$$

R_3 stands for a phenyl or alkyl group which makes the Schiff base a stable imine. Schiff bases can be synthesized from an aromatic amine and a carbonyl compound by nucleophilic addition forming a

1 + 2 → 3

hemiaminal, followed by dehydration to generate the imine. In a typical reaction, 4,4'-diaminodiphenyl ether reacts with o-vanillin[1]

A mixture of 4,4'-diaminodiphenyl ether 1 (1.00 g, 5.00 mmol) and o-vanillin 2 (1.52 g, 10.00 mmol) in methanol (40.00 mL) is stirred at room temperature for one hour to give an orange precipitate and after filtration and washing with methanol to give the pure Schiff base 3 (2.27 g, 97.00%)

A mixture of 4,4'-diaminodiphenyl ether 1 (1.00 g, 5.00 mmol) and o-vanillin 2 (1.52 g, 10.00 mmol) in methanol (40.00 mL) is stirred at room temperature for one hour to give an orange precipitate and after filtration and washing with methanol to give the pure Schiff base 3 (2.27 g, 97.00%)

There is a Schiff base intermediate in the fructose 1,6-bisphosphate aldolase catalyzed reaction during glycolysis.(The detail discussion of this processes is out of syllabus of this book)

DIPHENYLAMINE

Introduction

Diphenyl amine, is one of the most important diamine.

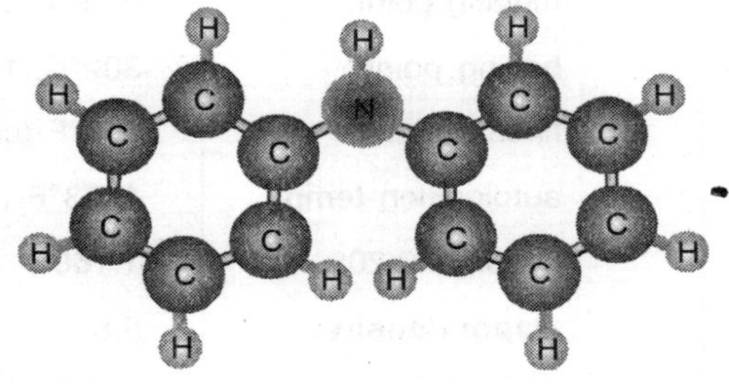

Preparation

(1) By heating acetanilide with bromo benzene and potassium carbonate in the presence of copper powder (Ullmann Reagent)

$$C_6H_5NHCOCH_3 + C_6H_5Br + K_2CO_3 \xrightarrow[\text{reflux}]{Cu} C_6H_5NHC_6H_5 + CH_3COOK + KBr$$

N-phenylacetamide + Bromobenzene → N-phenylaniline diphenylamine + Potassium Acetate + Potassium bromide

(2) By heating Phenol with aniline in the presence of zinc chloride at 260ºC

$$C_6H_5OH + C_6H_5NH_2 \xrightarrow[260^\circ C]{ZnCl_2} C_6H_5NHC_6H_5$$

phenol + aniline → N-phenylaniline diphenylamine

(3) Bu heating aniline and its hydrochloride at 140ºC under pressure (This is commercial process)

140 °C

H_2NHCl Aniline hydrogen chloride + H_2N → + Ammoniochloronium

Physical Properties

vapor pressure:	*1 mm Hg at 108.3°C*
solubility:	*Soluble in carbon disulfide, benzene, alcohol,. Insoluble in water.*
melting point:	52.9°C
boiling point:	302°C, 179°C at 22 mm Hg
flash point:	307°F (closed cup)
autoignition temp.:	1173°F (combustible)
density 22/20:	1.160
vapor density:	5.82

Chemcial Properties

(I) Diphenylamine ($K_b = 6.3 \times 10^{-14}$) is a weaker base than aniline ($Kb = 4.2 \times 10^{-10}$) since contains two phenyls groups against one in the latter. It forms salt which are completely hydrolyzed in aqueous solution.

(II) When heated with sodium of sodamide it forms sodio diphenyl derivatives more readily than aniline

Diphenyl amine + Na → Sodio diphenyl amine + $\frac{1}{2}$ H_2

(III) It reacts with nitrous acid to form the N - nitrosoamine

$$\underset{\text{Diphenyl amine}}{(C_6H_5)_2NH} + HNO_2 \longrightarrow \underset{N\text{-nitroso-}N\text{-phenylaniline}}{(C_6H_5)_2N{-}N{=}O} + H_2O$$

(IV) Its solutions in concentrated sulphuric acid give a light blue colour and an evolution of nitrogen dioxide fumes with a trace of nitrous acid or nitrite salt.

Uses

Diphenyl amine is used (i) as a reagent for testing nitrites (ii) as an redox indicator (with the combination of barium hydroxide and sulphonic acid, as barium diphenyl amine sulphonic acid B.D.A.S indicator.) (iii) In high - boiling baths (iv) stabilizers for smoke less powders and (v) in the manufacture of phenothiozine an intestinal disinfectant.

N-METHYLANILINE

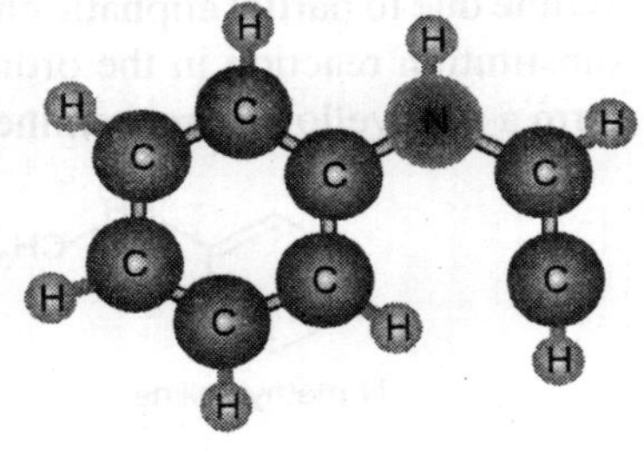

It is a mixed aliphatic -aromatic secondary amine, it may be obtained by the methylation of aniline with calculated amount of methyl iodide or dimethyl sulphate.

Industrially N-methylaniline is prepared by the following methods

(*i*) By heating aniline and methyl alcohol together with some sulphuric acid under pressure

$$\underset{\text{aniline}}{C_6H_5NH_2} + \underset{\text{methanol}}{H{-}\overset{H}{\underset{H}{C}}{-}O{-}H} \xrightarrow[\Delta]{H_2SO_4} \underset{\text{N-methylaniline}}{C_6H_5NH{-}CH_3}$$

(*ii*) By condensing aniline with formaldehyde (CH OH) and reducing the product with zinc and sodium hydroxide.

$$\underset{\text{Aniline}}{C_6H_5NH_2} + \underset{\text{Formaldehyde}}{H{-}\overset{H}{C}{=}O} \xrightarrow{-H_2O} \underset{\text{N-methyleneaniline}}{C_6H_5N{=}CH_2} \xrightarrow[Zn/NaOH]{+\,2H} \underset{\text{N-methylaniline}}{C_6H_5NH{-}CH_3}$$

Physical Properties

Boiling point:	194-196°C
Melting point:	–57°C
Relative density (water = 1):	0.99
Solubility in water:	none
Vapour pressure, Pa at 36°C:	133
Relative vapour density (air = 1):	3.7
Relative density of the vapour/air-mixture at 20°C (air = 1):	1.0
Flash point: (c.c.)	79.5°C
Octanol/water partition coefficient as log Pow:	1.7

Chemical Properties

N- methylaniline is a colourless liquid b.p 194°C. It is a some what stronger base ($Kb = 7.1 \times 10^{-10}$) than aniline due to partial aliphatic character. Its reactions are just like aliphatic secondary amines except the substitution reaction in the ortho and para position of the benzene ring. It reacts with nitrous acid to form a pale yellow nitrosoamine which gives *Liebermann Nitroso reaction*

N-methylaniline + HNO_2 —Rearrangement, Δ→ p - toluedine hydrochloride (H_3C–C_6H_4–NH_2HCl)

The may be purely aromatic or mixed aromatic - aliphatic regarding the hydrocarbon group arrangement with the N - atom, Aryl or Alkyl.

AR–N(AR)–AR: 3° arylamine (purely aromatic)

AR–N(AR)–AL: 3° alkyl diarylamine (mixed aliphatic aromatic)

AR–N(AL)–AL: 3° dialkyl arylamine (mixed aliphatic aromatic)

N,N DIPHENYLANILINE

Introduction

N,N diphenylaniline is a tertiary fully aromatic amine, with three benzene ring surrounding the amine group.

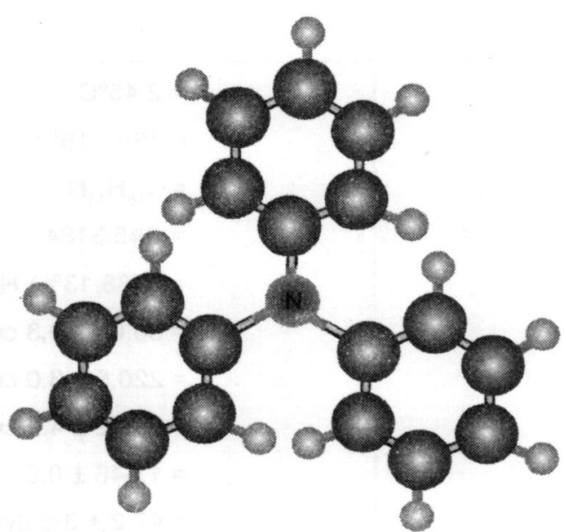

Preparation

It may be prepared

(1) By heating diphenylamine with iodobenzene and potassium carbonate in the presence of copper as a catalyst (Ullmann reaction)

N-phenylaniline diphenylamine + Iodobenzene $\xrightarrow[(K_2CO_3)]{Cu}$ N,N-diphenylaniline

(2) By heating sodio phenylamine and bromobenzene at 300°C

N-phenylaniline diphenylamine (Na–NH) + Bromobenzene $\xrightarrow{300°C}$ N,N-diphenylaniline

Physical Properties

Melting Point	= 2.45°C
Boiling Point	= 193 - 195°C
Molecular Formula	= $C_{18}H_{15}N$
Formula Weight	= 245.3184
Composition	= C(88.13%) H(6.16%) N(5.71%)
Molar Refractivity	= 80.09 ± 0.3 cm^3
Molar Volume	= 220.5 ± 3.0 cm^3
Parachor	= 578.2 ± 4.0 cm^3
Index of Refraction	= 1.646 ± 0.02
Surface Tension	= 47.2 ± 3.0 dyne/cm
Density	= 1.112 ± 0.06 g/cm^3
Polarizability	= 31.75 ± 0.5 10^{-24} cm^3
Monoisotopic Mass	= 245.120449 Da
Nominal Mass	= 245 Da
Average Mass	= 245.3184 Da

Chemical Properties

As it is seen from the formula the three benzene ring are attached with the amine group completely neutralizing the basic character, but still it is a weak base, however, its salts are with hydrofluoric acid, tetraflurobroric acid and perchloric acid. Have been prepared under special conditions.

$$(C_6H_5)_3N + HBF_4 \xrightarrow{\text{(benzene)}} (C_6H_5)_3NH + BF^-_4$$

N, N DIMETHYLANILINE

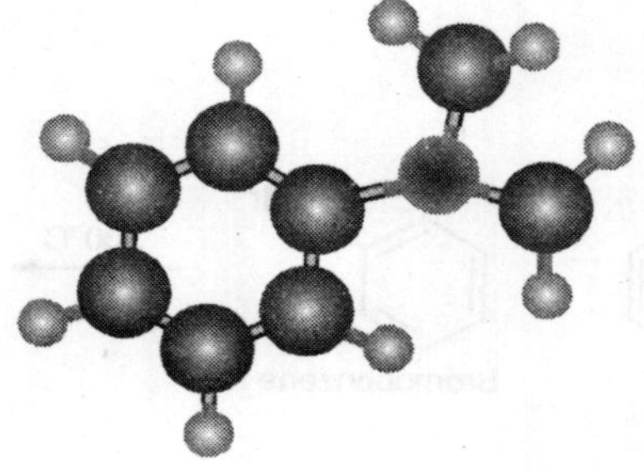

Preparation

N, N Dimethylaniline is prepared by the action of dimethyl sulphate and aniline

Aniline + Methyl sulphate → N,N-dimethylaniline

It can also be obtained by the reaction of sulphonated methanol on aniline

Aniline + Methanol —sulphuric acid→ N,N-dimethylaniline

> Methyl alcohol is used in excess to stop the formation of methylaniline

Physical Properties

Physical State	Pale Yellow to Brown, Oily Liquid, Amine-like Odor
Melting Point	2.45°c
Boiling Point	193-195°c
Specific Gravity	0.956
Solubility in Water	Negligible
Ph	Weak Base
Vapor Density	4.17
Autoignition	371°C
Nfpa Ratings	Health: 3 Flammability: 2 Reactivity: 0
Refractive Index	1.5580
Flash Point	63
Stability	Stable Under Ordinary Conditions

Chemical Properties

N, N dimethylaniline reacts with acids resulting salts and also reacts with alkyl halides to give ammonium compounds. The hydrogen atom present in para position of the benzene ring is very active and also easily replaceable by other atoms.

(I) Nitrous acid reacts with N, N dimethylaniline to produce nitrosodimethylaniline

$$C_6H_5N(CH_3)_2 + HNO_2 \longrightarrow ON{-}C_6H_4{-}N(CH_3)_2 + H_2O$$

N,N-dimethylaniline → *N,N*-dimethyl-4-nitrosoaniline (nitrosodimethyl aniline)

nitrosodimethylaniline when boiled with sodium hydroxide, the $-N(CH_3)_2$ group is separated forming p - nitrosophenol and diphenylamine.

$$ON{-}C_6H_4{-}N(CH_3)_2 + H_2O + NaOH \longrightarrow ON{-}C_6H_4{-}OH + (CH_3)_2NH$$

N,N-dimethyl-4-nitrosoaniline → 4-nitrosophenol + *N*-methylmethanamine

(2) It also reacts with carbonyl chloride to form diphenylmethane.

$$ON{-}C_6H_4{-}N(CH_3)_2 + COCl_2 \longrightarrow [(CH_3)_2N{-}C_6H_4]_2C{=}O + 2HCl$$

N,N-dimethyl-4-nitrosoaniline + phosgene → [3-(dimethylamino)phenyl][4-(dimethylamino) phenyl]methanone micher's ketone

Michler's ketone, a dye intermediate and derivative of dimethylaniline.

(3) When N, N dimethylaniline is heated at 290 - 310°C and under pressure its methyl groups gets attached in para and in ortho position.

H_3C CH_3 N — 290-310 °C, pressure → H_3C CH_3 N, CH_3 → H_2N, CH_3, CH_3

N,N-dimethyl-4-nitrosoaniline   N,N,4-trimethylaniline   2,4-dimethylaniline

Uses

N,N-Dimethylaniline is used as an intermediate to manufacture dyes, vanillin, tetramethyldiaminobenzophenone (white to greenish solid used as important intermediate for making dyes, pigments and photosensitizers) and other organic products. It is also used as a stabilizer for colorimetric peroxidase determination.

β-PHENYLELYLAMINE

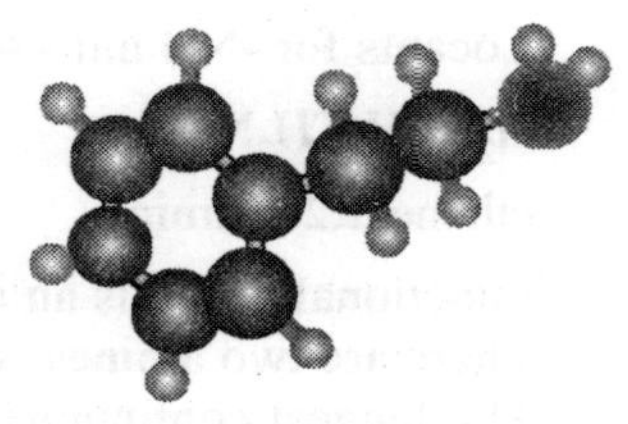

Phenethylamine (2-Phenylethylamine) is an alkaloid and monoamine. In the human brain, it is believed to function as a neuromodulator or neurotransmitter (trace amine). A colourless liquid that forms a solid carbonate salt with carbon dioxide (CO_2) upon exposure to air, phenethylamine in nature is synthesized from the amino acid phenylalanine by enzymatic decarboxylation. It is also found in many foods, especially after microbial fermentation, e.g., in chocolate. It has been suggested that phenethylamine from food (e.g., chocolate) may have psychoactive effects in sufficient quantities. However, it is quickly metabolized by the enzyme MAO-B, preventing significant concentrations from reaching the brain.

Substituted phenethylamines are a broad and diverse class of compounds that include neurotransmitters, hormones, stimulants, hallucinogens, entactogens, anorectics, bronchodilators, and antidepressants.

Physical Properties

β-Phenethylamine	
Chemical name	2-Phenyl-ethylamine or 2-phenylethanamine
Chemical formula	$C_8H_{11}N$
Molecular mass	121.18 g/mol
Density	0.965 g/ml
Melting point	–60 °C
Boiling point	200 °C

Chermical Properties

The term diamine simply implies the presence of two amines. Polyamines contain two or more amine groups.

The root name is based on the longest chain containing both the amine groups.

The chain is numbered so as to give the one of the amine groups the lowest possible number (i.e. first point of difference).

The appropriate multiplier (i.e. di- for two, tri for three etc.) is inserted before the -amine suffix or before the root.

For secondary or tertiary amines, the N locant is used in the same manner as for amines.

If more than one N is substituted, then use N', N" etc.

The N locant is listed before numerical locants, e.g. N,N',2-trimethyl....

Functional group is an amine, therefore suffix = -amine
There are two amines, so insert the multiplier di
The longest continuous chain is C2 therefore root = eth

Locants for -NH units are 1- and 2- 1,2-ethyldiamine or ethyl-1,2-diamine

$H_2NCH_2CH_2NH_2$
ethane-1,2-diamine

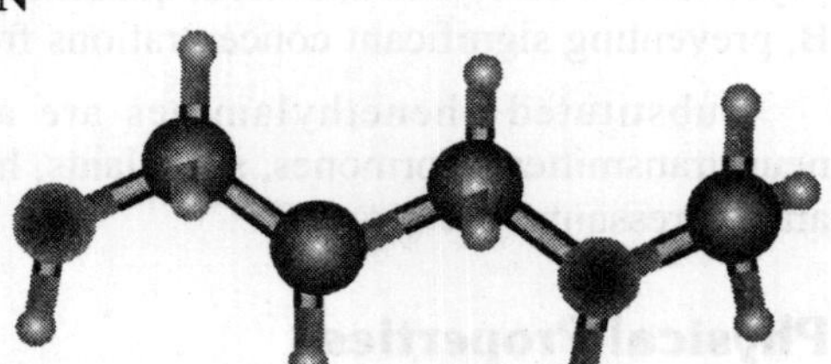

Functional group is an amine, therefore suffix = -amine
There are two amines, so insert the multiplier di
The longest continuous chain is C3 therefore root = prop

Locants for -NH units are 1- and 3-
There is a C1 substituent = methyl
The methyl group is located on the amine, so locant = N

N-methyl-1,3-propyldiamine
or
N-methylpropyl-1,3-diamine
diamine

$H_2NCH_2CH_2CH_2NHCH_3$

Functional group is an amine, therefore suffix = -amine
There are two amines, so insert the multiplier di
The longest continuous chain is C2 therefore root = eth

Locants for -NH units are 1- and 2-
There is are three C1 substituents = trimethyl
The methyl groups are located different amines, so locants = N-, N- and N'-

N,N,N'-trimethyl-1,2-ethyldiamine
or
N,N,N'-trimethylethyl-1,2-diamine
diamine

$(CH_3)_2NCH_2CH_2NHCH_3$

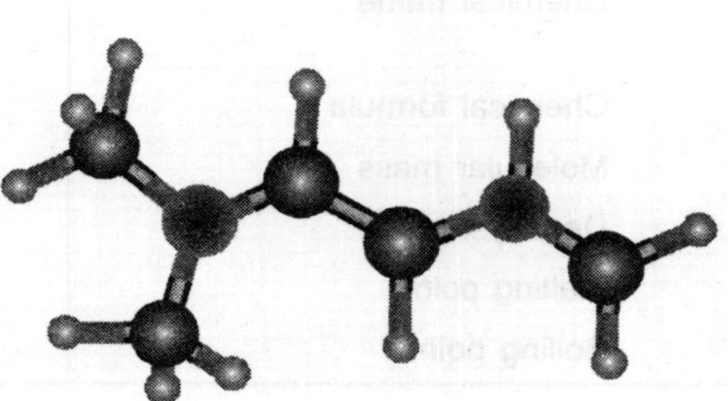

SOME MORE DIAMINES

Physical Properties

Molecular Weight	88.15
Boiling Point (°C)	139 at 101.3kPa (760 mmHg)
Melting Point (°C)	– 72
Specific Gravity (20/4°C)	0.851
Flash Point (°C)	42 (Tag Closed Cup)
Refractive Index (nD20)	1.4468
Solubility	100g/100g Water at 20°C

O-PHENYLENEDIAMINE (OPD)
1,2-BENZENEDIAMINE

Product Description and Uses

OPD is a chemical intermediate used in the synthesis of fungicides, corrosion inhibitors, pigments and pharmaceuticals. OPD is also used to remove elemental sulfur in mining ores, and to remove aldehyde colour formers in polymeric products.

Physical Properties

Property	Typical Value
CAS Registry Number	95-54-5
CAS Name	1,2 benzenediamine
Molecular Formula	$C_6H_4(NH_2)_2$
Molecular Weight	108.1
Amine Equivalent Weight	54
Bulk Density (flakes)	33 lb/ft3
Specific Gravity at 160°C (320°F)	1.031
Boiling Point (760 mmHg), °C (°F)	252 (486)
Melting Point Range, °C (°F)	102-104 (215-219)

Flash Point (closed cup), °C (°F)	156 (313)
Solubility Parameter	12.1
Solubility in Water, 35°C (95°F)	4 wt%
Heat of Vaporization, 280°C (536°F)	129 cal/g
Heat of Combustion	7,355 cal/g
Critical Temperature (Tc), °C (°F)	507 (945)
Critical Volume (Vc)	315 mL/(g mol)
Critical Pressure (Pc)	51.1 atm

P-PHENYLENEDIAMINE

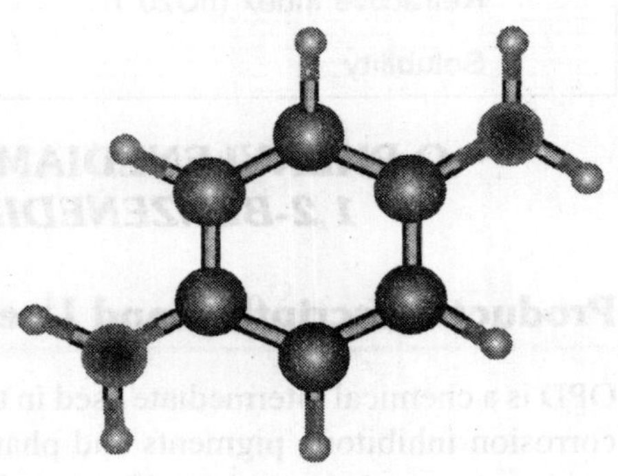

p-Phenylenediamine (PPD), also called 1,4-diaminobenzene or 1,4-phenylenediamine is an aromatic amine used as a component of engineering polymers and composites, aramid fibers, hair dyes, rubber chemicals, textile dyes and pigments. PPD is selected because of its low toxicity, high temperature stability, high strength, and chemical and electrical resistance.

Physical Properties

Molecular Formula	= $C_6H_8N_2$
Formula Weight	= 108.14112
Composition	= C(66.64%) H(7.46%) N(25.90%)
Molar Refractivity	= 34.72 ± 0.3 cm^3
Molar Volume	= 93.9 ± 3.0 cm^3
Parachor	= 258.9 ± 4.0 cm^3
Index of Refraction	= 1.660 ± 0.02
Surface Tension	= 57.5 ± 3.0 dyne/cm
Density	= 1.150 ± 0.06 g/cm^3
Polarizability	= 13.76 ± 0.5 10-24cm^3
Monoisotopic Mass	= 108.068748 Da
Nominal Mass	= 108 Da
Average Mass	= 108.1411 Da

Uses

This product is added to real henna or used alone to create black tattoo-like body art.

This compound is used in almost every hair dye on the market, regardless of brand. The darker the colour, usually, the higher the concentrations. Even the so-called "natural" and "herbal" hair colours, while ammonia-free, contain PPD. Some products sold as henna have PPD added, particularly "black henna". Using body art quality (BAQ) pure henna, or indigo is the only way to avoid PPD in hair dye.

The CDC lists p-phenylenediamine as being known as one of many contact allergens. The NIOSH Pocket Guide to Chemical Hazards lists exposure routes as through inhalation, skin absorption, ingestion, skin and/or eye contact. It lists: throat irritation (pharynx and larynx), bronchial asthma, and sensitization dermatitis as symptoms of exposure. Sensitization is a life long issue, what may lead to active sensitization to black clothing, printer's ink, fax ink, hair dye, fur dye, leather dye, and photographic products among many other products.

N, N-DIETHYL-1,3—PROPANEDIAMINE

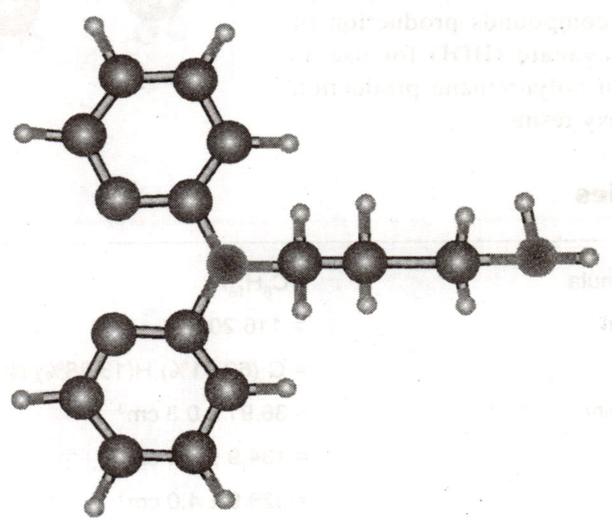

Physical Properties

Molecular Weight	130.23
Boiling Point (°C)	169 at 101.3kPa (760 mm Hg)
Melting Point (°C) –	
Specific Gravity (20/4°C)	0.826
Flash Point (°C)	60 (Tag Closed Cup)

Refractive Index (nD20)	1.4416
Solubility 100g/100g Water at 20°C	3.
Specifications Appearance Colourless clear liquid Purity (GC%) 99.0min.	
Moisture (%)	0.5 max.
Specific Gravity (20/4°C)	0.822~0.830

HEXAMETHYLENE DIAMINE

Hexamethylene diamine or 1,6-Hexanediamine is an organic compound with a hexamethylene hydrocarbon chain and amine functional groups at each end.

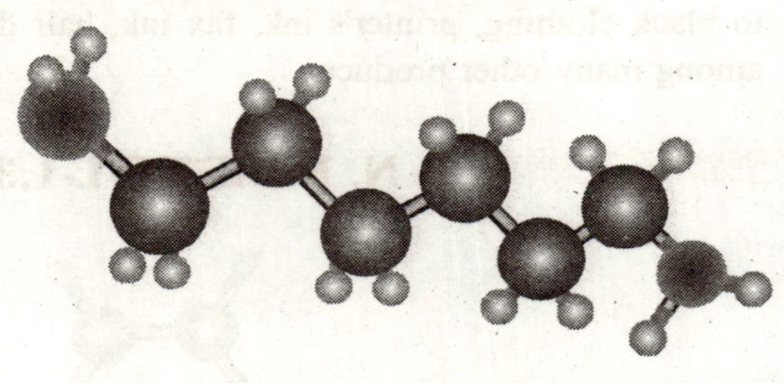

Its main uses are as a raw material in the production of nylon compounds production of hexamethylene diisocyanate (HDI) for use as monomer feedstock in polyurethane production cross-link agent in epoxy resins.

Physcial Properties

Molecular Formula	$= C_6H_{16}N_2$
Formula Weight	= 116.20464
Composition	= C (62.01%) H(13.88%) N(24.11%)
Molar Refractivity	$= 36.91 \pm 0.3$ cm^3
Molar Volume	$= 134.9 \pm 3.0$ cm^3
Parachor	$= 328.9 \pm 4.0$ cm^3
Index of Refraction	$= 1.459 \pm 0.02$
Surface Tension	$= 35.3 \pm 3.0$ dyne/cm
Density	$= 0.861 \pm 0.06$ g/cm^3
Polarizability	$= 14.63 \pm 0.5$ $10^{-24}cm^3$
Monoisotopic Mass	= 116.131349 Da
Nominal Mass	= 116 Da
Average Mass	= 116.2046 Da

CHAPTER

28

Dyes Indicators and Pigments

DEFINITION OF DYES INDICATORS AND PIGMENTS

Definition of Dyes

A dye can generally be described as a coloured substance that has an affinity to the substrate to which it is being applied. The dye is generally applied in an aqueous solution, and may require a mordant to improve the fastness of the dye on the fibre.

Definition of Indicator

Indicators are those which change from one colour to another when hydrogen ion concentration reaches to certain level, this hydrogen ion concentration is different for each indicator, indicators are all either soluble in water or in organic solvent.

Definition of Pigment

Any substance, usually in the form of a dry powder that imparts colour to another substance or mixture. Most pigments are insoluble in organic solvents as well as in water.

History and Introduction of Dyes

Humanity, from antiquity to modernity, has carried a wish to use dyes as a way to symbolize or beautify. From the ancient Egyptian use of the madder and saffron plants to dye cloth for burial wrappings to the modern employment of dies in the clothing industry, dyes have maintained their status as one of the most prevalent and widespread examples of applied chemical technology. Although some of the most ancient pigments were natural minerals, such as red lead oxide made by the Egyptians from the heating of lead with basic lead carbonate, this chapter will deal mainly with organic dyes. A brief discussion of the chemistry behind the colors ensues and will proceed to an overview of the development of dye chemistry throughout the ages.

The visible spectrum of light consists of radiation of wavelengths between approximately 350 to 780 nm. Absorptions of light by a molecule for these wavelengths require the high energy of electronic transitions in a molecule. While sigma bonds are generally not effected by light with wavelengths in the visible region, electrons in pi bonds are more susceptible to promotion into a higher-energy orbital. The typical case of promotion in the visible spectrum is from a p to p* orbital. Certain chromophores, or functional groups involved in absorption, such as conjugated bond systems, tend to have greater absorption in the visible region. The Woodward-Fieser rules summarize these tendencies and explain why dyes tend to be large conjugated molecules. With this very limited introduction to the chemistry of dyes, we are now ready to examine the rich history of dyes and dying. As was mentioned above, the ancient Egyptians used madder to dye wrappings for mummies. We now know that the dye from this plant is alizarin, which was synthesized in 1869 both by William Perkin and Heinrich Caro in independent laboratories. Until Perkin's breakthrough synthesis of the "coal-tar" dye mauve, which will be discussed in length below, most dyes came from natural sources. Some exceptions, though, are Runge's synthesis of aurine from phenol in 1834, and Laurent's successful conversion of phenol into picric acid in 1842. These natural dyes often utilized a mordant to improve color fastness. A mordant can simply be described as a "metallic salt with an affinity for both fibers and dyestuffs." The first conclusive evidence relating to the use of mordants suggests that they were used in Egypt's Middle Kingdom sometime between 2200 and 1500 B.C. At this same time, indigo was being grown for the purpose of dying not only in Egypt, but in Persia and Indochina as well. By this period in history, the Chinese had a dying tradition nearly 1000 years old.

The next great civilizations of Greece and Rome undoubtedly had dye chemistry intertwined in their culture, including their literature and mythology. Homer, the great Hellenic poet, tells in the Odyssey of using "coloured rugs and coverlets," (10) to honor a guest. The great Roman epic poet Virgil also writes of the importance of color as he describes "purple robes..." as a "hero's shrouding." It is fabled that Alexander the great deceived the Persians by using a red dye, possibly madder juice, to make it appear his army was wounded. Another Roman myth proposes that Hercules discovered Tyrian purple, the plum-colored dye which comes from the snail Murex. He supposedly noticed the dark purplish stain upon his dog's jaws after the animal had bitten a snail. Tyrian purple was popular for many centuries being worn only by royalty or those with high religious esteem. Some other dyes of antiquity include the "lac" dyes used in India or Cochineal dye in Mexico which came from the rather unique source of insects.

The next great leap forward for the dye chemistry industry occurred in the latter part of the 12th century when Norman dyers developed a Guild of Dyers in London. Although, dye chemistry remained a type of informal artisan's trade until 1429 when the first European book on dyeing was published in Italy. Later, in 1662, the first English text concerning this topic was published. Dyeing was slowly becoming an academic discipline.

It was not until the 19th century that the pace of advancement in this industry quickened to a gallop. With the advances of organic chemistry inevitably came the first widespread marketing of a synthetic dye. In 1856, 18 year-old William Perkin was studying beneath A.W. Hofmann at the Royal College of Chemistry in London. While on a break from his studies he tried to synthesize quinine after a lecture in which Hofmann described the need for a synthesis of this anti-malarial drug. Perkin attempted to oxidize allyl toluidine with potassium dichromate to achieve the product. With discouraging results, Perkin once again tried this synthesis by using aniline instead of toluidine. By luck, Perkin's aniline contained some *ortho-* and *para-*toluidine impurities. This reaction led to the production of a black sludge which he noticed contained a striking purple dye when he washed his glassware with ethanol or water. Perkin quickly devised a way of extracting the new dye and with the help of his father started the first synthetic dye works in 1858. The new dye was called mauve or mauveine by the French. Interestingly enough, though, it had also been called Tyrian purple. Thus, Perkin successfully ushered the history of dyeing from its "pre-aniline" period into the new "post-aniline" period.

The field of synthetic dye chemistry was now exploding as the increased textile production of the late 19th century industrial revolution increased the demand for inexpensive dyes. Soon Germany would become the leader in the synthetic dye industry as chemists educated by the likes of Liebig and Wohler dominated the synthesis of "coal-tar" dyes. These chemists discovered dozens of new dyes a year which were processed at and shipped from large-scale factories. These improvements in the making of dyes quickly eliminated the production of natural dyes due to their expense and inefficiency as compared to the new synthetic colours.

The synthetic dye industry today is vast and contains many groups of dying processes and dyes. From the synthesis of biological stains used in the preparation of microscope slides to the production of acetate rayon dyes and nylon dyes used in the preparation of commercial textiles, the industry continues to develop new processes and dyes to serve the needs and wants of humanity. One area of early synthetic dye chemistry though, azo dyes, remains one of the largest and most important to the industry. The birth of azo dyes came in 1858, the same year Perkin started his factory for the production of mauve, although their value was not appreciated until Bottiger produced congo red, the first direct cotton dye, in 1884. Johann Peter Griess had made the original discovery that a diazo compound could be derived from the reaction of nitrous acid with aromatic amines. Upon experimentation, he further concluded that this diazo compound could couple to another aromatic amine resulting in the formation of a dye. This area of chemistry has been greatly expanded and refined and now includes trisazo, tetrakisazo and polyazo dyes. The arenediazonium ion, containing the -N=N- chromophore, serves as a weak electrophile which may perform an electrophilic aromatic substitution on an aromatic ring to produce a vast and diverse array of different dyes. Upon referral to the above discussion of the chemistry behind the colours, one can see how these dyes with their great amounts of conjugated p bonds serve as excellent dyes.

The future of the synthetic dye chemistry appears certain. As the global market continues to expand and western culture proceeds to penetrate even the world's most isolated regions, the demand for inexpensive dyestuffs will continue to rise. It is promising that this demand will be well met as "the possibilities of further synthesis [of dyes] are unlimited." With these prospects in sight for the synthetic dye chemistry, one might say that this industry certainly promises a bright and colorful future.

Classification of Dyes

Structure	Color Index	Others
Acridine	Acid	Amphoteric
Anthraquinone	Basic	Neutral
Arylmethane	Direct	Indigoid
		Fluorine dyes
Azo	Mordant	
	Natural	
Nitro	Solvent	
Phthalocyanine		
Quinone Imine		
Tetrazolium		
Thiazole		
Xanthene		

Acridine dyes

Acridine dyes are derived from acridine

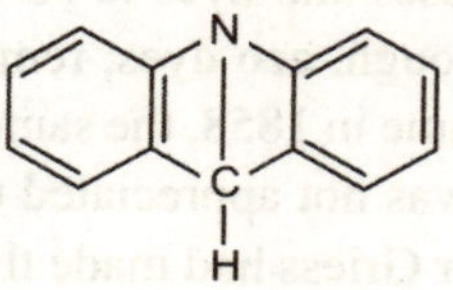

They have the general formula

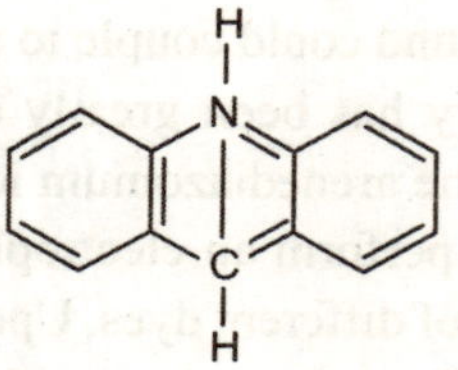

The general formula shows the chromophore to be a quinoid ring. It should also be noted that the molecule is positively charged, shown on the nitrogen atom attached to the quinoid ring, and the dyes are consequently basic. There are a few dyes in this class that are used in histotechnology, but they tend to be for fairly specialised purposes which take advantage of their fluorescent characteristics. The commonest examples would be acridine orange and acriflavine.

Acridine Orange

Property	Data
Common name	Acridine orange
Suggested name	Acridine orange
C.I. name	Basic orange 14
Class	Acridine
Ionisation	Basic
Solubility aqueous	Readily
Solubility ethanol	Moderate
Absorption maximum	497, 467 (Conn)
	489 (Aldrich)
	539 (Gurr)
Colour	Yellow
Empirical formula	$C_{17}H_{20}N_3Cl$
Formula weight	301.8

Acridine orange has been recommended as a fluorochrome to differentiate between DNA (green fluorescence) and RNA (red fluorescence). It also makes a good fluorescent Schiff reagent.

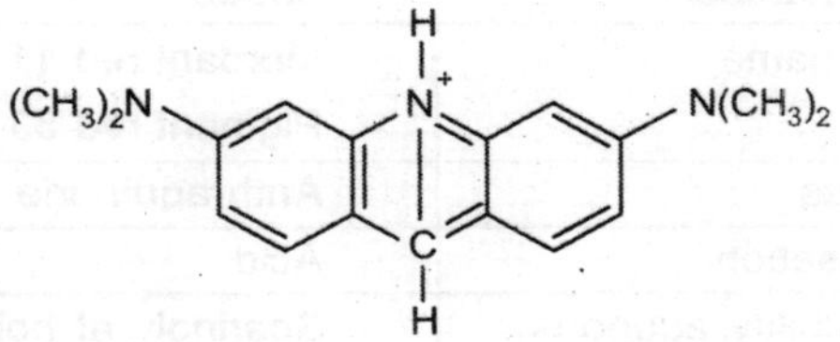

$(CH_3)_2N$ — N^+–H — $N(CH_3)_2$ — C–H

Anthaquinone Dyes

Anthaquinone dyes are derived from anthracene

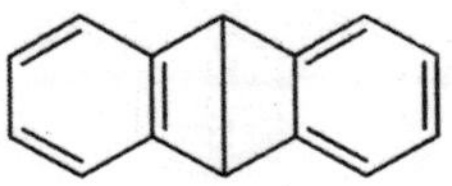

They have the general formula

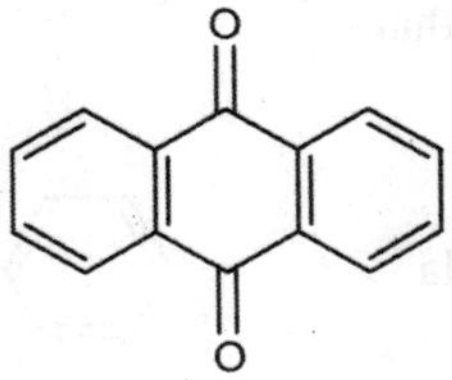

The general formula shows the chromophore to be a quinoid ring. This class of dyes may have either hydroxyl groups or amino groups attached to the general structure. Those that have hydroxyl groups (hydroxyanthraquinones) are the more numerous for histological purposes, and include the mordant dyes alizarin and alizarin red S, both used to demonstrate calcium. Also represented is anthracene blue SWR, a substitute for hematoxylin.

Dyes having amino groups (aminoanthraquinones) are less common, but include nuclear fast red, a mordant dye useful for staining nuclei red.

Nuclear Fast Red

Property	Data
Common name	Alizarin
Suggested name	Alizarin
Other names	—
C.I. number	58000
C.I. name	Mordant red 11 Pigment red 83
Class	Anthraquinone
Ionisation	Acid
Solubility aqueous	Sparingly at boiling Soluble in alkalis.
Solubility ethanol	0.125%
Absorption maximum	601, 567 (Conn) 609, 567 (Aldrich)
Empirical formula	$C_{14}H_8O_4$
Formula weight	240.2

O NH$_2$ OH SO$_3$Na O OH

According to Conn, Nuclear fast red has not been specifically identified, but is considered to be this dye. It forms bright red lakes with calcium, and a useful red nuclear staining lake with aluminum.

Arylmethane dyes

Arylmethane dyes are derived from methane

Diarylmethanes have the general formula

Triarylmethanes have the general formula

Arylmethane dyes are so called because they are derived from methane, but in which some of the hydrogen atoms are replaced with **aryl** rings. Since a synonym for aryl is **phenyl,** they may also legitimately be called **phenylmethane** dyes. Aryl rings are often referred to as benzene rings. However, benzene is a specific chemical compound and this usage is technically incorrect, although very common.

Diarylmethanes

Diarylmethane (or **diphenylmethane**) dyes have two aryl rings. Only one dye in this subgroup has gained much use, and that is auramine O, a fluorescent dye used for the demonstration of acid fast bacteria and to make a fluorescent Schiff reagent. Its structural formula is shown below. Note the alkylamino groups attached to the aryl rings. This dye is subclassified as an **aminodiarylmethane** as a consequence. The chromophore is the carbon to nitrogen double bond.

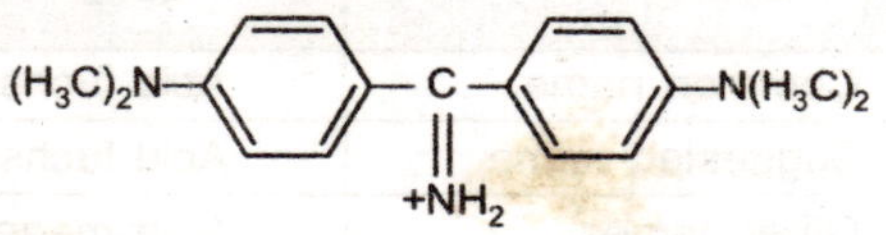

Triarylmethanes

Triarylmethane (or **triphenylmethane**) dyes contain three aryl rings. These are much more common and numerous examples can be given. The pararosanin homologues are good examples. These dyes also contain amino groups, and for that reason are sometimes identified as **aminotriarylmethane** dyes, and are often basic. In pararosanilin, shown below, each of the three aryl rings has an amino group. The quinoid ring is the chromophore.

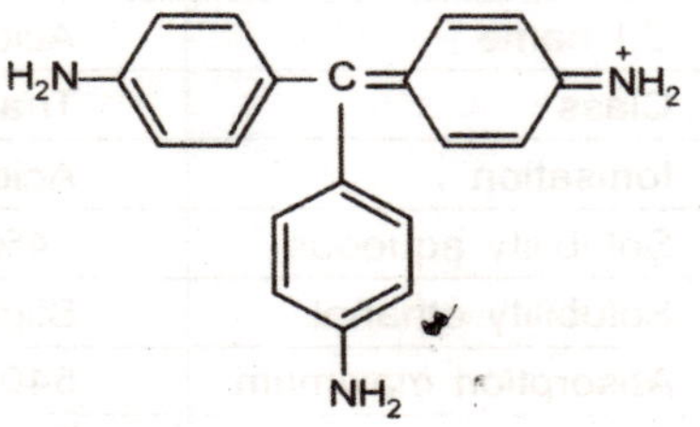

Not all aminotriarylmethane dyes are basic. Some are acid dyes. These have strongly anionic sulphonic groups as well, making the overall charge on the molecule negative. Examples are acid fuchsin and methyl blue.

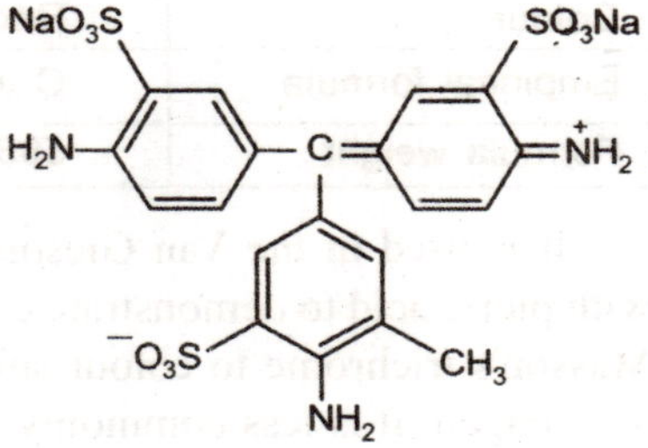

Another subgroup of the arylmethanes has hydroxyl groups instead of amino groups. These dyes are termed hydroxytriarylmethane dyes, and are acid dyes. An example is chromoxane cyanin R, shown below, which has gained some use as a hematoxylin substitute. Dyes in this subgroup are not widely used.

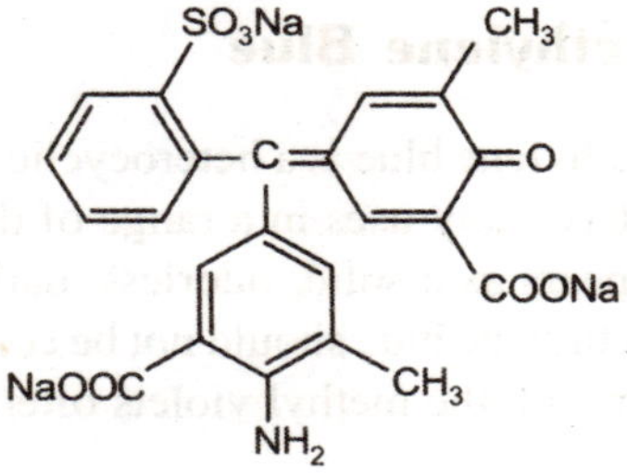

Auramine O

Auramine O, also called Basic yellow 2, Pyocatanium aureum, aizen auramine, Pyoktanin Yellow, Canary Yellow, Pyoktanin, or C.I. 41000, is a diarylmethane dye used as a fluorescent stain. It has the appearance of yellow needle crystals. It can be used to stain acid-fast bacteria (eg. Mycobacterium, where it binds to the mycolic acid in its cell wall) in a way similar to Ziehl-Neelsen stain. It is very soluble in water and soluble in ethanol. It can be used as a fluorescent version of Schiff reagent.

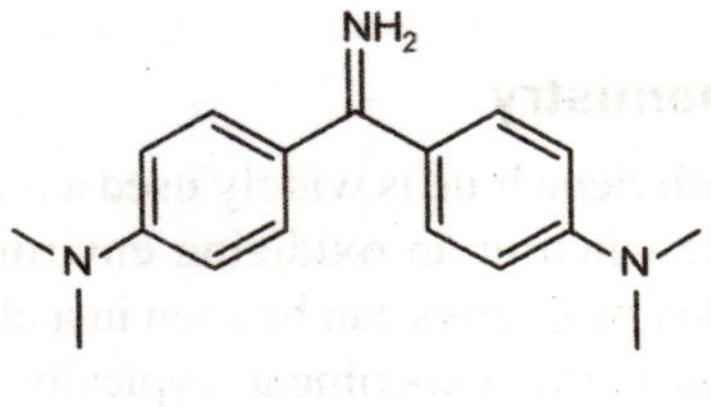

Auramine O can be used together with Rhodamine B as the Truant auramine-rhodamine stain for Mycobacterium tuberculosis. It can be also used as an antiseptic agent.

Acid Fuchsin

Property	Data
Common name	Acid fuchsin
Suggested name	Acid fuchsin
Other names	Acid magenta Acid rubin Acid roseine
C.I. number	42685
C.I. name	Acid violet 19
Class	Triarylmethane
Ionisation	Acid
Solubility aqueous	14%
Solubility ethanol	Slight
Absorption maximum	540-546
Colour	Red
Empirical formula	$C_{20}H_{17}N_3Na_2O_9S_3$
Formula weight	585.6

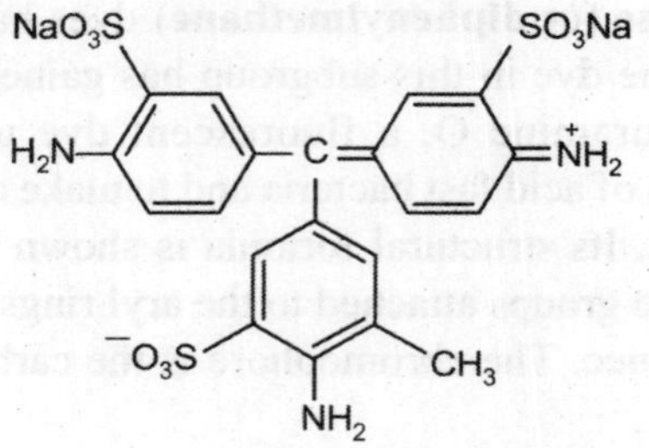

Acid fuchsin is made from basic fuchsin homologues by the addition of sulphonic groups. There are four of these compounds, each of which could have up to three sulphonic groups attached, making twelve possible chemicals. Although there may be slight differences in their properties they are all satisfactory. Commercial Acid fuchsin is probably a variable mixture.

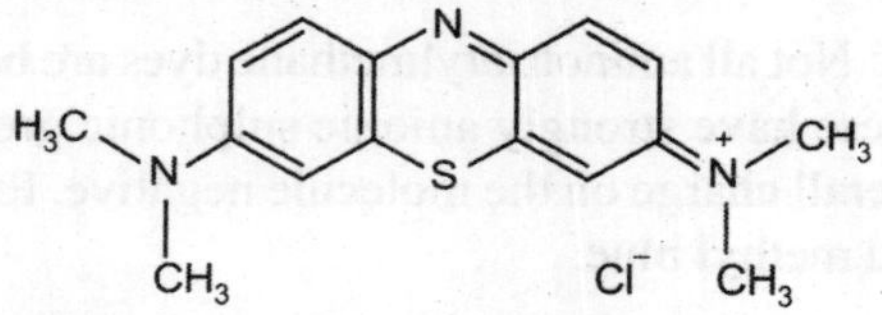

It is used in the Van Gieson method in conjunction with picric acid to demonstrate collagen fibres red, and in Masson's trichrome to colour smooth muscle in contrast to collagen. It is less commonly used in a method for mitochondria.

Methylene Blue

Methylene blue is a heterocyclic aromatic chemical compound with molecular formula: $C_{16}H_{18}ClN_3S$. It has many uses in a range of different fields, such as biology or chemistry. At room temperature it appears as a solid, odorless, dark green powder that yields a blue solution when dissolved in water. Methylene blue should not be confused with methyl blue, another histology stain, new methylene blue, nor with the methyl violets often used as pH indicators.

Uses

Chemistry

Methylene blue is widely used a redox indicator in analytical chemistry. Solutions of this substance are blue when in an oxidizing environment, but will turn colorless if exposed to a reducing agent. The redox properties can be seen in a classical demonstration of chemical kinetics in general chemistry, the "blue bottle" experiment. Typically, a solution is made of dextrose, methylene blue, and sodium hydroxide.

Upon shaking the bottle, oxygen oxidizes methylene blue, and the solution turns blue. The dextrose will gradually reduce the methylene blue to its colorless, reduced form. Hence, when the dissolved oxygen is entirely consumed, the solution will turn colourles. Methylene blue is also used to make the reaction between Fehling's solution and reducing sugars more visible.

Biology

In biology methylene blue is used as a dye for a number of different staining procedures, such as Gram's stain, Wright's stain, and Jenner's stain. Since it is a temporary staining technique, methylene blue can also be used to examine RNA or DNA under the microscope or in a gel: as an example, a solution of methylene blue can be used to stain RNA on hybridization membranes in northern blotting to verify the amount of nucleic acid present. While methylene blue is not as sensitive as ethidium bromide, it is less toxic and it does not intercalate in nucleic acid chains, thus avoiding interference with nucleic acid retention on hybridization membranes or with the hybridization process itself.

Medicine

Owing to its reducing agent properties, methylene blue is employed as a medication for the treatment of methemoglobinemia, which can arise from ingestion of certain pharmaceuticals or broad beans. Basically, methylene blue acts reducing the heme from methemoglobin to hemoglobin, however since methylene blue is toxic, any methemoglobinemia treatment with this substance should be strictly evaluated by a doctor. Methylene blue also blocks accumulation of cyclic guanosine monophosphate (cGMP) by inhibiting the enzyme guanylate cyclase: this action results in reduced responsiveness of vessels to cGMP-dependent vasodilators like nitric oxide and carbon monoxide. Methylene blue is used in endoscopic polypectomy as an adjunct to saline or epinephrine, and is used for injection into the submucosa around the polyp to be removed. This allows the submucosal tissue plane to be identified after the polyp is removed, which is useful in determining if more tissue needs to be removed, or if there has been a high risk for perforation.

Methylene blue is also used as a dye in chromoendoscopy, and is sprayed onto the mucosa of the gastrointestinal tract in order to identify dysplasia, or pre-cancerous lesions. Methylene blue was used at the end of the century as a successful treatment for malaria and the anti-malarial drug, chloroquine, is derived chemically from it. Interest in its use has recently been revived,especially because it is very cheap, but the result of a recent clinical trial was disappointing.

Since its reduction potential is similar to that of oxygen and can be reduced by components of the electron transport chain, large doses of methylene blue are sometimes used as an antidote to cyanide poisoning.

Warning

Methylene Blue is probably an MAOI and if infused intra-venously at doses exceeding 5mg/kg, may precipitate serious serotonin toxicity, serotonin syndrome, if combined with any selective serotonin reuptake inhibitors (SSRIs) or other serotonin reuptake inhibitor e.g. duloxetine, sibutramine, venlafaxine, clomipramine, imipramine.

Aquaculture

Methylene blue is used in aquaculture and by tropical fish hobbyists as a treatment for fungal infections. It can also be effective in treating fish infected by the parasitic protozoa Ichthyophthirius multifilius (ich).

Pranks

Methylene blue is highly stable in the human body, and if ingested, it resists the acidic environment of the stomach as well as the many hydrolytic enzymes present. It is not significantly metabolized by the liver, and is instead quickly filtered out by the kidneys. A common prank amongst biochemists is to add small amounts of methylene blue (generally a few drops of a stain solution will suffice) to coffee, cola, or another dark beverage. The stain's color will be masked by the beverage, and its taste is fairly faint. Within a few hours, the methylene blue will have been removed by the prank victim's kidneys, causing his urine to change color. The urine may become green if little methylene blue was added; larger amounts create a deep blue color. The prank is fairly harmless if small amounts of methylene blue are used, although allergies are possible and it is advisable to use pharmaceutical-grade stain which has been carefully protected from contamination. The main risk of methylene blue pranks is generating a sense of panic in the victim.

Azo Dyes

Aromatic azo compounds (R = R′ = aromatic) are usually stable and have vivid colors such as red, orange, and yellow. Therefore, they are used as dyes, which are called azo dyes. Some azo compounds, eg. Methyl orange, can also be used as acid-base indicators, due to their ability to function as weak acids, and the different colours of the acid and salt forms. Azobenzene is another typical aromatic azo compound. Their colour originates from absorbance in the visible region of the spectrum due to the delocalization of electrons in the benzene and azo groups forming a conjugated system, whereby the N=N group is the chromophore.

Classification

Azo dyes are subgrouped according to the number of azo chromophores in the molecule.

Monoazo

Monoazo dyes contain a single azo chromophore. Examples are Orange G and Ponceau 2R

Disazo

Disazo dyes contain two azo chromophores. Examples are Amido black 10B and Biebrich scarlet

Trisazo

Trisazo dyes contain three azo chromophores. These are not common in histotechnology.

Tetrakisazo

Tetrakisazo dyes have four azo chromophores. An example is Sirius red F3B

Lysochromes

Lysochromes, or fat dyes, are usually disazo dyes and frequently have a single hydroxyl group ortho to the azo nitrogen.

Lillie, in Conn, draws attention to the work of Michaelis in the early part of the century saying that when this is so intramolecular reconfiguration occurs which results in the loss of hydrogen from the hydroxyl, gaining of a hydrogen by one of the azo nitrogens, and formation of an ortho quinoid ring. While still intensely coloured, the resulting compound is insoluble in water, but is soluble in non-polar solvents, including triglycerides.

Different types of Azo Dyes are discussed below:

Alizarin

Alizarian is found in the roots of the madder plant (*Rubia tinctorum* L.), and was uswd as a dye in ancient times in Egypt, Persia and India. It is only rarely used in microtechnique, although it has been suggested for demonstrating some metals, with which it forms coloured lakes.

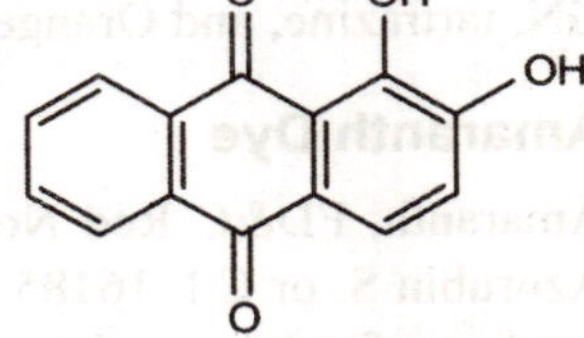

Alizarin Yellow R

Alizarin Yellow R (Chrome Orange, Mordant Orange 1) is a yellow coloured azo dye made by the diazo coupling reaction. It usually comes as a sodium salt. In its pure form it is a rust-colored solid. It is a pH indicator

The pH range changes from 10.1 to 12.0

Yellow below 10.1 and red above 12.0

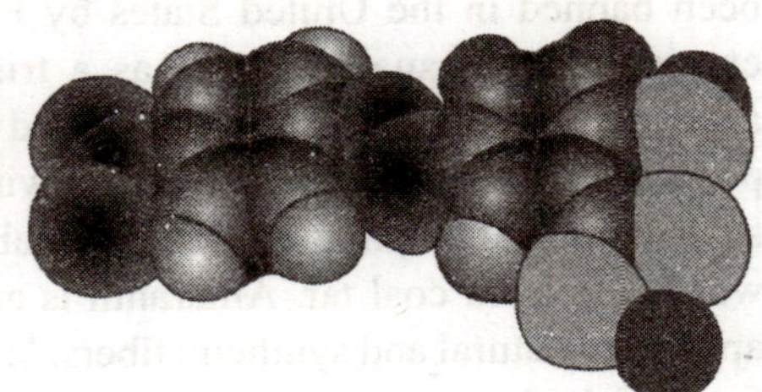

Allura Red AC

Allura Red AC is a red azo dye that goes by several names including: Allura Red, Food Red 17, C.I. 16035, and FD&C Red 40,2-Naphthalenesulfonic acid, 6-hydroxy-5-((2-methoxy-5-methyl-4-sulfophenyl)azo)-, disodium salt, and Disodium 6-hydroxy-5-((2-methoxy-5-methyl-4-sulfophenyl)azo)-2-naphthalene-sulfonate. It is used as a food dye and has the E number E129. Allura Red AC was originally introduced in the United States as a replacement for the use of E123 as a food coloring. Its CAS registry number is 025956-17-6.

It has the appearance of a dark red powder. It usually comes as a sodium salt, but can be also be used as both calcium and potassium salts. It is soluble in water. In water solution, its maximum absorbance lies at about 504 nm. Its melting point is at >300°C. Allura Red AC is approved by the Food and Drug Administration for use in cosmetics, drugs, and food. It is used in some tattoo inks and is used in many products, such as orange soda.

Despite the popular misconception, Allura Red AC is not derived from the cochineal insect. (The red coloring carmine is derived from the female Dactylopius coccus, a South and Central American beetle). Allura Red AC has few health risks associated with it in comparison to other azo dyes. Upon its introduction onto the market, there were fears that Allura Red AC was carcinogenic; however, studies have since shown that this is not the case. The initial reports of its consumption causing tumors have since been shown to have been caused by the presence of para-cresidine.

Although para-cresidine is an important reactant in the manufacture of Allura Red AC and is a known carcinogen, further studies conducted since have found no trace of para-cresidine to be present in food-grade Allura Red AC. Allura Red AC may also cause an adverse reaction in the small number of people with an aspirin intolerance or allergy, manifesting itself as a skin rash (urticaria). It has also been implicated in cases of hyperactivity in young children. Related dyes include Sunset Yellow FCF, Scarlet GN, tartrazine, and Orange B.

Amaranth Dye

Amaranth, FD&C Red No. 2, C.I. Food Red 9, Acid Red 27, Azorubin S, or C.I. 16185, is a dark red to purple azo dye once used as a food dye and to color cosmetics, but since 1976 it has been banned in the United States by FDA as it is a suspected carcinogen. It usually comes as a trisodium salt. It has the appearance of reddish-brown, dark red to purple water-soluble powder that decomposes at 120 °C without melting. Its water solution has absorption maximum at about 520 nm. Amaranth was made from coal tar. Amaranth is an anionic dye. It can be applied to natural and synthetic fibers, leather, paper, and phenol-formaldehyde resins.

HO SO_3^- ^-O_3S N=N SO_3^-

Amido Black 10B

In conjunction with picric acid, in a van Gieson type procedure, this dye demonstrates collagen and reticulin.

NaO_3S SO_3Na N=N N=N NO_2 OH NH_2

Aniline Yellow

Aniline Yellow, also known as para-aminoazobenzene, 4-phenylazoaniline, AAB, Brasilazina oil Yellow G, Ceres Yellow R, Fast spirit Yellow, Induline R, Oil Yellow AAB, Oil Yellow AN, Oil Yellow B, Oil

Yellow 2G, Oil Yellow R, Organol Yellow, Organol Yellow 2A, Solvent Yellow 1, Somalia Yellow 2G, Stearix Brown 4R, Sudan Yellow R, Sudan Yellow RA, and C.I. 11000, is a yellow azo dye and an aromatic amine. It is a derivate of azobenzene. It has the appearance of an orange powder. Aniline Yellow is used in microscopy for vital staining, in pyrotechnics for yellow colored smokes, in yellow pigments and inks including inks for inkjet printers. It is also used in insecticides, lacquers, varnishes, waxes, oil stains, and styrene resins. It is also an intermediate in synthesis of other dyes, e.g. chrysoidine, induline, Solid Yellow, and Acid Yellow. Aniline Yellow was the first azo dye. It was first produced in 1861 by C. Mene. The second azo dye was Bismarck Brown in 1863. Aniline Yellow was commercialized in 1864 as the first commercial azo dye, a year after Aniline Black. It is manufactured from aniline. Aniline Yellow was involved in the 1981 Spanish Toxic Oil Syndrome (TOS). A Madrid-based company imported denaturated rapeseed oil, dyed by aniline yellow to mark it as unsuitable for human consumption, to be used as a fuel in steel mills. However, the company distilled the oil to remove the dye, and sold it as a much more valuable olive oil for cooking. The result was a rash of pneumonia-type illnesses, with a second stage with lesions, weight loss, paralysis, and muscle wasting. The net result was over 20,000 sick and 400 dead. The chemistry of the poisonous reaction is still subject of a debate.

Azorubine

Azorubine, carmoisine, Food Red 3, Azorubin S, Brillantcarmoisin O, Acid Red 14, or C.I. 14720 is a synthetic red food dye from the azo dye group. It usually comes as a disodium salt. It is a red to maroon powder. It is used for the purposes where the food is heat-treated after fermentation. It has E number E122. Some of the foods it can be present in are blancmange, marzipan, Swiss roll, jams, preserves, yoghurts, jellies, breadcrumbs, and cheesecake mixes. It appears to cause allergic or intolerance reactions, particularly amongst those with an aspirin intolerance. Other reactions can include a rash similar to nettle rash and skin swelling. Asthmatics sometimes react badly to it.

Bismarck Brown Y

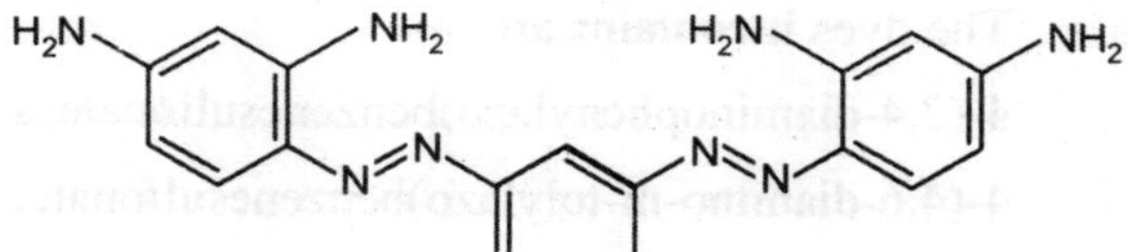

Bismarck brown Y, or Bismarck brown, Manchester brown, Phenylene brown, Basic Brown 1, or C.I. 21000, is a diazo dye. It is used in histology for staining tissues. It stains acid mucins to yellow color. It can be used with live cells. It is also used to stain cartilage in bone specimens. It is also used as one of Kasten's Shiff-type reagents in the periodic acid-Schiff stain to stain cellulose, and in Feulgen stain to stain DNA. It was more common in the past; today it is partially replaced by other stains.

Bismarck brown Y is a constituent of Papanicolaou stains. It can also be used as a counterstain for Victoria blue R for staining of acid-fast microorganisms.

Black 7984

Black 7984, Food Black 2, or C.I. 27755, is a brown-to-black synthetic diazo dye. It usually comes as a tetrasodium salt. When used as a food dye, it has E number E152. It is also used in cosmetics.

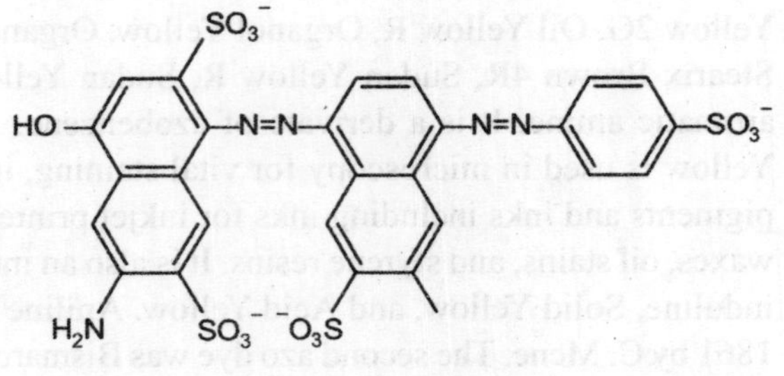

Its use is discontinued in USA and EU since 1984. It is currently delisted and not used anymore both in European Union and USA. It is also not permitted in Australia and Japan. It appears to cause allergic or intolerance reactions, particularly amongst those with an aspirin intolerance. It is a histamine liberator, and may worsen the symptoms of asthma.

Brilliant Black BN

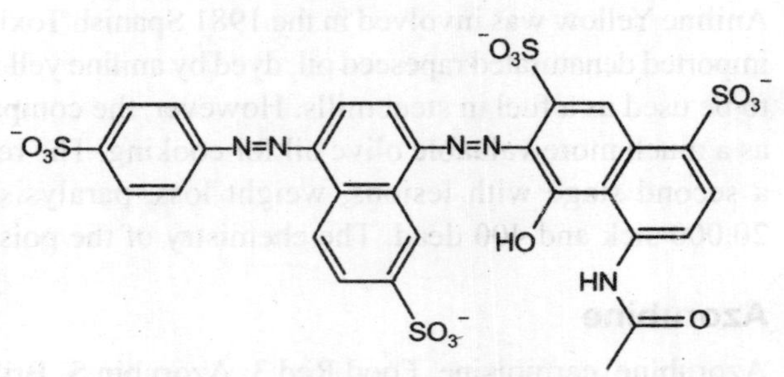

Brilliant Black BN, Brilliant Black PN, Brilliant Black A, Black PN, Food Black 1, Naphthol Black, C.I. Food Brown 1, or C.I. 28440, is a synthetic black diazo dye. It is soluble in water. It usually comes as tetrasodium salt. It has the appearance of solid, fine powder or granules. Calcium and potassium salts are allowed as well.

When used as a food dye, its E number is E151. It is used in food decorations and coatings, desserts, sweets, ice cream, mustard, red fruit jams, soft drinks, flavored milk drinks, fish paste, and other foods.

Brown FK

Brown FK, also called Kipper Brown, Chocolate Brown FK, and C.I. Food Brown 1, is a brown mixture of six synthetic azo dyes, with addition of sodium chloride, and/or sodium sulphate. It is very soluble in water.

The dyes it contains are:

4-(2,4-diaminophenylazo)benzenesulfonate, sodium salt

4-(4,6-diamino-m-tolylazo)benzenesulfonate, sodium salt

4,4'-(4,6-diamino-1,3-phenylenebisazo)-di(benzenesulfonate), disodium salt

4,4'-(2,4-diamino-1,3-phenylenebisazo)-di(benzenesulfonate), disodium salt

4,4'-(2,4-diamino-5-methyl-1,3-phenylenebisazo)-di(benzenesulfonate), disodium salt

4,4',4-(2,4-diaminobenzene-1,3,5-trisazo)-tri(benzenesulfonate), trisodium salt

When used as a food dye, its E number is E154. It is used in smoked and cured mackerels and other fish, but also in some cooked hams and other meats, and crisps. It yields healthy color that does not fade during cooking, nor tends to leach.

It is one of the colourants that the Hyperactive Children's Support Group recommends be eliminated from the diet of children, especially when in combination with benzoates.[citation needed] It can provoke allergic reactions in people sensitive to salicylates, and can intensify the symptoms of asthma. It is banned in the European Union (with exception of the United Kingdom), Australia, Austria, Canada, United States, Finland, Japan, Ireland, Sweden, Switzerland, New Zealand, and Norway.

Brown HT

Brown HT, also called Chocolate Brown HT, Food Brown 3, and C.I. 20285, is a brown synthetic coal tar diazo dye. When used as a food dye, its E number is E155. It is used to substitute cocoa or caramel as a colorant. It is used mainly in chocolate cakes, but also in milk and cheeses, yoghurts, jams, fruit products, fish, and other products. It may provoke allergic reactions in asthmatics, people sensitive to aspirin, and other sensitive individuals, and may induce skin sensitivity.

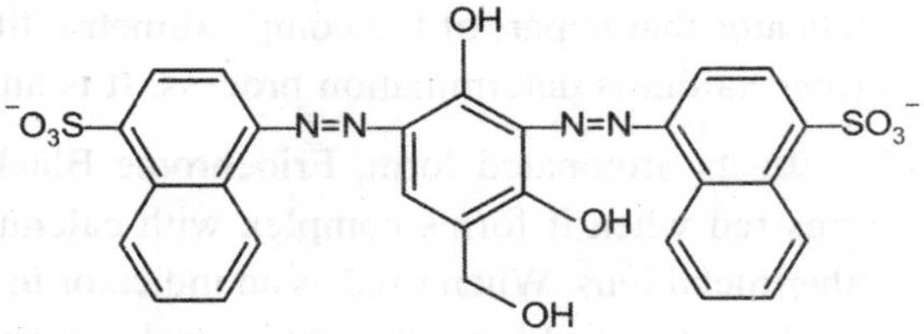

Chrysoine Resorcinol

Chrysoine resorcinol (Chrysoine; Resorcinol Yellow; Gold Yellow; C.I. Food Yellow 8; C.I. Acid Orange 6; C.I. 14270; Yellow T; Tropaeolin O; Tropaeolin R; sodium p-(2,4-dihydroxyphenylazo) benzenesulfonate) is a colourant which is used as a food additive.

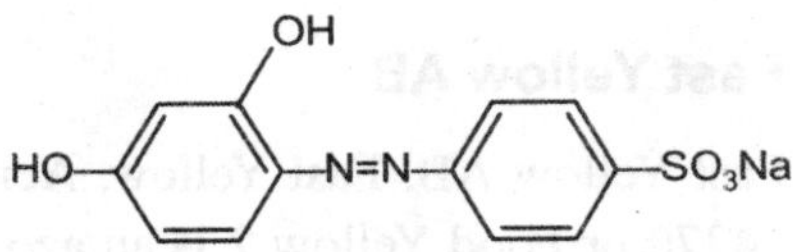

Citrus Red 2

Citrus Red 2, Citrus Red No. 2, C.I. Solvent Red 80, or C.I. 12156 is an artificial dye. As a food dye, it is permitted by Food and Drug Administration (FDA) since 1956 only for use on skin on some Florida oranges. While the dye is a carcinogen, it does not penetrate the orange peel into the pulp. Its chemical formula is $C_{18}H_{16}N_2O_3$. It is an orange to yellow solid or a dark red powder with melting point is 156°C.

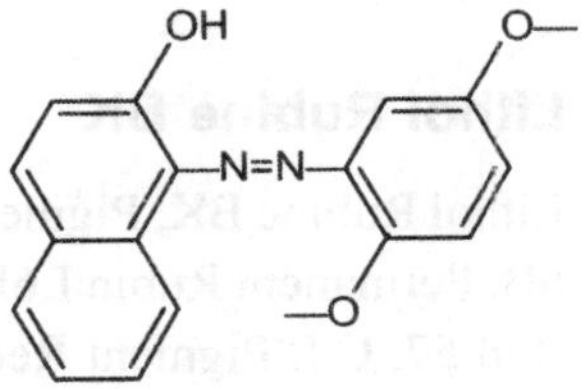

Congo red

Congo red is the sodium salt of benzidinediazo-bis-1-naphtylamine-4-sulfonic acid (formula: $C_{32}H_{22}N_6Na_2O_6S_2$; molecular weight: 696.66 g/mol). It is a secondary diazo dye. Congo red is water soluble, yielding a red colloidal solution; its solubility is better in organic solvents such as ethanol.

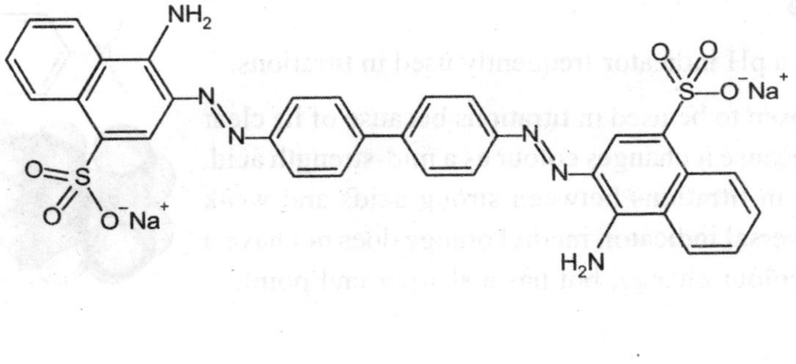

It has a strong, though apparently non-covalent affinity to cellulose fibres. However, the use of congo red in the cellulose industries (cotton textile, wood pulp & paper) has long been abandoned, mainly because of its toxicity.

Eriochrome Black T or Solochrome Black T

Eriochrome Black T or Solochrome Black T is a complexometric indicator that is part of the complexometric titrations, eg. in the water hardness determination process. It is an azo dye.

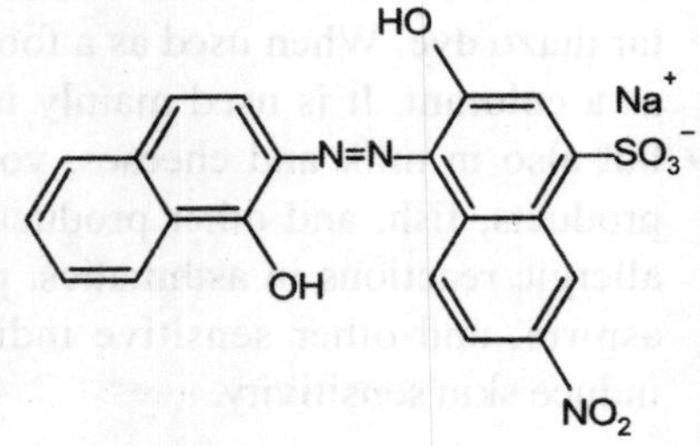

In its protonated form, Eriochrome Black T is blue and it turns red when it forms complex with calcium, magnesium, or other metal ions. When used as an indicator in an EDTA titration the characteristic blue end-point is reached when sufficient EDTA is added and metal ions form complexes with the EDTA instead of the eriochrome.

Fast Yellow AB

Fast Yellow AB, Fast Yellow, Acid Yellow, C.I. 13015, C.I. 14270 or Food Yellow 2 is an azo dye denoted by E number E105. It is used as a food dye. It is now delisted in both Europe and USA and is forbidden if used in foods and drinks; as toxicological data shown it is harmful.

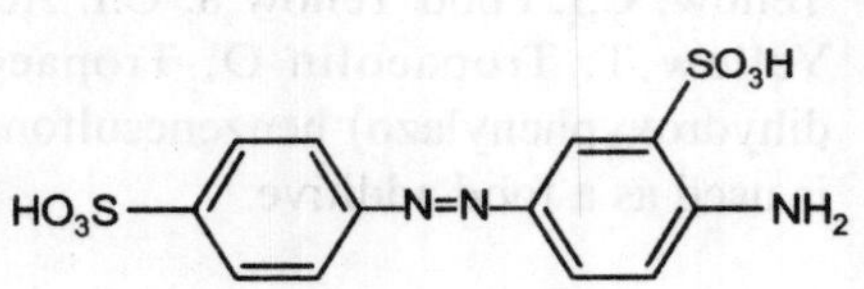

Lithol Rubine BK

Lithol Rubine BK, Pigment Rubine, Carmine 6B, Brilliant Carmine 6B, Permanent Rubin L6B, Litholrubine, Latolrubine, C.I. Pigment Red 57, C.I. Pigment Red 57:1, D&C Red No. 7, or C.I. 15850:1, is a reddish synthetic azo dye. It has the appearance of a red powder, slightly soluble in hot water, insoluble in cold water, and insoluble in ethanol. When dissolved in dimethylformamide, its absorption maximum lies at about 442 nm. It usually comes as a calcium salt.

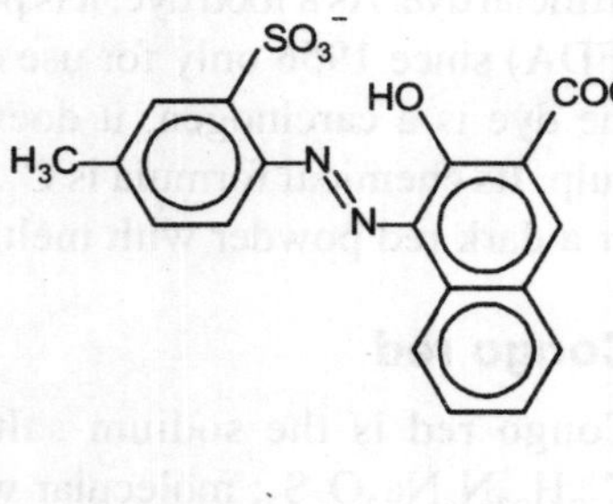

Methyl Orange

Methyl Orange is a pH indicator frequently used in titrations.

It is often chosen to be used in titrations because of its clear colour change. Because it changes colour as a mid-strength acid, it is usually used in titrations between strong acids and weak bases. Unlike universal indicator, methyl orange does not have a full spectrum of colour change, but has a sharper end point.

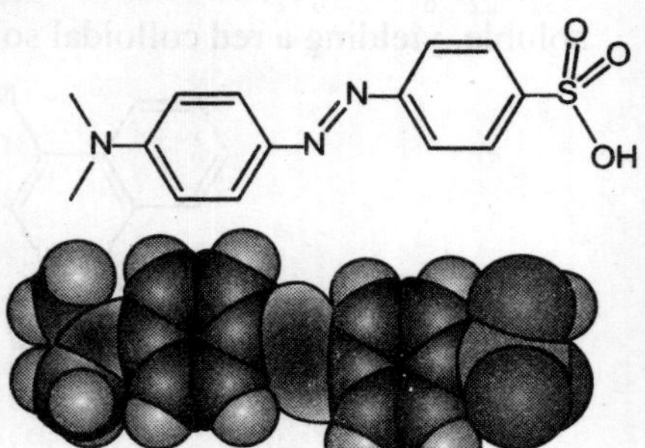

Methyl orange (pH indicator)		
below pH 3.1		*above pH 4.4*
3.1	↔	4.4
Methyl orange in xylene cyanole solution (pH indicator)		
below pH		*above pH*
3.2		4.2
3.2	↔	4.2

In a solution becoming less acidic, methyl orange moves from red.

Synthesis of Methyl Orange

Experimental Procedure - Methyl Orange Synthesis

1. Combine 4.35 g of anhydrous sulfanilic acid, 1.32 g of anhydrous sodium carbonate, and 50 mL of distilled water in a 125-mL conical flask. Warm the mixture until you obtain a homogeneous solution. Cool to room temperature. Make a solution of 1.85 g of sodium nitrite in 5 mL of water. Add this to the reaction mixture in the Erlenmeyer. Place 30 g of crushed ice in a 400-mL beaker, and add 5.25 mL of concentrated hydrochloric acid to it. Then pour the reaction mixture into the beaker with stirring. Finely divided crystals of the diazobenzene sulfonate will appear, but DO NOT isolate them. Keep the mixture in an ice bath and go on to the next step.

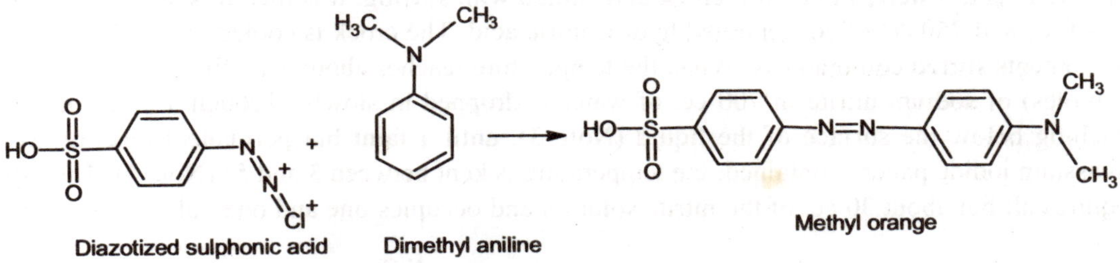

Diazotized sulphonic acid + Dimethyl aniline → Methyl orange

2. Make a solution of 3.0 g of N,N-dimethylaniline in 1.5 mL of acetic acid. Add this to the cold suspension of diazobenzene sulfonate while stirring vigorously. Then let the mixture stand for 15 minutes. (You should observe the formation of red crystals during this time). Add 17 mL of 20% NaOH solution to the cooled mixture. The color of the mixture will become orange, and a suspension of fine orange particles will appear. Heat the mixture, while stirring, to almost boiling. (Heat source = bunsen burner OR steam bath). Add 5 g of NaCl and warm the mixture until all of the salt has dissolved. Remove beaker from heat source and cool to room temperature. Then cool further in an ice water bath. Isolate crystals by suction filtration.

Methyl Red

Methyl Red, also called C.I. Acid Red 2, is an indicator dye that turns red in acidic solutions. It is an azo dye, and is a dark red crystalline powder. Methyl red is a pH indicator; it is red in pH under 4.4, yellow in pH over 6.2, and orange in between.

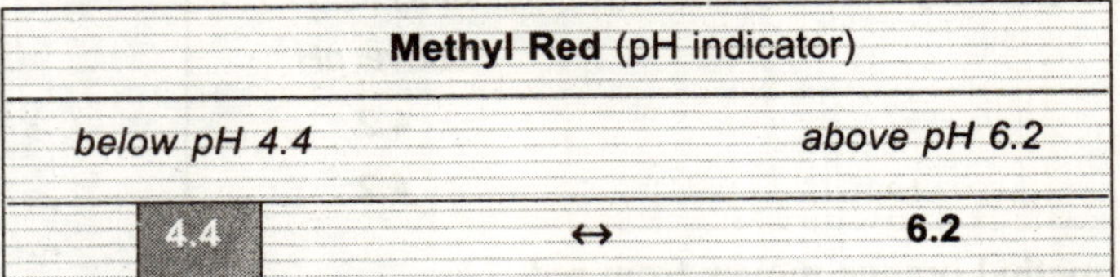

Methyl Red (pH indicator)		
below pH 4.4		*above pH 6.2*
4.4	↔	6.2

In microbiology, methyl red is used in the Methyl Red (MR) Test, used to identify bacteria producing stable acids by mechanisms of mixed acid fermentation of glucose. Cf. Voges-Proskauer (VP) test. Murexide and methyl red are investigated as promising enhancers of sonochemical destruction of chlorinated hydrocarbon pollutants.

Methyl Red Synthesis

Procedure

Six hundred and eighty-five grams (4.75 moles) of technical anthranilic acid (generally about 95 per cent pure) (Note 1) is dissolved in 1.5 l. of water and 500 cc. of concentrated hydrochloric acid (sp. gr. 1.17) (Note 2), by heating. The insoluble dark impurity present in small amounts is filtered off, and the filtrate is transferred to a 16-l. crock and chilled with stirring. It is then mixed with a mush of 2.5 kg. of ice and 750 cc. of concentrated hydrochloric acid. The crock is cooled externally with ice, and the contents stirred continuously. When the temperature reaches about 3°, a filtered solution of 360 g. (5 moles) of sodium nitrite in 700 cc. of water is dropped in slowly, through a long capillary tube reaching below the surface of the liquid (Note 3), until a faint but permanent reaction to starch-potassium iodide paper is obtained; the temperature is kept between 3 and 5° (Note 4). This operation requires all but about 30 cc. of the nitrite solution and occupies one and one-half to two hours.

To the solution of the diazonium salt is now added 848 g. (885 cc., 7 moles) of dimethylaniline; this may be done rapidly, as the temperature does not rise appreciably. Stirring is continued for one hour, the temperature being kept at 5°. Five hundred cubic centimetres of a filtered solution of 680 g (8.3

moles) of crystallized sodium acetate diluted to 1200 cc. is then added, and the stirring continued for four hours. If a foamy solid rises to the surface during this time and refuses to become incorporated by the stirrer, a few drops of ethyl acetate may be added to reduce the foam. The mixture is allowed to stand overnight in an ice bath which is well insulated by several thicknesses of burlap; the temperature must be kept below 7° to get a good yield of product. The remainder of the sodium acetate solution is then added with stirring, and, after the mixture has been stirred for an additional period of one to three hours, the temperature is allowed to rise slowly to 20-25° in the course of 24 hours. Just enough sodium hydroxide solution is then added, with stirring, to cause the mixture to have a distinct odor of dimethylaniline (about 240 cc. of a 40 per cent solution is generally required), and the mixture is allowed to stand for 48 hours or longer at room temperature (20-25°) (Note 5).

The solid is then filtered off, washed first with water, then with 400 cc. of 10 per cent acetic acid (to remove the dimethylaniline), and finally with distilled water. The last filtrate is generally pale pink. The solid is sucked as dry as possible, spread out on a tray in order to allow most of the water to evaporate (fifteen to twenty hours), and then suspended in 4 l. of methyl alcohol in a 12-litre. flask. This mixture is stirred on the steam bath under a reflux condenser for one to two hours, allowed to cool slowly, and then chilled in an ice bath and filtered. The solid product is washed with a second 4 litre. of cold methyl alcohol (Note 6). After being dried in air, it varies in weight from 820 to 870 g.

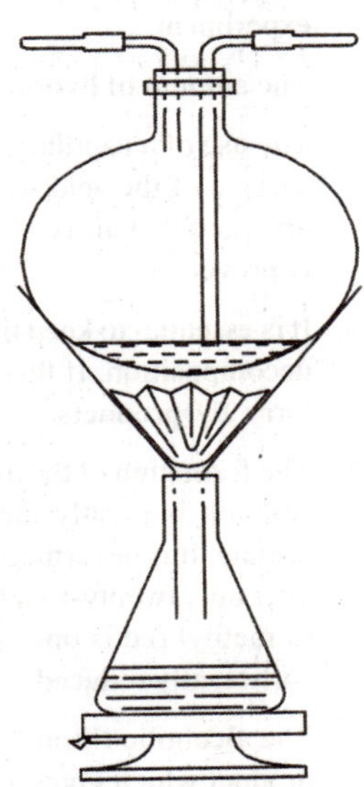

Fig. 28.1 Preparation of Methyl Rod

The product is extracted with boiling toluene in the following manner: 150 g. is placed in a fluted filter paper of 29-cm. diameter in a 25-cm glass funnel which passes through the cork of a 2-litre. flat-bottomed conical flask containing 1250 cc. of toluene (Fig. 28.1).

The flask is heated on an electric stove, and a 12-litre. round-bottomed flask, clamped to a ring stand, is placed on the funnel to act as a condenser, cold water being run through the flask. The toluene is boiled until the condensed liquid runs through almost colorless (this requires from four to ten hours). The heating is then discontinued, and, as soon as the liquid ceases to boil, the flask is removed to a bath containing water at 90-100°; the level of the water should be slightly above the level of the liquid in the flask. This arrangement permits the temperature to fall slowly so that large crystals are obtained. In the meantime a second conical flask containing 1250 cc. of toluene is attached to the funnel, and a new charge of 150 g. of crude methyl red is placed in the paper. The process is repeated until all the crude material has been so treated. When extraction is complete it is found that a certain amount of black amorphous insoluble matter remains on the filter; this is discarded. The crystals of methyl red (Note 7) are filtered off and washed with a little toluene. The weight of pure material is 755-805 g. The mother liquors are concentrated to one-fourth of their volume, and the crystals which separate on cooling are recrystallized from fresh toluene. The recovered toluene can, of course, be employed again. The total yield of pure methyl red is 790-840 g (62-66% of the theoretical amount) (Note 8). It melts at 181-182°.

The watery mother liquors from the crude methyl red are rendered alkaline with sodium hydroxide and distilled until no more dimethylaniline passes over. In this way 250 g to 400 g of moist dimethylaniline is recovered.

Notes

1. It is advisable to titrate the crude anthranilic acid with standard alkali and phenolphthalein before starting the experiment.
2. The amount of hydrochloric acid indicated must not be reduced; otherwise, diazoamino compounds are formed.
3. The use of a capillary tube for the addition of sodium nitrite prevents loss of nitrous acid by local reaction at the surface of the acid solution. The tube should not be tightly connected to the dropping funnel, but should be so arranged that air is sucked through with every drop. In this way, the entrance of the acid liquor into the capillary is prevented.
4. It is essential to keep the temperature low while unreacted diazobenzoic acid remains in solution, in order to avoid decomposition. If this precaution is not taken, the yields are considerably diminished, through the formation of tarry by-products.
5. The formation of the azo compound takes place slowly on the addition of the dimethylaniline, but the speed of the reaction is greatly increased when the hydrogen-ion concentration is lowered by the addition of the sodium acetate. It is nevertheless necessary to allow the reaction mixture to stand a long time; if the product is filtered off after only twenty-four hours, a further quantity of dye will separate from the filtrate on standing. The hydrochloride of methyl red is only sparingly soluble in cold water, and is apt to separate in blue needles if the acidity is not sufficiently reduced.
6. The alcoholic filtrate, obtained on digesting and washing the crude methyl red, contains a more soluble red by-product which gives a brownish-yellow solution in alkali. The methyl alcohol may be recovered with very little loss by distillation; it is, however, impracticable to attempt to recover any methyl red from the residue, owing to the tarry nature of the by-product. The proportion of this by-product is greatly increased if the temperature of the mixture is allowed to rise too soon after the addition of the sodium acetate.
7. Methyl red is described as crystallizing in needles from glacial acetic acid; on recrystallization from toluene it separates in plates.

   When the methyl red is crystallized from toluene, it sometimes separates in the form of bright-red lumps, probably on account of too rapid crystallization. Under these conditions it is advisable to crystallize again, using a somewhat larger amount of toluene. The sodium salt may be prepared by dissolving the crude product in an equal weight of 35 per cent sodium hydroxide which has been diluted to 3 or 4 litres, filtering, and evaporating under diminished pressure. The isolation of methyl red as the sodium salt avoids the toluene extraction, and yields a water-soluble product (orange leaflets) which is very convenient for use as an indicator.
8. In checking these directions, an 80 per cent anthranilic acid was used; it gave a correspondingly lower yield of methyl red (650-700 g.).

Discussion

Methyl red can be prepared by diazotization of anthranilic acid in alcoholic solution, the product being allowed to react with dimethylaniline in the same solvent, or both operations can be carried out in aqueous solution.

Methyl Yellow

Methyl Yellow, or **C.I. 11020**, is a chemical compound which may be used as a pH indicator.

Methyl yellow (pH indicator)		
below pH 2.9		*above pH 4.0*
2.9	↔	4.0

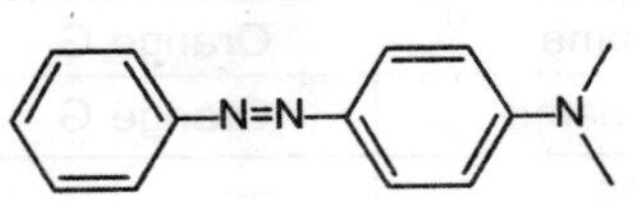

Synonyms:

- 4-dimethylaminoazobenzene
- N,N-dimethyl-4-aminoazobenzene
- methyl yellow
- *p*-dimethylaminoazobenzene
- butter yellow
- aniline yellow

In aqueous solution at low pH, methyl yellow appears red. Between pH 2.9 and 4.0, methyl yellow undergoes a transition, to become yellow above pH 4.0.

Oil Red O

Property	Data
Common name	Oil red O
Suggested name	Oil red O
Other names	—
C.I. number	26125
C.I. name	Solvent red 27
Class	Azo
Ionisation	Lysochrome
Solubility aqueous	Insoluble
Solubility ethanol	Moderate
Absorption maximum	518 (359)
Colour	Red
Empirical formula	$C_{26}H_{24}N_4O$
Formula weight	408.51

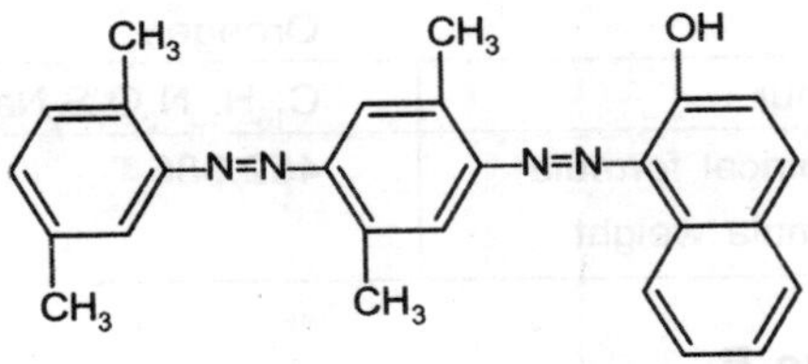

Oil red O is a lysochrome (fat soluble dye) predominantly used for demonstrating triglycerides in frozen sections, but which may also stain some protein bound lipids in paraffin sections. It has largely replaced sudan III and sudan IV as it is much deeper red in colour, and consequently more clearly visible.

Compare this dye to the other common lysochromes sudan III, sudan IV and sudan black B.

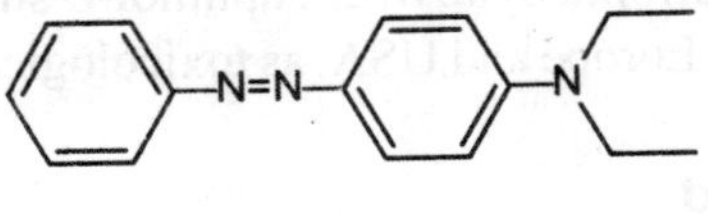

Oil Yellow DE

Oil Yellow DE, also known as Solvent Yellow 56 and C.I. 11021, is a synthetic greenish-yellow azo dye. It has the appearance of reddish yeloow powder with melting point of 168°C. Chemically it is N, N-diethyl-p=(phenylazo) aniline, $C_{16}H_{19}N_3$.

It is used to dye hydrocarbon solvents, oils, fats and waxes, notably petrol, mineral oil and shoe polishes and polystyrene resings. In pyrotechnics, it is used in some yellow colored smokes.

Orange G

Property	Data
Common name	Orange G
Suggested name	Orange G
Other names	—
C.I. number	16230
C.I. name	Acid orange 10
Class	Azo
Ionisation	Acid
Solubility	10.86%
aqueous	0.22%
Solubility ethanol	472 (Gurr), 475
Absorption maximum	(Aldrich), 476-480 (Conn) Orange
Colour	$C_{16}H_{10}N_2O_7S_2Na_2$
Empirical formula	452.386
Formula weight	

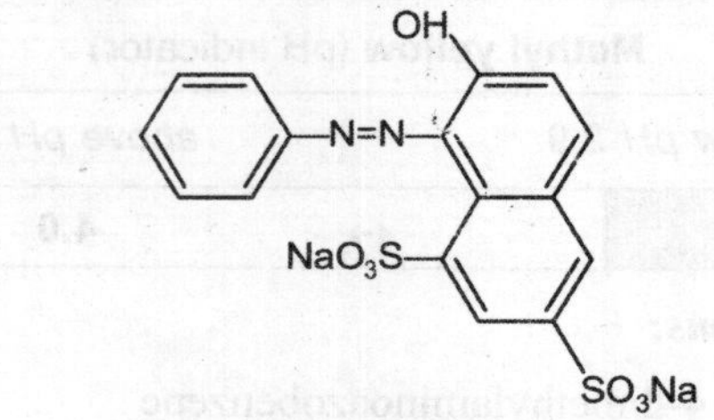

Orange G is a valuable acid dye used in many staining methods, including Papanicolauo's OG6 stain. It is often combined with other yellow dyes in alcoholic solution to stain erythrocytes in trichrome methods, and is used for demonstrating cells in the pancreas and pituitary.

Orange B

Orange B is a food dye from the azo dye group. It is approved by Food and Drug Administration (FDA) for use only in hot dog and sausage casings or surfaces, only up to 150 ppm of the finished food weight. It usually comes as disodium salt.

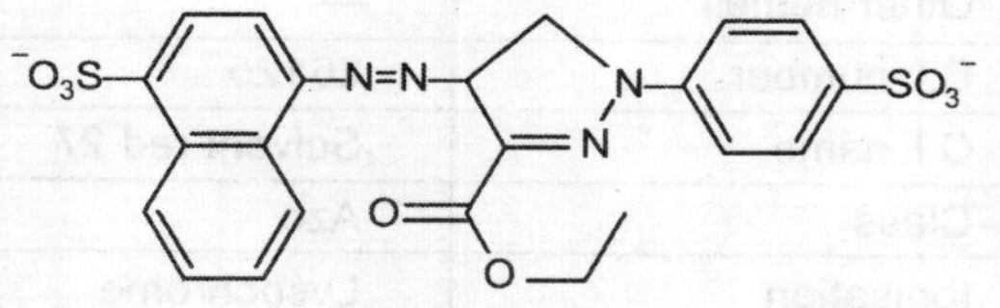

Orange GGN

Orange GGN, also known as alpha-naphthol orange, is a food dye. In Europe it is denoted by the E Number E111. It is the disodium salt of 1-(m-sulfophenylazo)-2-naphthol-6-sulfonic acid. It is currently delisted in Europe and USA, as toxicological data shown it is harmful.

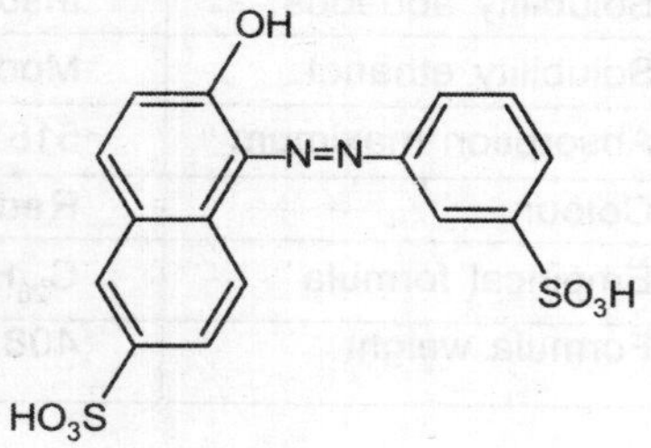

Para Red

Para Red (paranitraniline red, Pigment Red 1, C.I. 12070, Recolite Para Red B, Carnelio Para Red BS) is a chemical dye. Chemically, the dye is similar to Sudan I. The dye was discovered in 1880 by von Gallois and Ullrich, and was the first azo dye. It dyes cellulose fabrics a brilliant red, but is not very fast.

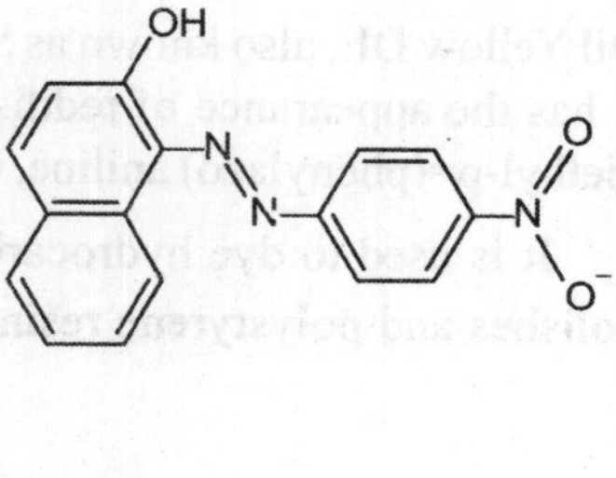

Ponceau 2R

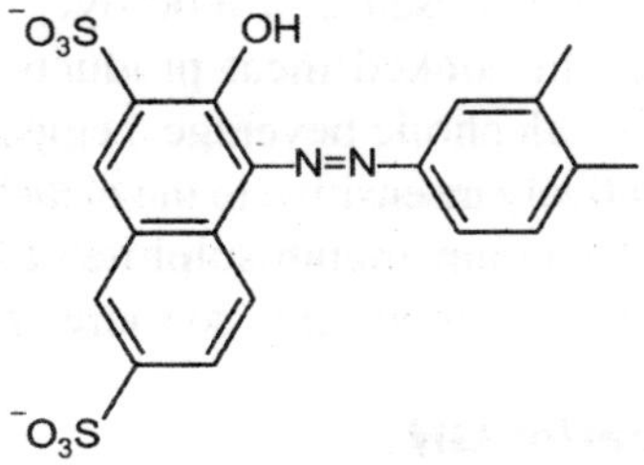

Ponceau 2R, Xylidine ponceau, Ponceau G, Red R, Acid Red 26, Food Red 5, or C.I. 16150 is a red azo dye used in histology for staining. It is easily soluble in water and slightly in ethanol. It usually comes as a disodium salt.

Xylidine ponceau is mostly used only in the Masson's trichrome stain as a red counterstain, where it imparts orange hue to the red-stained cytoplasmic structures.

Ponceau 4R

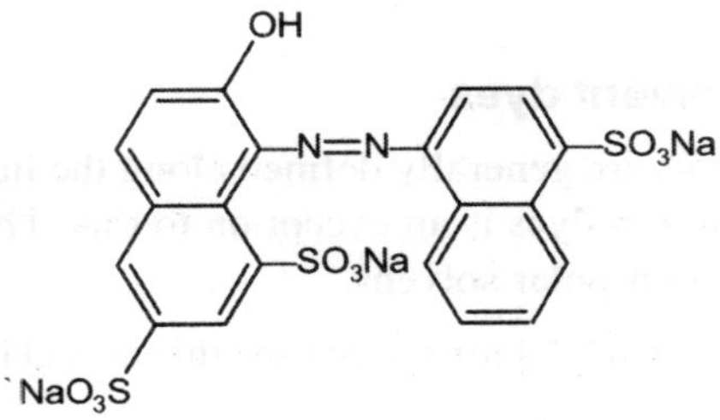

Ponceau 4R (also known as Food Red 7, C.I. 16255, Cochineal Red A, New Coccine, Acid Red 18, SX purple) is a colourant that may be added to foods to induce a colour change. It is denoted by E Number E124, and has the capacity for inducing an allergic reaction. Its chemical name is trisodium salt of 1-(4-sulpho-1-napthylazo)- 2-napthol- 6,8-disulphonic acid.

Ponceau 4R is a synthetic coal tar and red azo dye which can be used in a variety of food products.

Since it is an azo dye, it may elicit intolerance in people allergic to salicylates (aspirin). Additionally, it is a histamine liberator, and may intensify symptoms of asthma.

Ponceau 6R

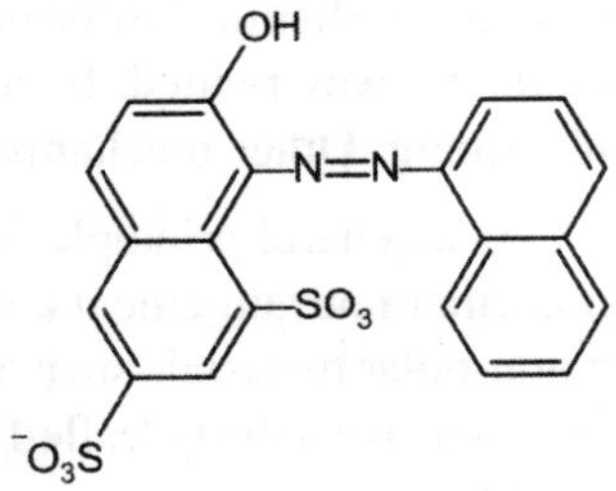

Ponceau 6R, or Crystal ponceau 6R, Crystal scarlet, Brilliant crystal scarlet 6R, Acid Red 44, or C.I. 16250, is a red azo dye. It is soluble in water and slightly soluble in ethanol. It is used as a food dye, with E number E126. It is also used in histology, for staining fibrin with the MSB Trichrome stain. It usually comes as disodium salt.

Ponceau S

Ponceau S, Acid Red 112, or C.I. 27195 is a sodium salt of a diazo dye that may be used to prepare a stain for rapid reversible detection of protein bands on nitrocellulose or PVDF membranes (Western blotting), as well as on cellulose acetate membranes. A Ponceau S stain is easily reversed with water washes, facilitating subsequent immunological detection. Common stain formulations include 0.1% (w/v) Ponceau S in 5% acetic acid or 2% (w/v) Ponceau S in 30% TCA and 30% sulfosalicylic acid.

Red G or Red 2G

Red 2G, Acid Red 1, Food Red 10, Amidonaphthol red G, azogeranine, azophloxine, azofloxin, or C.I. 18050, is a synthetic red azo dye. It is soluble in water and slightly soluble in ethanol. It usually comes as a disodium salt of 8-actamido-1-hydroxy-2-phenylazonaphthalene-3,6 disulphonate.

When used as a food dye, its E number is E128. It is mainly used in cooked meat products and sausages, and also in some jams, alcoholic beverages, eggs, and milk and dairy products. It is relatively insensitive to the bleaching effect of sulfur dioxide (E220) and sodium metabisulphite (E223). In the intestines it can be converted to aniline, so there are concerns Red 2G

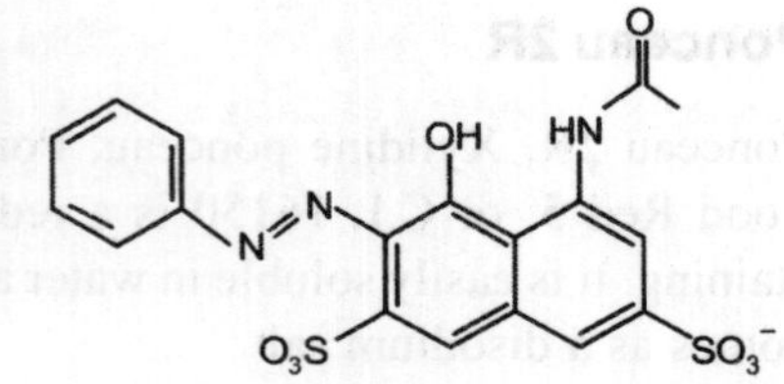

Scarlet GN

Scarlet GN, or C.I. Food Red 2, Ponceau SX, FD&C Red No. 4, or C.I. 14700 is a red azo dye used as a food dye. When used as a food additive, it has the E number E125. It usually comes as a disodium salt.

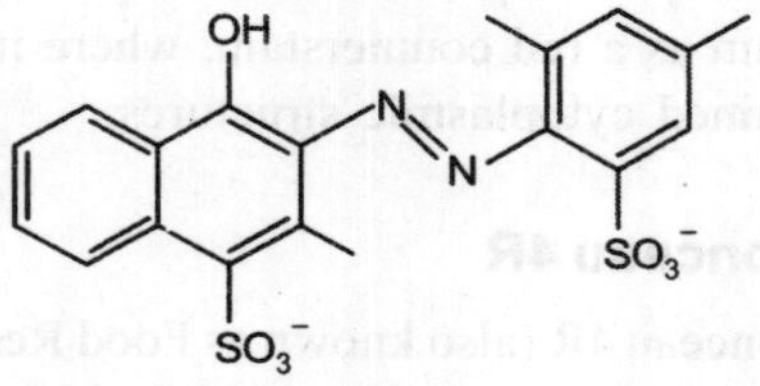

Solvent dyes

Dyes are generally defined along the lines of being coloured, aromatic compounds that can ionise. One class of dyes is an exception to this. This colour by dissolving in the target, which is invariably a lipid or non-polar solvent.

The Colour Index uses this as a classification and naming system. Each dye is named according to the pattern:

solvent + base colour + number

These dyes are thereby specifically identified as dyes of the stated colour, and whose primary mechanism of staining is by dissolving in the target. Note that this is a functional and colour classification. It contains no chemical information, neither does it imply that dyes with similar names but unique numbers are in any way related. It should also be noted that the classification refers to the primary mechanism of staining. Other mechanisms may also be possible, but are rare.

As a general principle, solvent dyes do not ionise. Many are azo dyes which have undergone some molecular rearrangement and lost the ability to ionise. In the process they gained the ability to dissolve in non-polar materials such as triglycerides. They are commonly used to stain such materials in sections. They are frequently called lysochrome dyes. The prefix lyso means dissolve, and chrome means colour.

Sudan III (solvent red 23), sudan IV (solvent red 24) and oil red O (solvent red 27) are commonly used for demonstrating fat in sections. Sudan black B (solvent black 3) is also very effective, but can also stain ionically under some circumstances.

Solvent Red 26

Solvent Red 26, also known as C.I. 26120, is a purplish red synthetic azo dye. Its chemical formula is $C_{25}H_{22}N_4O$, or 1-[[2,5-dimethyl-4-[(2-methylphenyl)azo]-phenyl]azo]-2-naphthol. It is soluble in oils and insoluble in water.

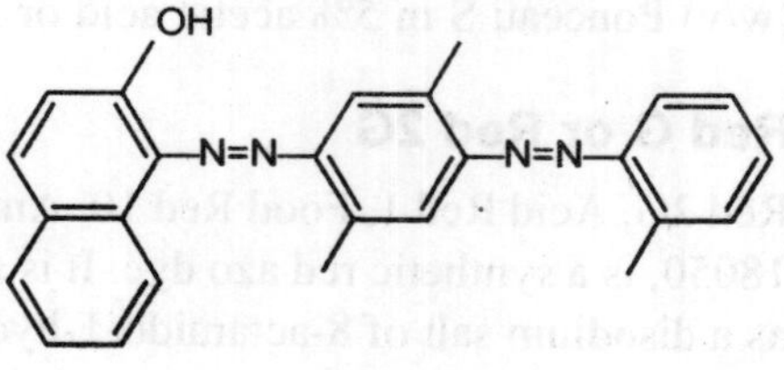

Its main use is as a standard fuel dye in the United States of America mandated by the IRS to distinguish low-taxed heating oil from automotive diesel fuel, and by the EPA to mark fuels with higher sulfur content; it is however increasingly replaced with Solvent Red 164, a similar dye with longer alkyl chains, which is better soluble in hydrocarbons. The concentration required by IRS is a spectral equivalent of 3.9 pounds per 1000 barrels, or 11.13 mg/l, of Solvent Red 26 in solid form; the concentrations required by EPA are roughly 5 times lower.

Solvent Red 164

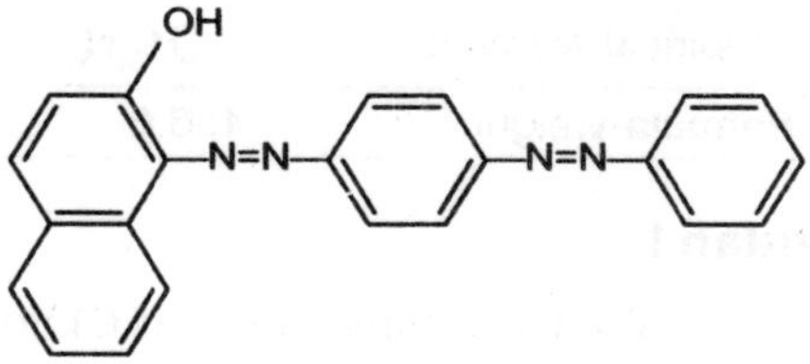

Solvent Red 164, also called Oil Red B, is a synthetic red diazo dye. Its chemical structure is 1-[[4-[phenylazo]-phenyl]azo]-2-naphthol. The inventors of the product were Morton International under the commercial name Automate Red B.

Its main use is as a fuel dye in the United States of America mandated by the IRS to distinguish low-taxed heating oil from automotive diesel fuel, and by the EPA to mark fuels with higher sulfur content; it is a replacement for Solvent Red 26 with better solubility in hydrocarbons. The concentration required by IRS is a spectral equivalent of 3.9 pounds per 1000 barrels, or 11.13 mg/l, of Solvent Red 26 in solid form; the concentrations required by EPA are roughly 5 times lower.

Solvent Yellow 124

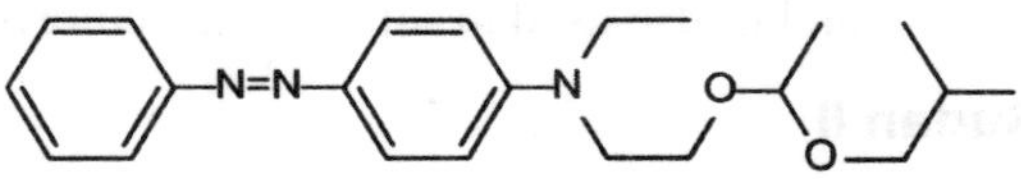

Solvent Yellow 124, also called SY124, Sudan 455, Somalia Yellow, T10 Yellow LBN, is a yellow azo dye used in European Union as a fuel dye. It is a marker used since August 2002 to distinguish diesel fuel intended for heating from a higher-taxed motor diesel fuel. It is added to fuels not intended for motor vehicles in amounts of 6 mg/l, or 7 mg/kg, under the name Euromarker.

Solvent Yellow 124 is a dye with structure similar to Solvent Yellow 56. This dye can be easily hydrolyzed with acids, splitting off the acetal group responsible for its solubility in nonpolar solvents, and yielding a water-soluble form which is easy to extract to water. Like a similar methyl orange dye, it changes color to red in acidic pH. It can be easily detected in the fuel at levels as low as 0.3 ppm by extraction to a diluted hydrochloric acid, allowing detection of the red diesel added into motor diesel in amounts as low as 2-3%.

Sudan Black B

Property	Data
Common name	Sudan black B
Suggested name	Sudan black B
Other names	Fat black HB
C.I. number	26150
C.I. name	Solvent black 3
Class	Azo

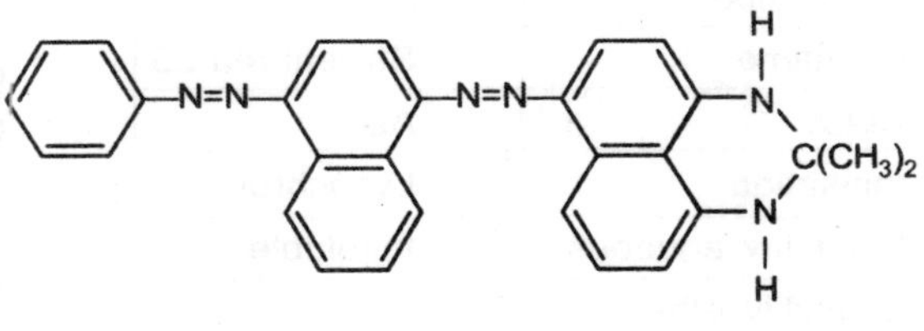

Ionisation	Lysochrome
Solubility aqueous	Insoluble
Solubility ethanol	Moderate
Absorption maximum	596-605 (Conn)
	598, 415 (Aldrich)
Colour	Blue-black
Empirical formula	$C_{29}H_{24}N_6$
Formula weight	456.6

Sudan black B is a lysochrome (ft soluble dye) predominantly used for demonstrating triglycerides in frozen sections. It is also valuable for demonstrating someprotein bound lipids in parffin sections. It may also stain other materials, not being completely restricted to lipids as the other dyes used.

Compare this dye to the other common lysochromes sudan III, sudan IV and oil red O.

Sudan I

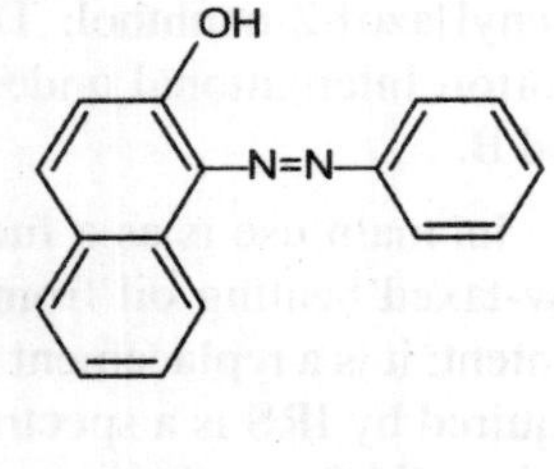

Sudan I (also commonly known as CI Solvent Yellow 14 and Solvent Orange R), is a lysochrome, an azo dye with a chemical formula of 1-phenylazo-2-naphthol. Sudan I is a powdered substance with an orange-red appearance. The additive is mainly used to colour waxes, oils, petrol, solvents and polishes. Sudan I has also been adopted for colouring various foodstuffs, including particular brands of curry powder and chili powder, although the use of Sudan I in foods is now banned in many countries due to inconclusive reports on its possible health risks.

Sudan I is also used in some orange colored smoke formulations.

Sudan II

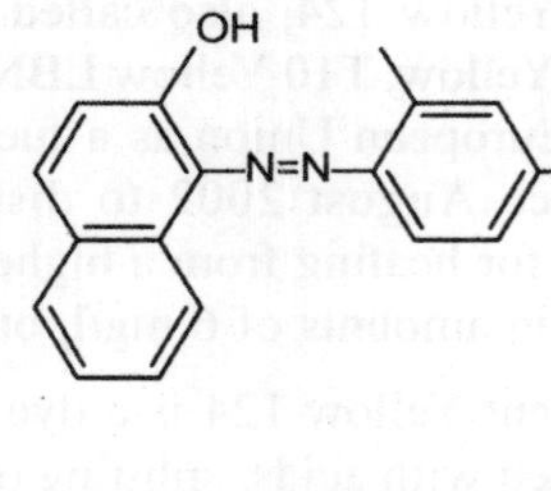

Sudan II ($C_{18}H_{16}N_2O$) is a lysochrome (fat-soluble dye) azo dye used for staining of triglycerides in frozen sections, and some protein bound lipids and lipoproteins on paraffin sections. It has the appearance of red powder with melting point 156-158°C and maximum absorption at 493(420) nm.

Sudan III

Property	Data
Common name	Sudan III
Suggested name	Sudan III
Other names	Sudan red BK
C.I. number	26100
C.I. name	Solvent red 23
Class	Azo
Ionisation	Lysochrome
Solubility aqueous	Insoluble
Solubility ethanol	0.15%

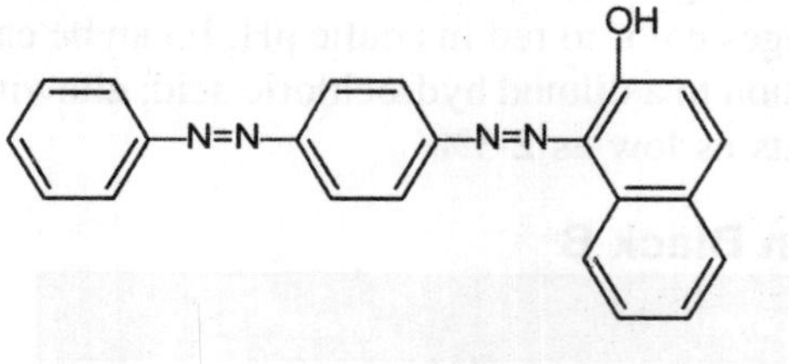

Sudan III is lysochrome (fat soluble dye) predominantly used for demonstrating triglycerides in frozen sections, but which may also stain some protein bound lipids in paraffin sections. It is less popular than oil red O as it has a more orange shade.

Absorption maximum	508-510 (Conn) 507,354 (Aldrich) 503 (Gurr)
Colour	Red
Empirical formula	$C_{22}H_{16}N_4O$
Formula weight	352.4

Compare this dye to the other common lysochoromes sudan IV, oil red O and sudan black B.

Sudan IV

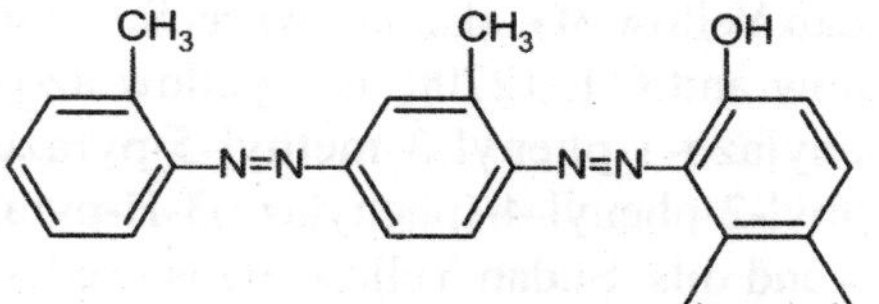

Property	Data
Common name	Sudan IV
Suggested name	Sudan IV
Other names	Scarlet R or red Scharlach R Biebrich scarlet R
C.I. number	26105
C.I. name	Solvent red 24
Class	Azo
Ionisation	Lysochrome
Solubility aqueous	Insoluble
Solubility ethanol	0.09%
Absorption maximum	522-529 (Conn) 513-515 (Gurr) 520, 357 (Aldrich)
Colour	Red
Empirical formula	$C_{24}H_{20}N_4O$
Formula weight	380.5

CH3 CH3 OH
N=N N=N

Sudan IV is a lysochrome (fat soluble dye) predominantly used for demonstrating triglycerides in frozen sections, but which may also stain some protein bound lipids inparaffin sections. It is les popular than oil red O as it has a more orange shade. Compare this dye to the other common lysochromes sudan III, oil red O and sudan black B. Gurr notes that a purified form of this dye hs the name "Biebrich scarlet R". It should not be confused with the water soluble dye which is much more commonly referred to as Biebrich scarlet. This latter dye is not a lysochrome, but is used as a plasma stain in trichrome techniques.

Sudan Red 7B

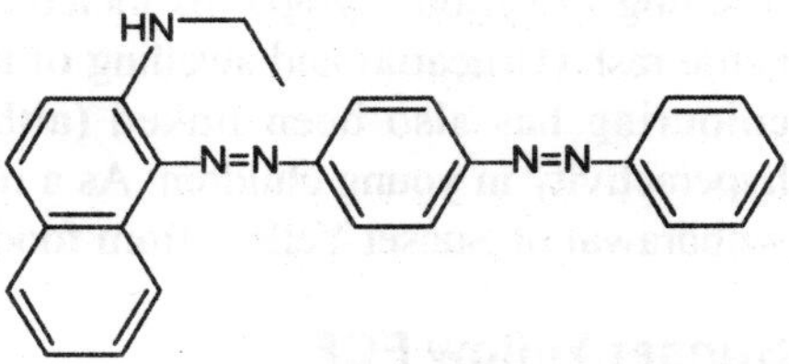

Sudan Red 7B, also known as Solvent Red 19, Ceres Red 7B, Fat Red 7B, Hexatype carmine B, Lacquer red V3B, Oil violet, Organol bordeaux B, Sudanrot 7B, Typogen carmine, and C.I. 26050, is a red diazo dye. Chemically it is N-ethyl-1-[[p-(phenylazo)phenyl]azo]-2-naphthalenamine. It is soluble in oils and insoluble in water.

HN
N=N N=N

Sudan Red G

OH
N=N
O

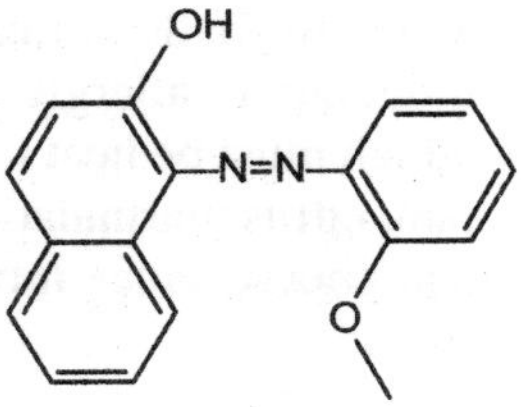

Sudan Red G is a yellowish red lysochrome azo dye. It has the appearance of an odorless reddish-orange powder with melting point 225°C. It is soluble in fats and used for coloring of fats, oils, and waxes, including the waxes used in turpentine-based polishes. It is also used in polystyrene, cellulose, and synthetic lacquers. It is insoluble in water. It is stable to temperatures of about 100-110 °C. It was used as a food dye. It is used in some temporary tattoos, where it can cause contact dermatitis. It is also used in hair dyes. It is a component of some newer formulas for red smoke signals and smoke-screens, together with Disperse Red 11.

Sudan Yellow 3G

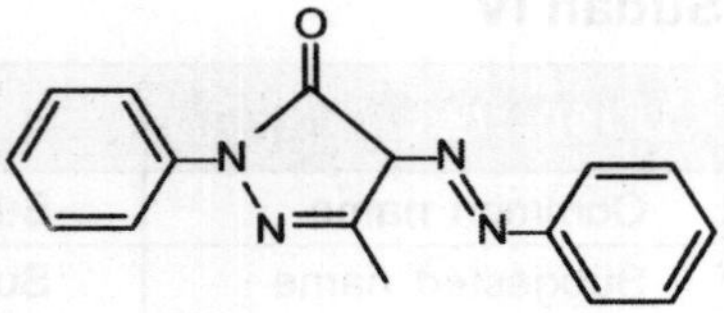

Sudan Yellow 3G, also known as Solvent Yellow 16, C.I. disperse yellow and C.I. 12700, is a yellow azo dye. Chemically it is 4-phenylazo-1-phenyl-3-methyl-5-pyrazolone or 2,4-dihydro-5-methyl-2-phenyl-4-(phenylazo)3-H-pyrazol-3-on. It is soluble in fats and oils. Sudan Yellow 3G is used as a pigment in cosmetics and printer toners, and as a dye in inks, including inks for inkjet printers. In pyrotechnics, it is used in some yellow coloured smokes.

Sunset Yellow is an azo dye used as artificial food colouring with E number E110, and in drugs and cosmetics. It is also known as A.F. Yellow #5, Acid Yellow tra, Sunset Yellow FCF, Certolake Sunset Yellow, C.I. Food Yellow 3, Edicol Supra Yellow FC, C.I. 53907, Orange Pal, Sun Yellow, 1351 Yellow, 1899 Yellow, FD&C Yellow #6, and Yellow Sun. It has the appearance of orange-red crystallic powder, decomposing at about 390°C, well-soluble in water (50-100 mg/ml at 24°C).

Sunset Yellow

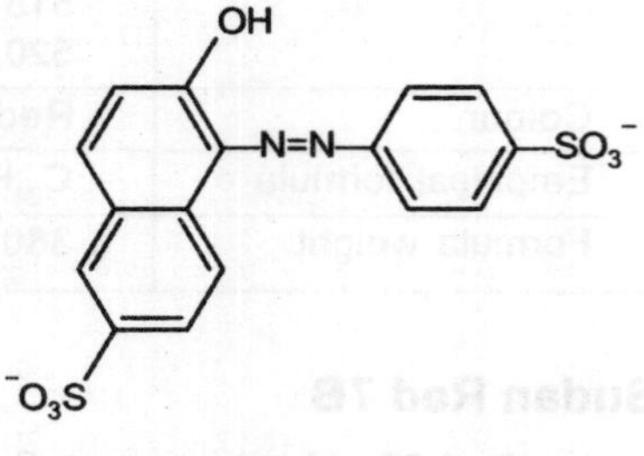

Sunset Yellow is most commonly used as a disodium salt of 1-p-sulfophenylazo-2-naphthol-6-sulfonic acid, however both the potassium and calcium salts are also used, if to a much lesser degree. Sunset Yellow is a sulfonated version of Sudan I, which is frequently present in it as an impurity. Sunset Yellow itself may be responsible for causing an allergic reaction in people with an aspirin intolerance, resulting in various symptoms including gastric upset, vomiting, nettle rash (Urticaria) and swelling of the skin (Angiodema). The colouring has also been linked (although inconclusively) to hyperactivity in young children. As a result of these problems, there have been repeated calls for the total withdrawal of Sunset Yellow from food use.

Sunset Yellow FCF

Sunset Yellow FCF (walso known as Orange Yellow S, and FD&C Yellow 6) is a colourant that may be added to foods to induce a colour change. It is denoted by E Number E 110, and has the capacity for inducing an allergic reaction. It is a synthetic coal tar and azo yellow dye useful in fermented foods which must be heat treated. It may be found in orage squash, orange jelly, marzipan, Swiss roll, apricot jam, citrus marmalade, lemon curd, sweets, hot chocolate mix and packet soups, trifle mix, bredcrumbs and cheese sauce mix and soft drinks.

Sunset Yellow FCF	
Chemical name	disodium salt of 6-hydroxy-5 -[(4-sulfophenyl)azo]-2 -naphthalenesulfonic acid
Chemical formula	$C_{16}H_{10}Na_2O_7S_2$
Molecular mass	424.35 g/mol
Melting point	300 °C

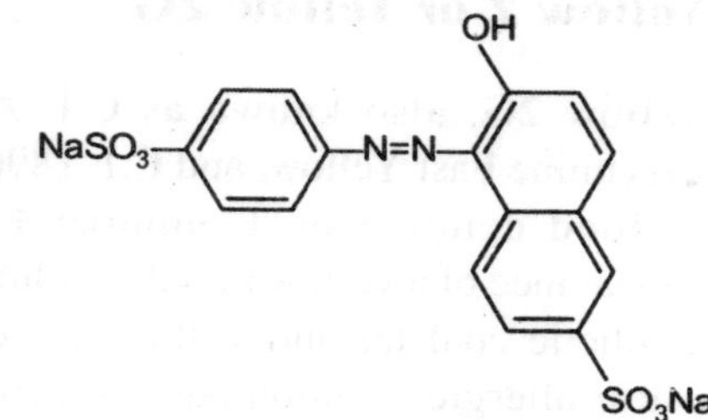

Tartrazine

Tartrazine (otherwise known as E102 or FD&C Yellow 5) is a synthetic lemon yellow azo dye used as a food colouring. It is derived from coal-tar. It is found in certain brands of fruit squash, fruit cordial, coloured soft drinks, instant puddings, cake mixes, custard powder, soups, sauces, kool-aid (medicinal), ice cream, ice lollies, sweets, chewing gum, marzipan, jam, jelly, marmalade, mustard, yogurt and many convenience foods together with glycerin, lemon and honey products.It may be found in the shells of medicinal capsules. It can also be used with E133 Brilliant Blue FCF or E142 Green S to produce various green shades (e.g. for tinned processed peas). It appears to cause the most allergic and intolerance reactions of all the azo dyes, particularly amongst those with an aspirin intolerance and asthmatics. Reactions can include migraine, blurred vision, itching, rhinitis, urticaria and purple skin patches (because of this more use is now being made of E160b Annatto).

Trypan Blue

Trypan blue is a vital stain that colours dead tissues or cells blue. It is a diazo dye.Liver cells or tissues with intact cell membranes will not be coloured. Since cells are very selective in the compounds that pass through the membrane, in a viable cell trypan blue is not absorbed; however, it traverses the membrane in a dead cell. Hence, dead cells are shown as a distinctive blue colour under a microscope.

Yellow 2 or Yellow 2G

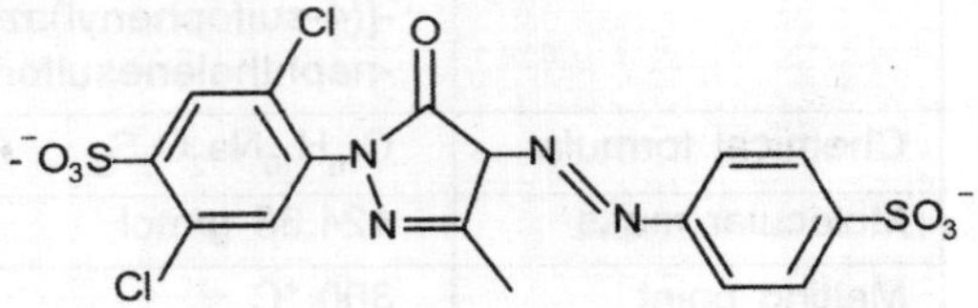

Yellow 2G, also known as C.I. Acid Yellow 17, Lissamine Fast Yellow, and C.I. 18965, is a colourant in food denoted by E number E107. It has the appearance of a yellow powder, soluble in water.It is a synthetic coal tar and yellow azo dye.It appears to cause allergic or intolerance reactions, particularly amongst those with an aspirin intolerance and asthma sufferers.It is one of the colours that the Hyperactive Children's Support Group recommends be eliminated from the diet of children.

Metanil Yellow

Property	Data
Common name	Metanil yellow
Suggested name	Metanil yellow
Other names	Tropaeolin G
C.I. number	13065
C.I. name	Acid yellow 36
Class	Azo
Ionisation	Acid
Solubility aqueous	5.4%
Solubility ethanol	1.4%
Absorption maximum	536 (Conn), 435 (Gurr), 414 (Aldrich)
Colour	Yellow
Empirical formula	$C_{18}H_{14}N_3O_3SNa$
Formula weight	375.391

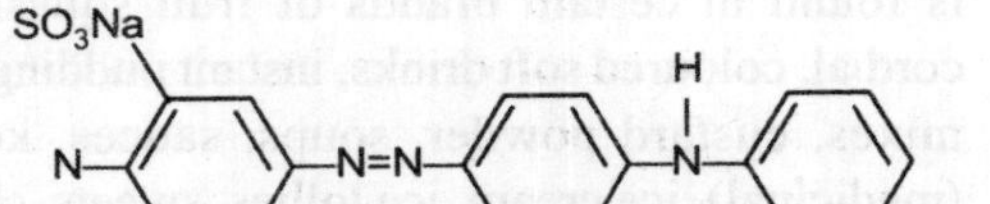

Metanil yellow has been used in some trichrome methods, including one by Masson, to stain collagen yellow in contrast to the red plasma stain. It has been used in solution with other small molecular weight, yellow, acid dyes to increase the intensity of staining.

Alcian Yellow

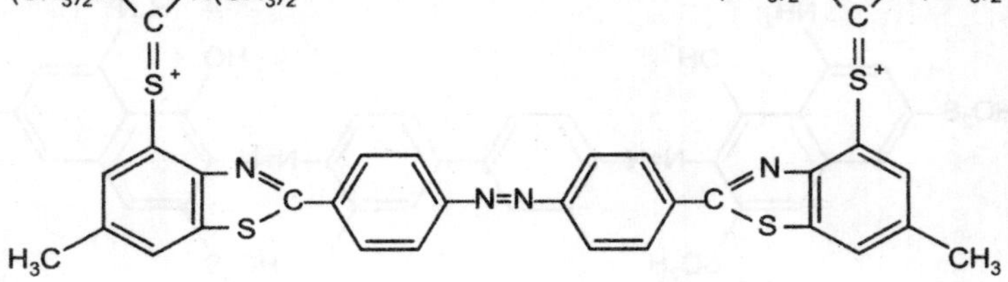

Property	Data
Common name	Alcian yellow
Suggested name	Alcian yellow
Other names	Ingrain yellow
C.I. number	12840
C.I. name	Ingrain yellow 1
Class	Azo
Ionisation	Basic
Solubility aqueous	—
Solubility ethanol	—
Absorption maximum	388
Colour	Yellow
Empirical formula	$C_{40}H_{46}N_8S_4Cl_2$
Formula weight	838.03

Alcian yellow is not a commonly used dye, but is similar in staining characteristics to alcian blue, and can stain acid mucins yellow. It is recommended in the Leung & Gibbon method for helicobacter, where it contrasts well with the blue organisms. It has been difficult to obtain of late, although it is still manufactured.

Azophloxine

Property	Data
Common name	Azophloxine
Suggested name	Azophloxine
Other names	Amidonaphthol red G
C.I. number	18050
C.I. name	Acid red 1
Class	Azo
Ionisation	Acid
Solubility aqueous	Soluble
Solubility ethanol	Slightly
Absorption maximum	540,502(Conn), 532,506 (Aldrich)
Colour	Red
Empirical formula	$C_{18}H_{13}N_3O_8S_2Na_2$
Formula weight	509.44

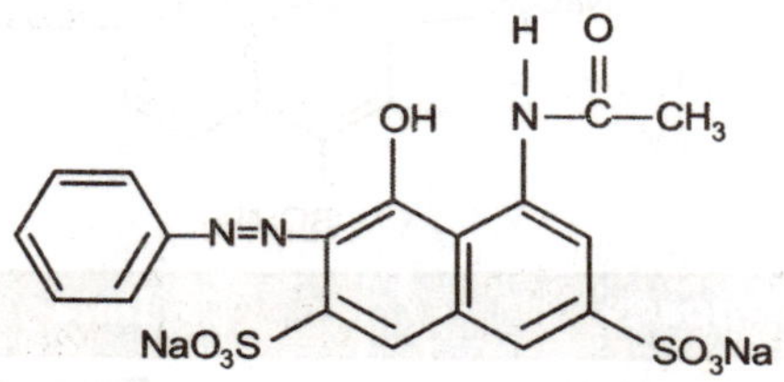

H O
OH N—C—CH$_3$
—N=N—
NaO$_3$S SO$_3$Na

This dye is rarely called for, although Goldner suggested it as an alternate plasma stain in his trichrome.

Azo-eosin

Property	Data
Common name	Azo-eosin
Suggested name	Azo-eosin
Other names	Azoeosin G
C.I. number	14710
C.I. name	Acid red 4
Class	Azo
Ionisation	Acid
Solubility aqueous	Moderate
Solubility ethanol	Slight
Absorption maximum	509 (Gurr) 508, 376 (Aldrich)
Colour	Red
Empirical formula	$C_{17}H_{13}N_2O_5SNa$
Formula weight	380.36

OCH$_3$ OH N=N SO$_3$Na

This dye has been used in histology only rarely. Lendrum, Slidders and Fraser recommended it for their standardised trichrome method, considering it to be a suitable plasma stain after silver impregnations.

It should be noted that it is not related to eosin, although it is an azo dye. The name presumably reflects a similarity in colour.

Evans Blue

NH$_2$ OH NaO$_3$S N=N CH$_3$ SO$_3$Na OH NH$_2$ N=N SO$_3$Na CH$_3$ SO$_3$Na

Property	Data
Common name	Evans blue
Suggested name	Evans blue
Other names	Azovan blue
C.I. number	23860
C.I. name	Direct blue 53
Class	Azo
Ionisation	Acid
Solubility aqueous	Soluble
Solubility ethanol	Soluble
Absorption maximum	611
Colour	Blue
Empirical formula	$C_{34}H_{24}N_6O_{14}S_4Na_4$
Formula weight	960.842

This dye is listed as a teratogen by Aldrich. It is sometimes used as a counterstain, especially in fluorescent methods to suppress background autofluorescence.

Xylidine Ponceau

Property	Data
Common name	Xylidine ponceau
Suggested name	Xylidine ponceau
Other names	Ponceau de xylidine Ponceau 2R
C.I. number	16150
C.I. name	Acid red 26
Class	Azo
Ionisation	Acid
Solubility aqueous	Readily
Solubility ethanol	Slightly
Absorption maximum	499 (Conn), 502 (Gurr) 503, 388 (Aldrich)
Colour	Red
Empirical formula	$C_{18}H_{14}N_2O_7S_2Na_2$
Formula weight	480.44

CH3 HO SO3Na H3C N=N SO3Na

Masson specified a dye he called Ponceau de xylidine for a variant of his trichrome staining method. Although this dye has not been absolutely identified as the dye he used, it is almost certainly the one, and is accepted as such by general consensus. Its use is largely confined to Masson's trichrome, where it gives a slight orange shading to the red of the cytoplasmic structures.

Chromotrope 2R

Property	Data
Common name	Chromotrope 2R
Suggested name	Chromotrope 2R
Other names	Carmoisine 6R
C.I. number	16570
C.I. name	Acid red 29
Class	Azo
Ionisation	Acid
Solubility aqueous	19.3%
Solubility ethanol	0.17%
Absorption maximum	503,542 (Conn) 510,530 (Aldrich)
Colour	Red
Empirical formula	$C_{16}H_{10}N_2O_8S_2Na_2$
Formula weight	468.386

OH OH N=N NaO3S SO3Na

Chromotrope 2R may be used as the plasma stain in trichrome techniques, most notably in Gomori's one-step method. It can replace xylidine ponceau in Masson's trichrome.

Biebrich Scarlet

Property	Data
Common name	Biebrich scarlet
Suggested name	Biebrich scarlet
Other names	Croceine scarlet
	Ponceau B or BS
C.I. number	26905
C.I. name	Acid red 66
Class	Azo
Ionisation	Acid
Solubility aqueous	Soluble
Solubility ethanol	0.05%
Absorption maximum	504 (Conn), 506 (Gurr)
	505 (Aldrich)
Colour	Red
Empirical formula	$C_{22}H_{14}N_4O_7S_2Na_2$
Formula weight	556.5

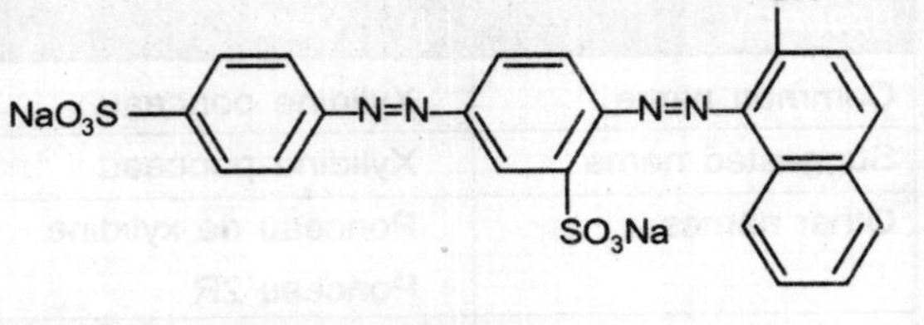

Biebrich scarlet is sometimes used as a plasma stain instead of Acid fuchsin in Masson's trichrome.

This dye should not be confused with the purified form of sudan III, referred to as biebrich scarlet R.

Woodstain Scarlet

Property	Data
Common name	Woodstain scarlet
Suggested name	Woodstain scarlet
Other names	Crocein scarlet 3B or MOO
C.I. number	Brilliant crocein MOO
C.I. name	27290
Class	Acid red 73
Ionisation	Azo
Solubility aqueous	Acid
Solubility ethanol	Moderate
Absorption	Slight
maximum	510 (Aldrich), 513 (Gurr)
Colour	Red
Empirical formula	$C_{22}H_{14}N_4O_7S_2Na_2$
Formula weight	556.5

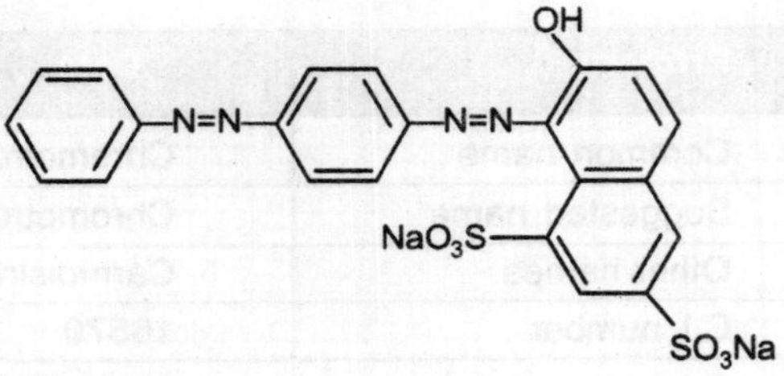

Woodstain scarlet has been recommended for the demonstration of muscle fibres in conjunction with, or as a replacement for, acid fuchsin. It is not commonly used. Do not confuse this dye with crocein scarlet, a synonym for Biebrich scarlet

Congo Corinth G

Property	Data
Common name	Erie garnet B
Suggested name	Congo corinth
Other names	—
C.I. number	22145
C.I. name	Direct red 10
Class	Azo
Ionisation	Acid
Solubility aqueous	Moderately
Solubility ethanol	Slightly
Absorption maximum	527
Colour	Red
Empirical formula	$C_{32}H_{21}N_5O_7S_2Na_2$
Formula weight	697.7

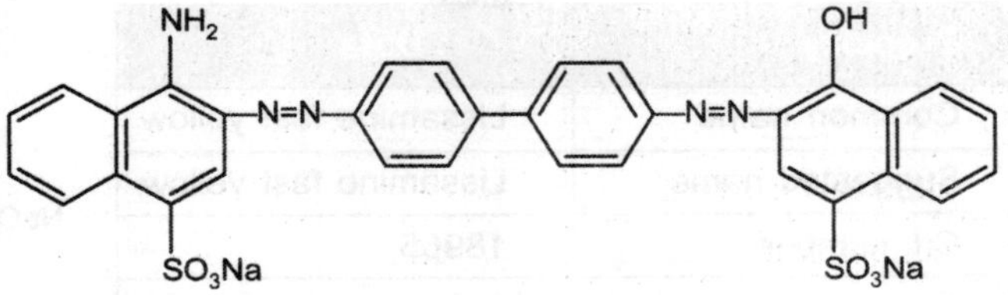

Congo corinth G is one of the dyes in Geschickter's solution, with Azure A. This solution is used to stain frozen sections for rapid diagnosis. The sections look very similar to those stained by polychromed methylene blue. Its structure is very similar to Congo red, and has been used to demonstrate amuloid.

Durazol Blue 4R

Property	Data
Common name	Durazol blue
Suggested name	Durazol blue 4R
Other names	Sirius supra blue F3R
Class	Azo
Ionisation	Acid
Solubility aqueous	Very
Solubility ethanol	Slight
Absorption maximum	573 (Gurr)
Colour	Blue
Empirical formula	$C_{34}H_{24}N_5O_{12}S_3Na_3$
Formula weight	860

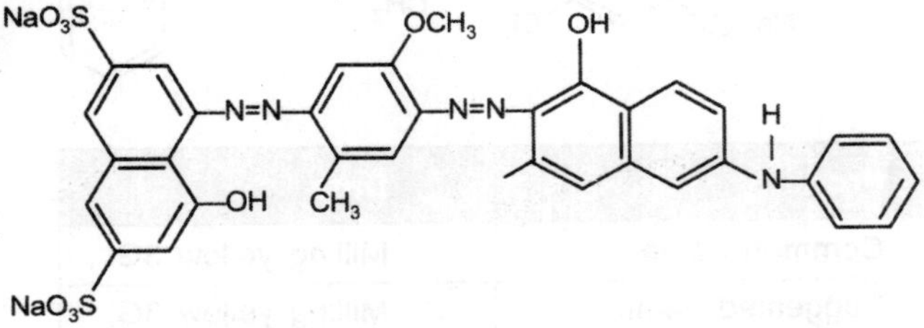

A dye named durazol blue has been suggested as a replacement for methyl blue in Lendrum's MSB. It is most likely to be this dye, durazol blue 4R. There is another dye named durazol blue 8G, but it is a copper phthalocyanine and is unlikely to be the dye meant. Unfortunately, the dye was not specifically identified and it is not clear what dye is meant.

Lissamine Fast Yellow

Property	Data
Common name	Lissamine fast yellow
Suggested name	Lissamine fast yellow
C.I. number	18965
Class	Azo
Ionisation	Acid
Solubility aqueous	Readily
Solubility ethanol	Poor
Absorption maximum	396 (Gurr)
Colour	Yellow
Empirical formula	$C_{16}H_{10}N_4O_7S_2Cl_2Na_2$
Formula weight	551

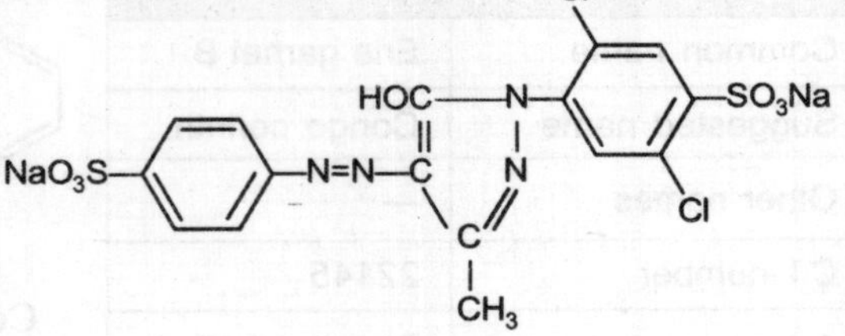

Lissamine fast yellow is rarely called for in medical histology. It is recommended as the erythrocyte stain in Garvey's method for elastic, fibrin and collagen. It has also been recommended as a substitute for martius yellow in Lendrum's MSB.

Milling Yellow

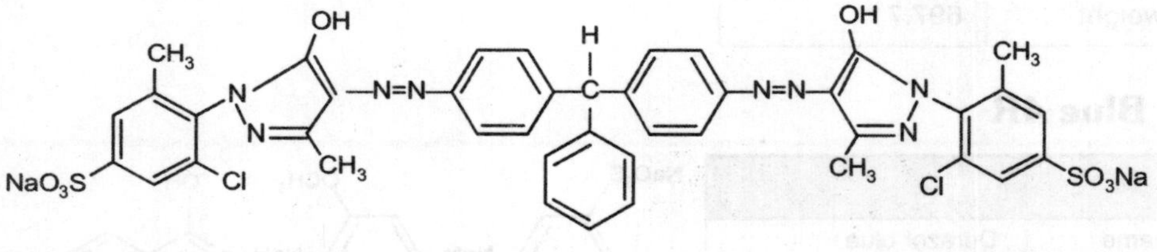

Property	Data
Common name	Milling yellow 3G
Suggested name	Milling yellow 3G
Classs	Azo
Ionisation	Acid
Solubility aqueous	Moderate
Solubility ethanol	Moderate
Absorption maximum	391 (Gurr)
Colour	Yellow
Empirical formula	$C_{41}H_{32}N_8O_8S_2Cl_2Na_2$
Formula weight	946

A large molecular weight dye named milling yellow 3G was specified in Slidders' fuchsin-miller technique for fibrin. It is likely this dye, but positive identification is not possible as neither

Naphthalene Blue Black

Property	Data
Common name	Naphthalene blue black
Suggested name	CS
Other names	Naphthalene blue black C
Class	Azo
Ionisation	Acid
Solubility aqueous	Freely
Solubility ethanol	Slightly
Absorption maximum	610-614
Colour	Blue black
Empirical formula	$C_{22}H_{13}N_6O_{12}S_3Na_3$
Formula weight	719

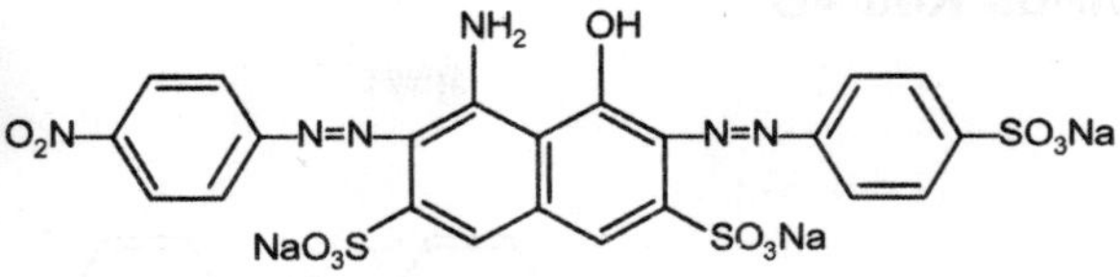

Lendrum used this dye for the demonstration of fibrin in his Masson 44/41 and Obadiah methods. He considered that it demonstrated the oldest fibrin at a stage when it was not distinguishable from collagen by any other staining method.

Pontamine Sky Blue 5B

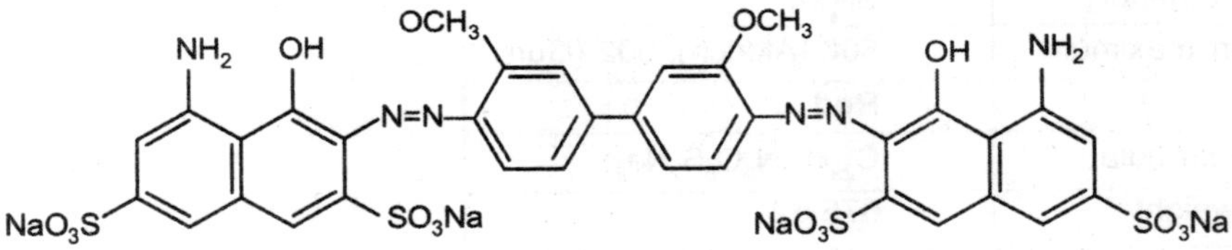

Property	Data
Common name	Pontamine sky blue
Suggested name	Pontamine sky blue 5B
Other names	Chicago blue 4B
Class	Azo
Ionisation	Acid
Solubility aqueous	Readily
Solubility ethanol	Poorly
Absorption maximum	588 (Gurr)
Colour	Blue
Empirical formula	$C_{34}H_{24}N_6O_{16}S_4Na_4$
Formula weight	993

A large molecular weight dye named pontamine sky blue was suggested as a substitute for methyl blue in Lendrum's MSB technique for fibrin. It is likely this dye, but positive identification is not possible as a CI number was not given. It was recommended with a comment that its use reduced the tendency of some fibrin to stain with the blue dye.

Sirius Red 4B

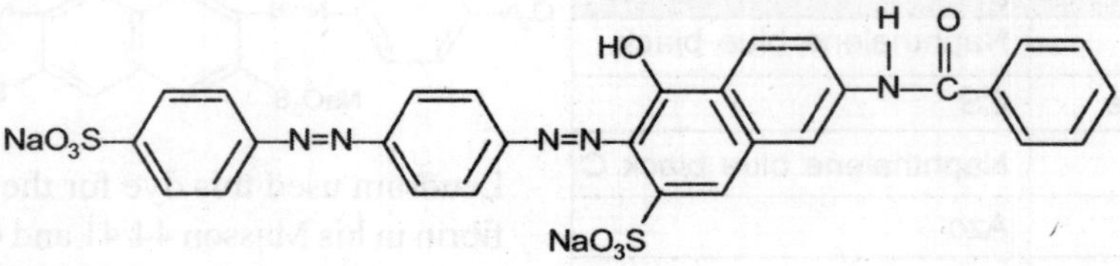

Property	Data
Common name	Sirius red 4B
Suggested name	Sirius red 4B
Other names	Chlorantine fast red 5B
C.I. number	28160
C.I. name	Direct red 81
Class	Azo
Ionisation	Acid
Solubility aqueous	Moderate
Solubility ethanol	Slight
Absorption maximum	508 (Aldrich), 502 (Gurr)
Colour	Red
Empirical formula	$C_{29}H_{19}N_5O_8S_2Na_2$
Formula weight	675.62

Sirius red 4B has been recommended as the plasma stain in some trichrome staining methods, or as a substitute for acid fuchsin in the van Gieson. It is sometimes confused with the similrly named dye, sirius red F3B, which may be used for demonstrating amyloid. Sirius red 4B is not used for that purpose and is not a substitute for congo red.

Sirius Red F3B

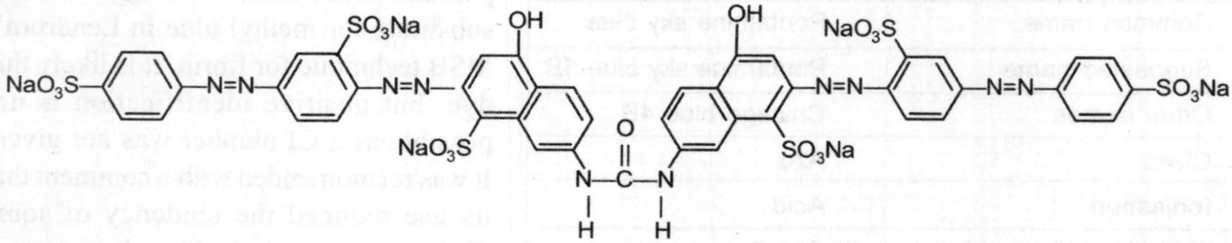

Property	Data
Common name	Sirius red F3B
Suggested name	Sirius red F3B
C.I. number	35780
C.I. name	Direct red 80

Sirius red F3B has been recommen- ded as a substitute for acid fuchsin in van Gieson type procedures. It is also used for the demonstration of amyloid in techniques similar to those using congo red, depending on hydrogen bonding and formation of a

Class	Azo
Ionisation	Acid
Solubility aqueous	Moderate
Solubility ethanol	Poorly
Colour	Red
Empirical formula	$C_{45}H_{26}N_{10}O_{21}S_6Na_6$
Formula weight	1373.1

"pseudocrystal." Like congo red it shows green birefringence, but it is a little darker.

Do not confuse this dye with sirius red 4B, which is not used for the demonstration of amyloid.

Nitro Dyes

Martius Yellow

Property	Data
Common name	Martius yellow
Suggested name	Martius yellow
Other names	Naphthol yellow
C.I. number	10315
C.I. name	Acid yellow 24
Class	Nitro
Ionisation	Acid
Solubility aqueous	4.6%
Solubility ethanol	0.15%
Absorption maximum	445 (Conn), 420 (Gurr) 435 (Aldrich)
Colour	Yellow
Empirical formula	$(O_2N)_2C_{10}H_5OH$
Formula weight	234.174

OH NO2 NO2

Ethanol solutions of martius yellow have been used to stain erythrocytes yellow, so they contrast well with red fibrin, in trichrome methods such as Lendrum's Picro Mallory and Slidder's Martius, Scrlet and Blue (MSB). It can be combined with other small molecular weight yellow dyes to increase stain intensity.

The sodium salt of martius yellow is the dye Manchester yellow with an absorptionj maximum of 432 (Aldrich).

Lithol Rubine BK

Lithol Rubine BK, Pigment Rubine, Carmine 6B, Brilliant Carmine 6B, Permanent Rubin L6B, Litholrubine, Latolrubine, C.I. Pigment Red 57, C.I. Pigment Red 57:1, D&C Red No. 7, or C.I. 15850:1, is a reddish synthetic azo dye. It has the appearance of a red powder, slightly soluble in hot water, insoluble in cold water, and insoluble in ethanol. When dissolved in dimethylformamide, its absorption maximum lies at about 442 nm. It usually comes as a calcium salt. It is used to dye plastics, paints, printing inks, and for textile printing.

SO_3^- HO COO^- H_3C N N

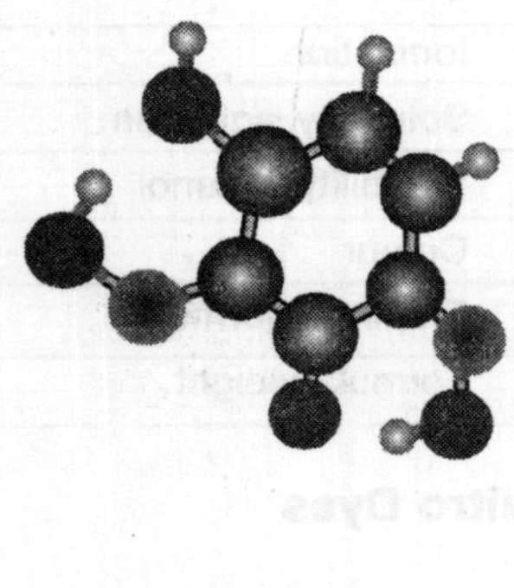

Nitroso Dye

Fast Green O

This dye is prepared by the action of resorcinol and nitrous acid.

Resorcinol $\xrightarrow{NaNO_2/HCl}$ Fast green O

Fast Green O forms a lake like substance with ferrous sulphate. This lake is fast to light and washing.

Phthalocyanine Dyes

History

The first known appearance of an unknown blue by-product (which we now know was metal-free phthalocyanine) was reported in 1907. In 1927, two researchers from Switzerland accidentally synthesized copper phthalocyanine, copper naphthalocyanine and copper octamethylphthalocyanine when they tried to convert o-dibromobenzene into phthalonitrile. They remarked the enormous stability of these complexes but failed to appreciate their discovery and to fully characterize these blue complexes. The real discovery also started as an accident, when a blue product was found in a reaction flask where only white product was expected.

A phthalocyanine is a macrocyclic compound having an alternating nitrogen atom-carbon atom ring structure. The molecule is able to coordinate hydrogen and metal cations in its center by coordinate bonds with the four isoindole nitrogen atoms. The central atoms can carry additional ligands. Most of the elements have been found to be able to coordinate to the phthalocyanine macrocycle. Therefore, a variety of phthalocyanine complexes exist.

Structure

The structure of a phthalocyanine molecule is closely related to that of the naturally occurring porphyrin systems (Fig. 28.2).

The phthalocyanine macrocycle is also related to some other macrocyclic complexes, as e.g., the subphthalocyanine, superphthalocyanine or hemiporphyrazine.

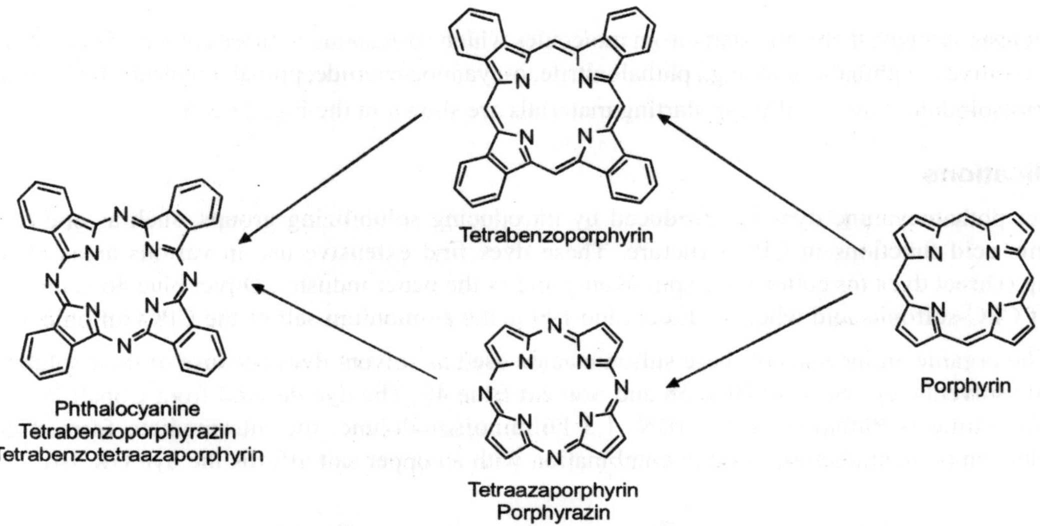

Fig 28.2. Relationship of the phthalocyanine with the porphyrin macrocycle.

Synthesis

The figure below (Fig. 28.3) shows that a phthalocyanine macrocycle consists of four identical corners.

Fig. 29.3.

Phthalocyanine

Fig. 29.3. A

Fig. 28.3 and 28.3 A. Strategy for the synthesis of phthalocyanines and Typical starting materials for the synthesis of phthalocyanines

A synthesis strategy, therefore, starts from molecules which correspond to these corners. Such molecules are derivatives of phthalic acid: e.g., phthalonitrile, o-cyanobenzamide, phthalanhydride, phthalimide or diiminoisoindole. Several of these starting materials are shown in the Fig. 28.3 A

Applications

Copper phthalocyanine dyes are produced by introducing solubilizing groups, such as one or more sulfonic acid functions in CPC structure. These dyes find extensive use in various areas of textile dyeing (Direct dyes for cotton), for spin dyeing and in the paper industry. Direct blue 86 is the sodium salt of CPC-sulfonic acid whereas direct blue 199 is the ammonium salt of the CPC-sulfonic acid.

The organic amine salts of these sulfonic acids used as solvent dyes because of their solubility in organic solvents, e.g. Solvent Blue 38 and Solvent Blue 48. The dye derived from Cobalt PC and an organic amine is Phthalogen Dye IBN. 1,3 Diiminoisoindolene, the intermediate formed during phthalocyanine manufacture, used in combination with a copper salt affords the dye GK 161.

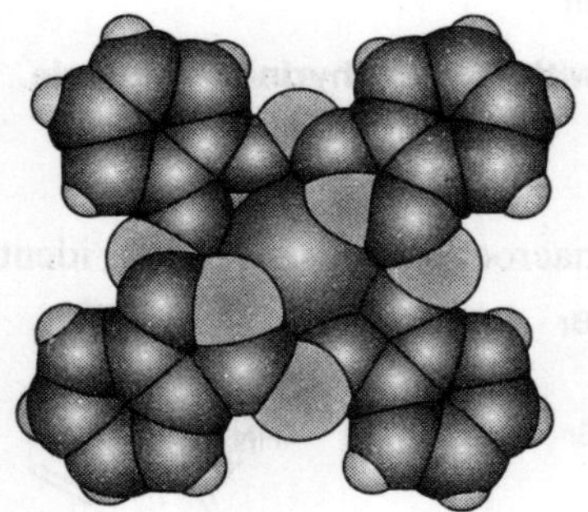

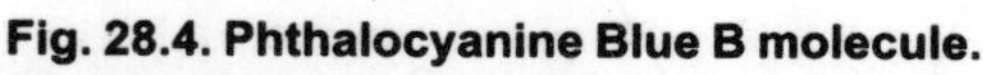

Fig. 28.4. Phthalocyanine Blue B molecule.

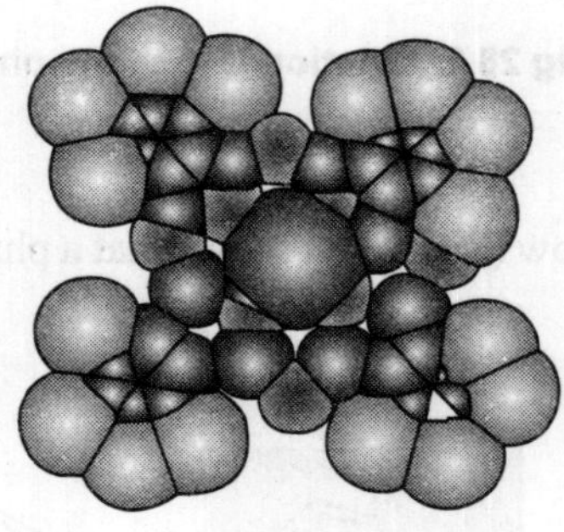

Fig. 28.4. A. Phthalocyanine Green G molecule.

Phthalocyanine Dyes

Phthalocyanin dyes have the general formula to the right. At the point marked X various groups are attached. These vary with each dye.

The phthalocyanine dyes are few in number, but those few are commonly used. Alcian blue is employed in one of the standard methods for acid mucopolysaccharides. Luxol fast blue is used to demonstrate myelin.

Phthalocyanine Blue BN

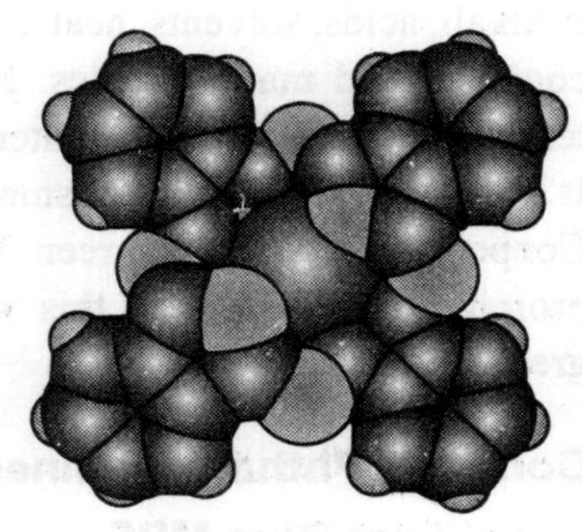

Phthalocyanine Blue BN, also called phthalo blue, helio blue, thalo blue, Winsor blue, phthalocyanine blue, C.I. Pigment Blue 15:2, Copper phthalocyanine blue, Copper tetrabenzoporphyrazine, Cu-Phthaloblue, PB-15, PB-36, and C.I. 74160, is a bright, greenish-blue crystalline synthetic blue pigment from the group of pinthalocyanine dyes. It was developed as a pigment in the mid-1930s. Its brilliant blue is used in paints and dyes. It is highly valued for its superior properties such as light fastness, tinting strength, covering power and resistance to the effects of alkalies and acids. It has the appearance of a blue powder, insoluble in water and most solvents.

Phthalocyanine blue is a complex of copper with phthalocyanine.

Phthalo Blue is a deep blue oil paint. When mixed with other colors, it quickly tends toward green. Also, with the addition of chlorine to it, it turns to a lighter green tone than the original prussian blue with the same chemical characteristics. It has to be used with great care, as it is a very powerful tinter and tends to overwhelm other pigments. When used alone, it should be mixed with some white, as it is transparent and almost black on its own. it was also generally a popular oil in Bob Ross's Joy Of Painting show where it is comonly used for sky ,water and other such components of a landscape.Due to its stability, phthalo blue is used in inks, coatings, and many plastics. In application it is transparent. The pigment is insoluble and has no tendency to migrate in the material. It is a standard pigment used in printing ink and the packaging industry.

Phthalocyanine Blue BN is also used as a source material for manufacture of Phthalocyanine Green G.

Phthalocyanine Green G

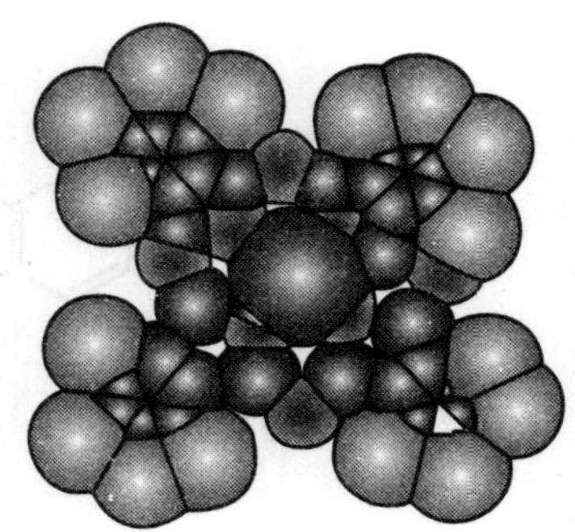

Phthalocyanine Green G, also called phthalo green, Pigment Green 7, Copper Phthalocyanine Green, C.I. Pigment Green 42, Non-flocculating Green G, Polychloro copper phthalocyanine, or C.I. 74260, is a synthetic green pigment from the group of phthalocyanine dyes, a complex of copper(II) with chlorinated phthalocyanine. It is a very soft green powder insoluble in water. Its CAS number is [1328-45-6] and [1328-53-6]. Its chemical formula ranges from $C_{32}H_3Cl_{13}CuN_8$ to $C_{32}HCl_{15}CuN_8$. It is a bright, high intensity colour used in oil and acrylic based artist's paints, and in other applications.Phthalocyanine green is a phthalocyanine blue pigment where most of the hydrogen atoms are replaced with chlorine. The strongly electronegative chlorine atoms influence the distribution of the electrons in the phthalocyanine structure, shifting its absorption spectrum. It is made by chlorination of the phthalocyanine blue as a melt of sodium chloride and aluminium chloride, to which chlorine is introduced at elevated temperature. Material from some manufacturers may be contaminated by trace amounts of polychlorinated biphenyls. The phthalo green molecules are highly stable. They are resistant to alkali, acids, solvents, heat, and ultraviolet radiation. Due to its stability, phthalo green is used in inks, coatings, and many plastics. In application it is transparent. The pigment is insoluble and has no tendency to migrate in the material. It is a standard pigment used in printing ink and packaging industry. It is also allowed in all cosmetics except those used around the eyes, and is used in some tattoos. Copper phthalocyanine green 36 is another variant where part of the chlorine atoms is replaced with bromine; the mixture of this with phthalocyanine green can reach a wide tonal range of greens to green-yellows.

Compare Phthalocyanines

Luxol Fast Blue MBS

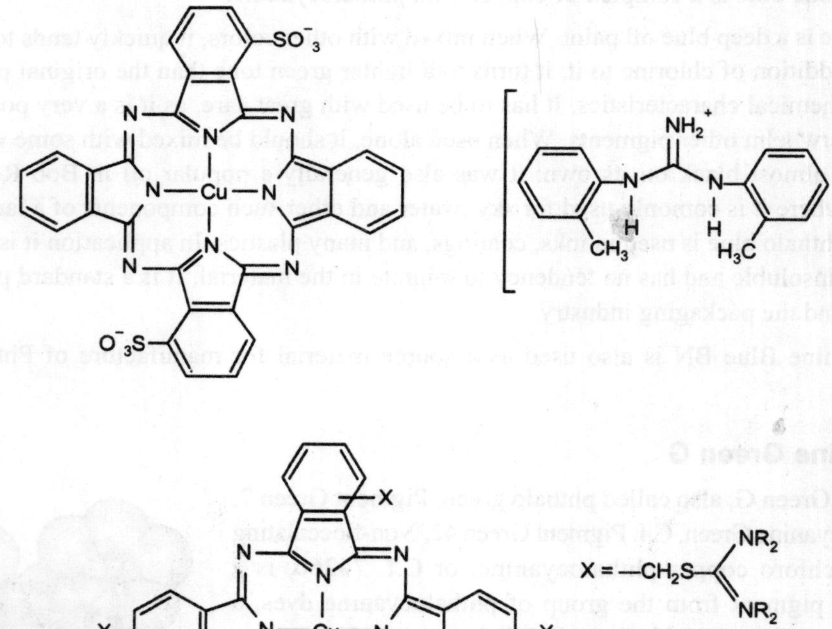

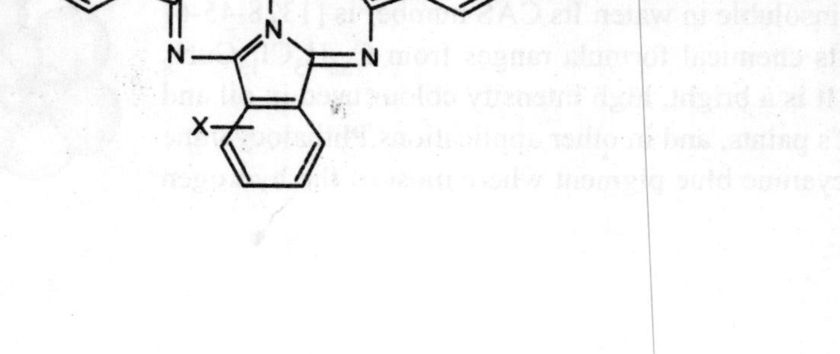

$X = -CH_2S-C(=NR_2)NR_2$

R = alkyl or aryl group

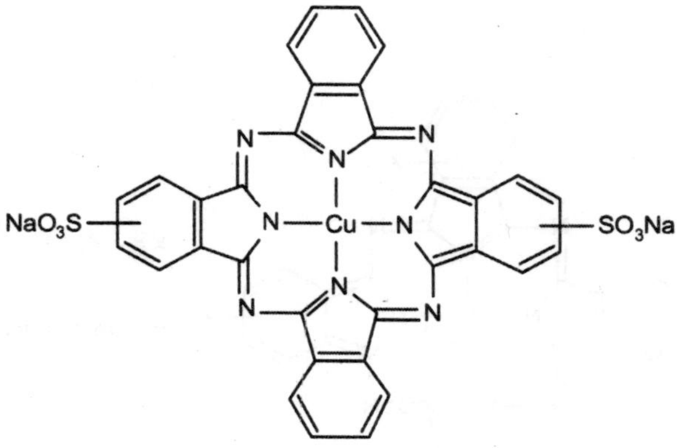

Durazol blue 8G

Luxol Fast Blue MBS

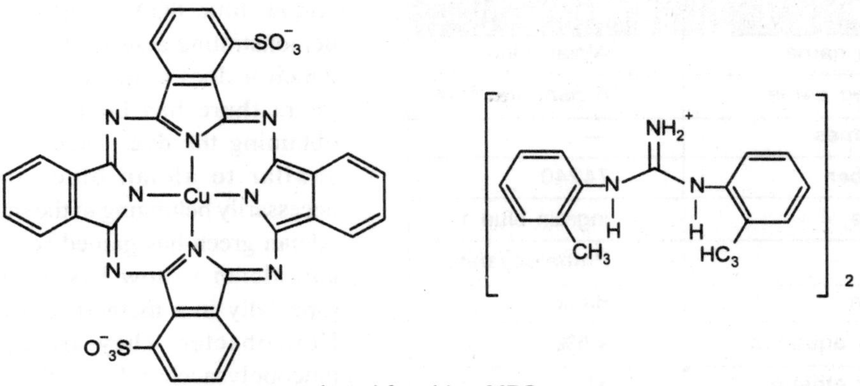

Luxol fast blue MBS

Property	Data
Common name	Luxol fast blue
Suggested name	Luxol fast blue MBS
Other names	Solvent blue 38
Class	Phthalocyanine
Solubility aqueous	Slight
Solubility ethanol	Soluble
Absorption maximum	666 (Aldrich)
Colour	Blue

Luxol fast blue is described as being a diarylguanidine salt of a sulphonated copper phthalocyanine. The diarylguanidine being on the right. The dye is soluble in alcohols and phospholipids, which may explain how it demonstrates myelin.

Compare the formula with other phthalocyanines.

Alcian Blue

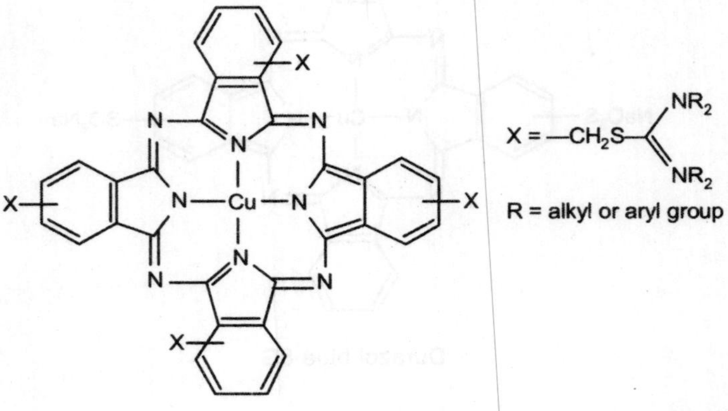

Alcian blue 8GX

Property	Data
Common name	Alcian blue
Suggested name	Alcian blue 8GX
Other names	—
C.I. number	74240
C.I. name	Ingrain blue 1
Class	Phthalocyanine
Ionisation	Basic
Solubility aqueous	9.5%
Solubility ethanol	6%
Absorption maximum	615 (Aldrich)
	629 (Gurr)
Colour	Blue
Empirical formula	—
Formula weight	1300

Alcian blue 8GX is primarily used for demonstrating acid mucopolysaccharides, which it does quite selectively. In recent years there has been some difficulty obtaining the dye. There are other dyes similar to alcian blue, although not necessarily belonging to the same dye class. Alcian green has gained some popularity, and alcian yellow has been suggested, especially in a method for demonstrating Helicobacter. They are used for acid mucopolysaccharides in the same way as alcian blue. They have no particular advantage for demonstrating mucin. According to the Merck Index the groups marked X split off during staining.Compare the formula with other phthalocyanines.

PHTHALEIN DYES

These dyes are the derivatives of dihydroxytriphenyl methane, these compounds respond to the pH change generally towards the higher side these compounds are generally very less but some of then are listed below.

Phenolphthalein

Systematic name	Phenolphthalein
Chemical formula	$C_{20}H_{14}O_4$
Molecular mass	318.33 g mol^{-1}
Density	1.299 g cm^{-3}, solid
Melting point	263-265°C
Boiling point	258°C

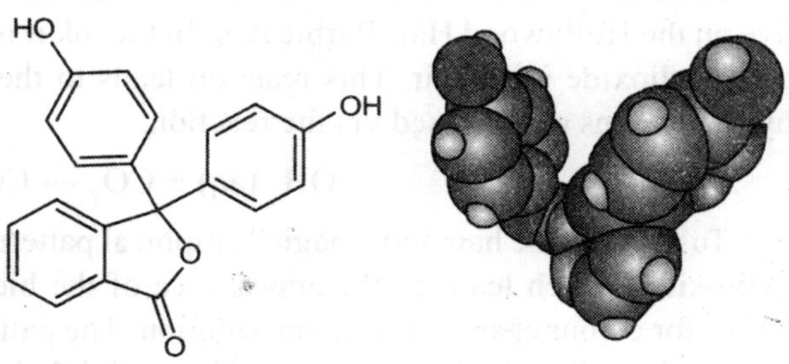

Phenolphthalein (pH indicator)		
below pH 8.2		*above pH 10.0*
colorless	→	pink

Phenolphthalein is a sensitive chemical with the formula $C_{20}H_{14}O_4$ (often written as "HIn" in chemistry shorthand notation). Often used in titrations, it turns from colorless in acidic solutions to pink in basic solutions. If the concentration of indicator is particularly strong, it can appear purple. In solutions containing a pH below 0, phenolphthalein turns a bright orange colour.In strongly basic solutions, phenolphthalein's pink colour undergoes a rather slow fading reaction and becomes colourless again. In other words, the molecule has four forms:

Phenolphthalein		H_2Phenolphthalein		Phenolphthalein^{2-}		Phenolphthalein(OH)$^{3-}$
strongly acidic ($pH < 0$)	→	acidic	→	basic	→	strong alkaline
orange		no colour		pink		no colour

The rather slow fading reaction that produces the colorless $InOH^{3-}$ ion is sometimes used in classes for the study of reaction kinetics. Phenolphthalein is insoluble in water, and is usually dissolved in alcohols for use in experiments. It is itself a weak acid, which can lose H^+ ions in solution. The phenolphthalein molecule is colorless. However, the phenolphthalein ion is pink. When a base is added to the phenolphthalein, the molecule $\rightleftharpoons$ ions equilibrium shifts to the right, leading to more ionization as H^+ ions are removed. This is predicted by Le Chatelier's principle.

Uses

Phenolphthalein has been used for over a century as a laxative, but is now being removed from the market because of concerns over carcinogenicity. However, the small amounts usually used in experiments are harmless. Phenolphthalein is also commonly used in a mixture, primarily by forensic scientists, to test for the presence of blood. Phenolphthalein is used to perform a presumptive blood test, and is commonly known as the Kastle-Meyer test. A dry sample is collected with a swab or filter paper. First a few drops of alcohol, then a few drops of phenolphthalein and finally a few drops of hydrogen peroxide are dripped onto the sample. If the sample turns pink then it is a positive test. This test is non-destructive to the sample; it can be kept and used in further tests at the lab. This test has the same reaction with blood from any animal, so further testing would be required to determine whether it originates from a human.

Phenolphthalein is used in toys, for example as a component of disappearing inks, or disappearing dye on the Hollywood Hair Barbie hair. In the ink it is mixed with sodium hydroxide, which reacts with carbon dioxide in the air. This reaction leads to the pH falling below the color change threshold as hydrogen ions are released via the reaction:

$$OH^{-}(aq) + CO_2 \rightarrow CO_3^{2-}(aq) + H^{+}(aq)$$

To develop the hair and "magic" graphical patterns, the ink is sprayed with a solution of ammonium hydroxide, which leads to the appearance of the hidden graphics by the same mechanism described above for colour change in alkaline solution. The pattern will eventually disappear by the same reaction with carbon dioxide detailed above. Thymolphthalein is used for the same purpose and in the same way, when blue color is desired.

Preparation of Phenolphthalein

Phenolphthalein is prepared by the actionof phenol and phthalic anhydride in the presence of conc. Sulphuric acid. The colour shifts from transparent to pink in the presence of alkali.

Phenol (2 molecules) + Phthalic anhydride —conc H_2SO_4, $-H_2O$→ Phenolphthalein (colourless)

If excess of strong alkali solution is added to the phenolphthalein solution the pink colour dissappears due to the loss of the quinoid structure and resonance.

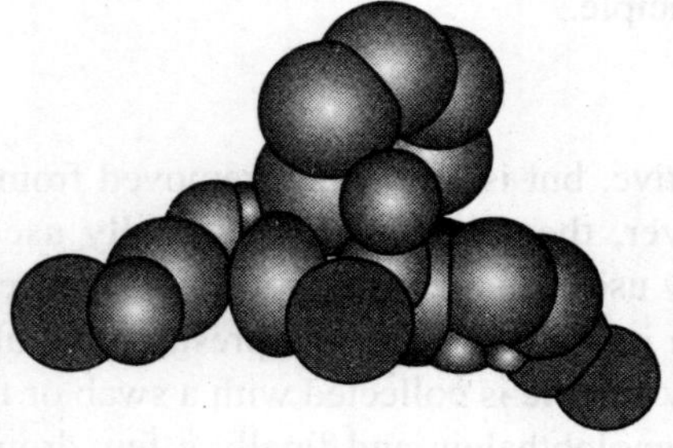

Phenolphthalein (colourless)

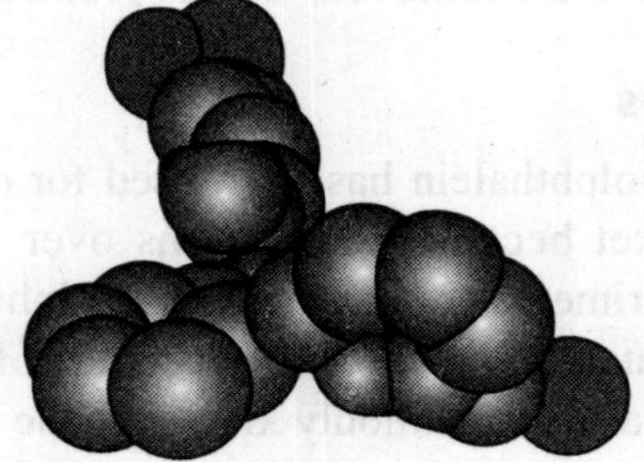

Phenolphthalein (pink)

Phenol Red

Phenol red (also known as phenolsulfonphthalein or PSP) is a pH indicator that is frequently used in cell biology laboratories.

Chemical structure and properties

Phenol red exists as a red crystal that is stable in air. Its solubility is 0.77 grams per liter (g/L) in water and 2.9 g/L in ethanol. It is a weak acid with $pK_a = 8.00$ at 20°C.

A solution of phenol red is used as a pH indicator: its colour exhibits a gradual transition from yellow to red over the pH range 6.6 to 8.0. Above pH 8.1, phenol red turns a bright pink (fuschia) colour.

Phenol red (pH indicator)		
below pH 6.6		*above pH 8.0*
6.6	↔	8.0

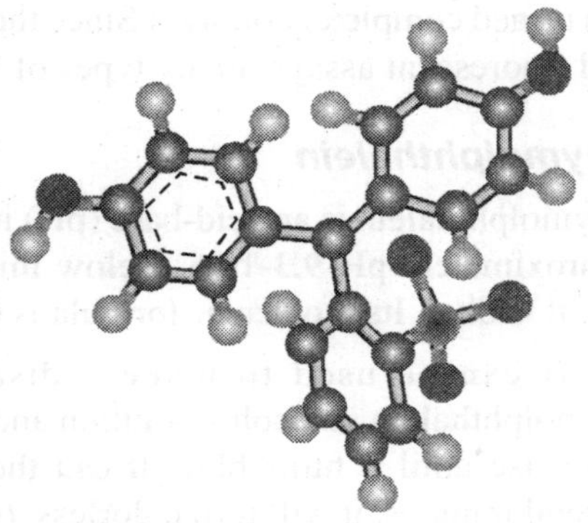

This observed color change is due to the fact that phenol red loses protons (and changes colour) as the pH increases. In crystalline form, and in solution under very acidic conditions (low pH), the compound exists as a zwitterion as in the structure shown above, with the sulfate group negatively charged, and the ketone group carrying an additional proton. This form is sometimes symbolically written as $H_2^+PS^-$ and is orange-red. If the pH is increased (pKa = 1.2), the proton from the ketone group is lost, resulting in the yellow negatively charged ion HPS. At still higher pH (pKa = 7.7), the phenol's hydroxide group loses its proton, resulting in the red ion PS^{2-}.

HO $\overset{+}{O}H$ SO_3^-

In several sources, the structure of phenol red is shown with the sulphur atom being part of a cyclic group, similar to the structure of phenolphthalein. However, this cyclic structure could not be confirmed by X-ray crystallography.

Several indicators share a similar structure to phenol red, including bromothymol blue, thymol blue, bromocresol purple, thymolphthalein, and phenolphthalein.

Phenolsulfonphthalein test

Phenol red is used in the phenolsulfonphthalein test, also known as the PSP test. This test was used to estimate the overall blood flow through the kidney and is now obsolete. The test is based on the fact that phenol red is excreted almost entirely in the urine. By measuring the amount of phenol red excreted colorimetrically, kidney function can be determined. Phenol red solution is administered intravenously, all of the urine produced is collected and the phenol red present determined.

Indicator for cell cultures

Most living tissues prosper at a near-neutral pH; that is, a pH close to 7. Blood's pH ranges from 7.35 to 7.45, for instance. When cells are grown in tissue culture, the medium in which they grow is held close to this physiological pH. A small amount of phenol red added to this growth medium will have a pink-red colour under normal conditions.In the event of problems, waste products produced by dying cells or overgrowth of contaminants will cause a change in pH, leading to a change in indicator colour. For example, a culture of relatively slowly-dividing mammalian cells can be quickly overgrown by bacterial contamination. This usually results in an acidification of the medium, turning it yellow. Many biologists find this a convenient way to rapidly check on the health of tissue cultures. In addition, the waste products produced by the mammalian cells themselves will slowly decrease the pH, gradually turning the solution orange and then yellow. This color change is an indication that even in the absence of contamination, the medium needs to be replaced (generally, this should be done before the medium has turned completely orange).Since the color of phenol red can interfere with some spectrophotometric and fluorescent assays, many types of tissue culture media are also available without phenol red.

Thymolphthalein

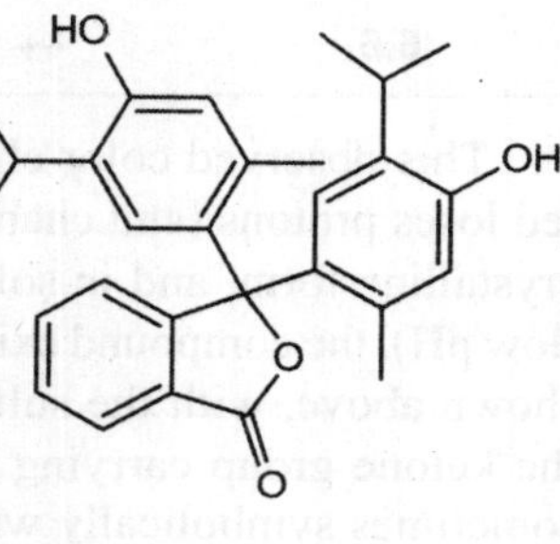

Thymolphthalein is an acid-base (pH) indicator. Its transition range is at approximately pH 9.3-10.5. Below this pH, it is colorless; above this pH, it is blue. Its molecular formula is $C_{28}H_{30}O_4$.

It can be used to make a disappearing ink. Dissolve the thymolphthalein in alcohol solution and add sodium hydroxide (NaOH) dropwise until it turns blue. It can then be sprayed onto cloth. After several minutes, it will turn colorless. (Carbon dioxide from the air will react with the basic solution. When the pH drops below 9.3, the thymolphthalein turns colorless.) Such inscription can be turned visible again by submerging it in liquid of pH higher than 10.5.

Thymolphthalein (pH indicator)		
below pH 9.3		*above pH 10.5*
9.3	↔	10.5

TRIPHENYLMETHANE DYES

Triphenylmethane, or triphenyl methane, is a hydrocarbon. It is a colourless to light brown powder. It is the basic skeleton of many synthetic dyes (Triarylmethane dyes); many of them are pH indicators, and some display fluorescence. It is insoluble in water and soluble in chloroform.The pKa of the hydrogen on the central carbon is around 31. Triphenylmethane is significantly more acidic than most other hydrocarbons because the planar trityl anion is stabilized by extensive delocalization over three phenyl rings. The trityl anion absorbs strongly in the visible region, making it red. This colour can be used as an indicator when maintaining anhydrous conditions with calcium hydride; the hydride reagent reacts with water to form solid calcium hydroxide, while it is also a strong enough base to generate the trityl anion. If the hydride is used up then the solution will turn colourless. Triphenylmethane can be synthesized by Friedel-Crafts reaction from benzene and chloroform with aluminium chloride catalyst. Synthesis from benzylidene chloride, prepared from benzaldehyde and phosphorus pentachloride, is used as well.

Physical Properties

Chemical name	Triphenylmethane
Chemical formula	$C_{19}H_{16}$
Molecular mass	244.33 g/mol
Density	1.014 g/cm3
Melting point	92-94 °C
Boiling point	359 °C

This compound is responsible for making many dyes which are used as indicators in the chemical laboratory. The different types of triarylmethane dyes are discussed below:

Aniline Blue WS

Aniline Blue WS, also called aniline blue, China blue, or Soluble blue, is a mixture of methyl blue and water blue. It may also be either one of them.Aniline blue or its constituents are used to stain collagen, as the fibre stain in Masson's trichrome. It can be used in the Mallory's connective tissue stain and Gomori's one-step trichrome stain. It is used in differential staining.

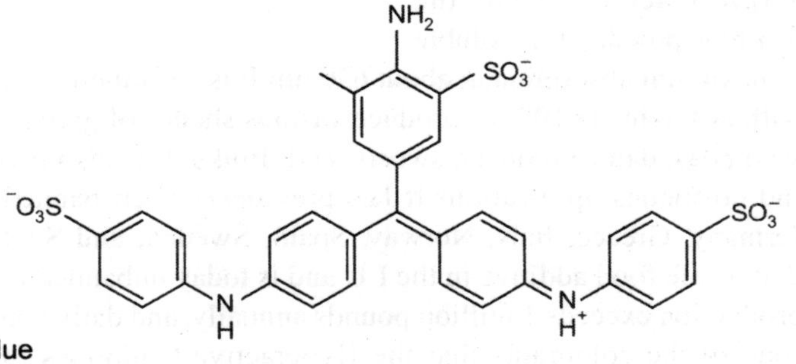

Water blue

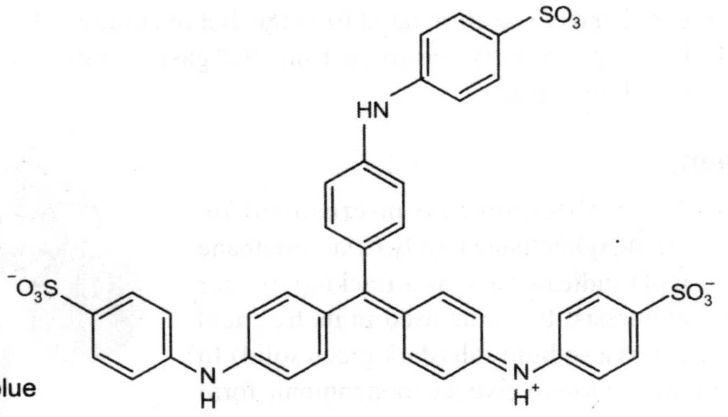

Methyl blue

Aurin (C.I. 43800)

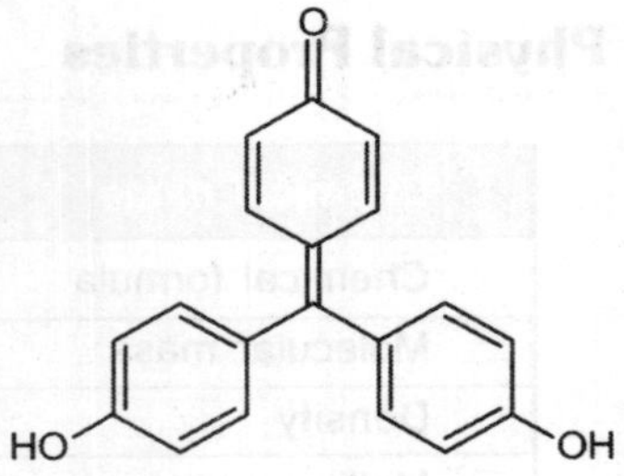

Aurin (C.I. 43800), sometimes named rosolic acid or corallin, chemically 4-[bis(p-hydroxyphenyl)methylene]-2,5-cyclohexadien-1-one, is an organic compound, forming yellowish or deep-red crystals with greenish metallic luster. It is practically insoluble in water, freely soluble in alcohol. It is soluble in strong acids to form yellow solution, or in aqueous alkalis to form carmine red solutions. Due to this behaviour it can be used as pH indicator with pH transition range 5.0 - 6.8. It used as intermediate in manufacturing of dyes.

Brilliant Blue FCF

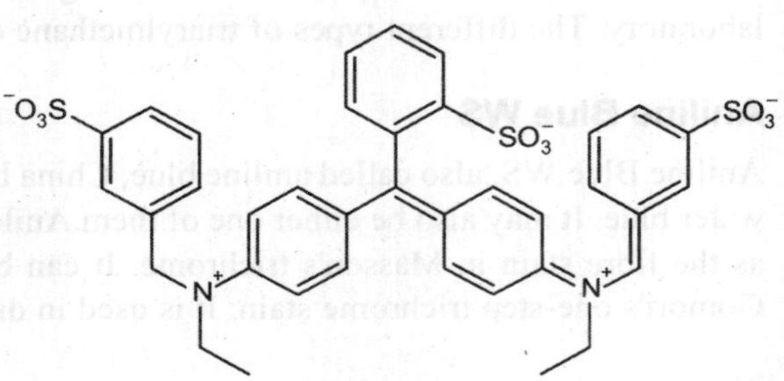

Brilliant Blue FCF (also known as FD&C Blue No.1, Food Blue 2, Acid Blue 9, D&C Blue No. 4, Alzen Food Blue No. 1, Alphazurine, Atracid Blue FG, Erioglaucine, Eriosky blue, Patent Blue AR, Xylene Blue VSG, and C.I. 42090) is a colorant that may be added to foods to induce a color change. It is denoted by E Number E133, and has the capacity for inducing an allergic reaction. It has the appearance of a reddish-blue powder. It is soluble in water; solution has maximum absorption at about 630 nm.It is a synthetic dye derived from coal tar. It can be combined with tartrazine (E102) to produce various shades of green. It is often found in ice cream, tinned processed peas, dairy products, sweets, and drinks. It is also used in soaps, shampoos, and other hygiene and cosmetics applications.It has previously been banned in Austria, Belgium, Denmark, France, Germany, Greece, Italy, Norway, Spain, Sweden, and Switzerland among others but has been certified as a safe food additive in the EU and is today unbanned in most of the countries. In the United States production exceeds 1 million pounds annually, and daily consumption is around 16 mg per person.It is one of the colourants that the Hyperactive Children's Support Group and the Feingold Association recommends to be eliminated from the diet of children. The chemical formation is C37H34N2Na2O9S3 .The dye is poorly absorbed from the gastro-intestinal tract and 95% of the adsorbed dye can be found in the urine.

Bromocresol Green

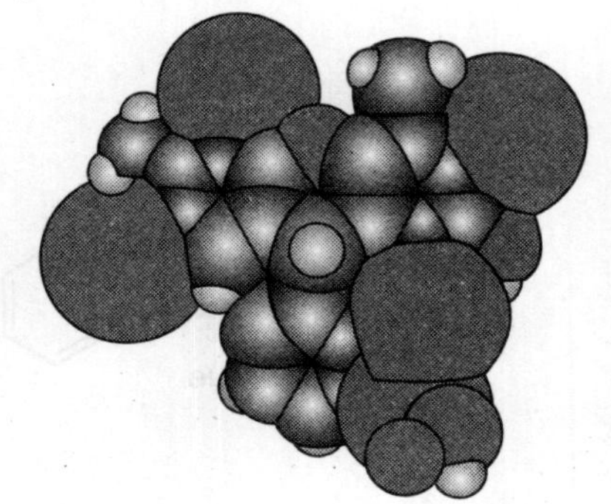

Bromocresol green (3,3',5,5'-tetrabromo-m-cresolsulfon phthalein) is a dye of the triphenylmethane family (triarylmethane dyes), which is used as a pH indicator and as a tracking dye for DNA agarose gel electrophoresis. It can be used in its free acid form (light brown solid), or as a sodium salt (dark green solid).In aqueous solution, both solids ionize to give the monoanionic form (yellow), that further deprotonates at higher pH to give the dianionic form (blue), which is stabilized by resonance:

Bromocresol green (pH indicator)		
below pH 3.8		*above pH* 5.4
3.8	↔	5.4

The pK (pKa) of this reaction is 4.9.

The acid and basic form of this dye have an isosbestic point in their spectra, around 515 nm.

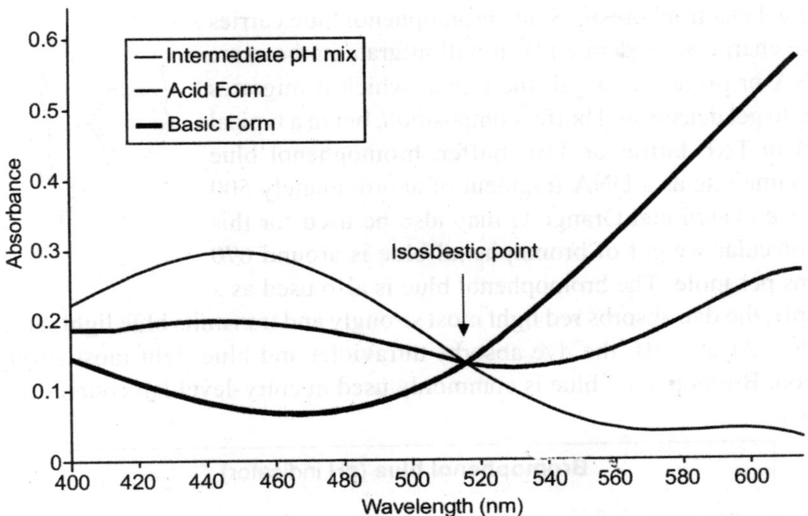

Preparation

This indicator is prepared by the action of bromine on m-cresol

It is used as an acid base indicator in the volumetric analysis.

Bromophenol Blue

Bromophenol blue, Tetrabromophenolsulfonephthalein, is an acid-base indicator whose useful range as an indicator lies between pH 3.0 and 4.6. It changes from yellow at pH 3.0 to purple at pH 4.6; this reaction is reversible. Bromophenol blue is also used as a colour marker to monitor the process of agarose gel electrophoresis and polyacrylamide gel electrophoresis. Since bromophenol blue carries a slight negative charge at moderate pH, it will migrate in the same direction as DNA or protein in a gel; the rate at which it migrates varies according to gel density and buffer composition, but in a typical 1% agarose gel in TAE buffer or TBE buffer, bromophenol blue migrates at the same rate as a DNA fragment of approximately 500 base pairs. Xylene cyanol and Orange G may also be used for this purpose.The molecular weight of bromophenol blue is around 670 Daltons or grams per mole. The bromophenol blue is also used as a dye. At neutral pH, the dye absorbs red light most strongly and transmits blue light. Solutions of the dye therefore are blue. At low pH, the dye absorbs ultraviolet and blue light most strongly and appears yellow in solution. Bromophenol blue is commonly used in entry-level lab courses to stain proteins in wet-mount slides.

Bromophenol blue (pH indicator)		
below pH 3.0		*above pH 4.6*
3.0	↔	4.6

Bromthymol Blue

Bromthymol Blue (also known as **Dibromothymolsulfonephthalein, Bromthymol Blue**, and **BTB**) is a chemical indicator for weak acids and bases. The chemical is also used for observing photosynthetic activities or respiratory indicators (turns yellow as CO_2 is added).Bromthymol Blue acts as a weak acid in solution and therefore can be in acid or base forms which appear yellow and blue respectively. It is green in neutral solution. It is typically sold in solid form as the sodium salt of the acid indicator. It also finds occasional use in the laboratory as a biological slide stain. At this point it is already blue, and a drop or two is used on a water slide. The cover slip is placed on top of the water droplet and the specimen in it, with the blue coloring mixed in. It is sometimes used to define cell walls or nuclei under the microscope.Bromothymol Blue is mostly used in measuring substances that would have relatively low acidic or basic levels (near a neutral pH). It's often used in pools, fish tanks, or measuring the presence of carbonic acid in a liquid.

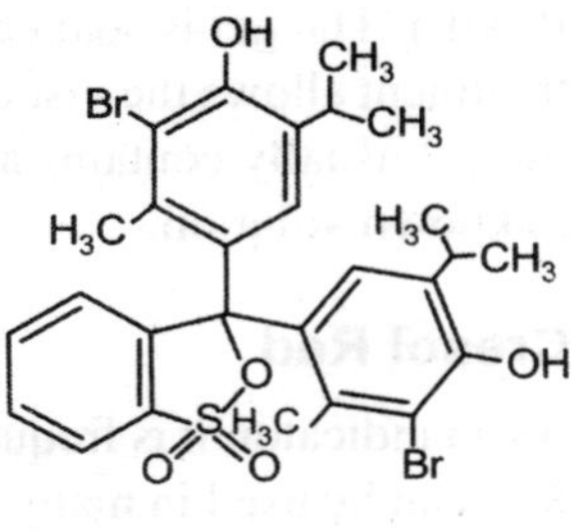

$$C_{27}H_{28}Br_2O_5S \leftrightarrow H^+ + C_{27}H_{27}Br_2O_5S^-$$

The pKa for bromothymol blue is 7.1.

Bromothylmol Blue (pH indicator)		
below pH 6.0		*above pH 7.6*
6.0	↔	7.6

A common demonstration of BTB's pH indicator properties involves exhaling through a tube into a neutral solution of BTB. As carbon dioxide is absorbed from the breath into the solution, forming carbonic acid, the solution changes color from green to yellow.

Coomassie

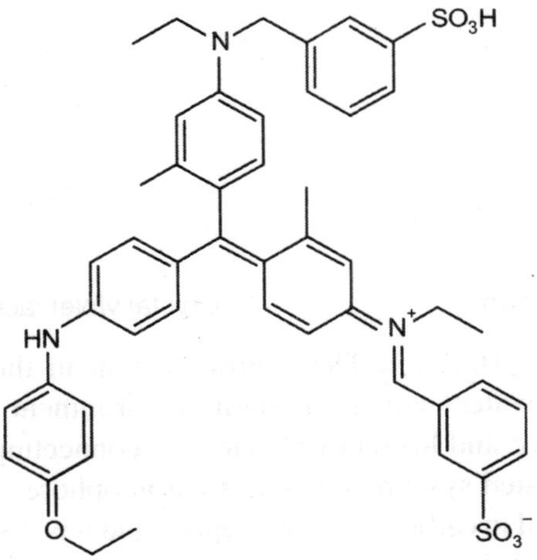

Coomassie (also known as Brilliant Blue, Brilliant Blue G, Acid Blue 90, C.I. 42655, or Brilliant Blue G 250) is a blue dye commonly used in sodium dodecyl sulfate polyacrylamide gel electrophoresis (SDS-

PAGE). The gel is soaked in dye for thirty minutes and then destained for thirty minutes or more. This treatment allows the visualization of bands indicating the protein content of the gel. The visualization on the gel usually contains a set of molecular weight marker so that protein MW can be determined in an unknown solution.

Cresol Red

As an indicator, it is frequently used for monitoring the pH in aquaria. Cresol Red can be used in many molecular reactions instead of water. For example restriction digestion or PCR.

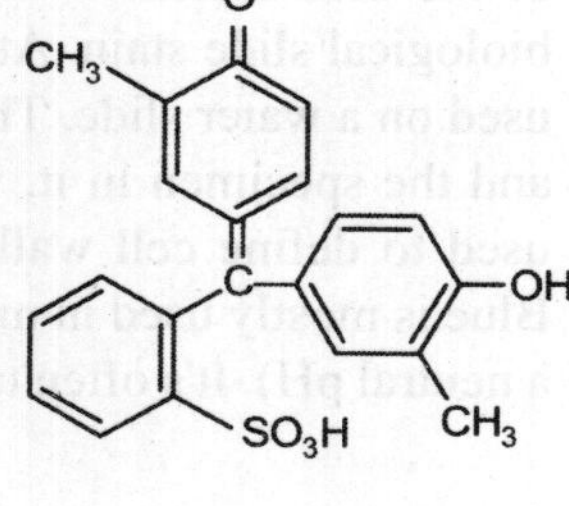

Cresol Red (pH indicator)		
below pH 7.2		*above pH 8.8*
7.2	↔	8.8

Cresol Red can also be used as a color marker to monitor the process of agarose gel electrophoresis and polyacrylamide gel electrophoresis. In a 1% agarose gel it runs approximately at the size of a 125 base pair (bp) DNA molecule. Bromophenol blue and xylene cyanol can also be used for this purpose.

Crystal violet lactone (CVL)

Crystal violet lactone (CVL) is a leuco dye, a lactone derivate of crystal violet 10B. In pure state it is a slightly yellowish crystalline powder, soluble in nonpolar or slightly polar organic solvents.

Crystal violet lactone, leuco form

Crystal violet lactone, protonated colored form

Its chemical formula is $C_{26}H_{29}N_3O_2$ The central carbon in the leuco form is in a tetraedric configuration, forming four covalent bonds. In acidic environment the lactone ring is broken, the central carbon loses one valence and becomes planar, interconnecting the π systems of the aromatic rings to form one large conjugated system acting as a chromophore with strong absorption in visible spectrum.It was the first dye used in carbonless copy papers, and it is still widely used in this application. It is also the leuco dye component in some thermochromic dyes, eg. in the Hypercolor line of clothing. One of its novel uses is a security marker for fuels. It is now being used in the purple type of zubbles.

It may cause allergic contact dermatitis in people handling the carbonless copy paper.

Ethyl Green

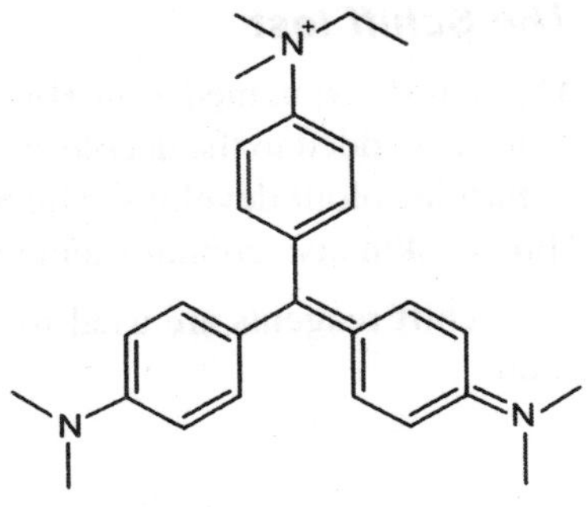

The dye Ethyl Green (Brilliant Green, C.I. 42590; $C_{27}H_{35}N_3ClBr$), due to its powerful bacteriostatic properties, is commonly used along with iodine tincture as a topical antiseptic in Eastern Europe / the former USSR (common name: Zelyonka), though not in the west due to its irritant properties and toxicity if ingested. It is soluble in wate. Ethyl green is made of crystal violet by adding an ethyl group; crystal violet is therefore a possible contaminant. Methyl green is a closely related dye used as a stain in histology. Methyl green and ethyl green are very similar and probably interchangeable.

Fast Green FCF

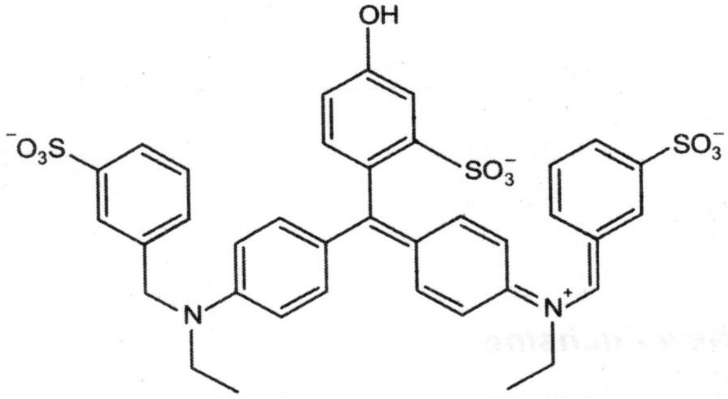

Fast Green FCF, also called Food green 3, FD&C Green No. 3, Green 1724, Solid Green FCF, and C.I. 42053, is a sea green triarylmethane food dye. Its E number is E143.Fast Green FCF is recommended as a replacement of Light Green SF yellowish in Masson's trichrome, as its color is more brilliant and less likely to fade. It is used as a quantitative stain for histones at alkaline pH after acid extraction of DNA. It is also used as a protein stain in electrophoresis. Its absorption maximum is at 625 nm.Fast Green FCF is poorly absorbed by the intestines. Its use as a food dye is prohibited in European Union and some other countries. It can be used for tinned green peas and other vegetables, jellies, sauces, fish, desserts, and dry bakery mixes at level of up to 100 mg/kg.

Fuchsine or Rosaniline Hydrochloride

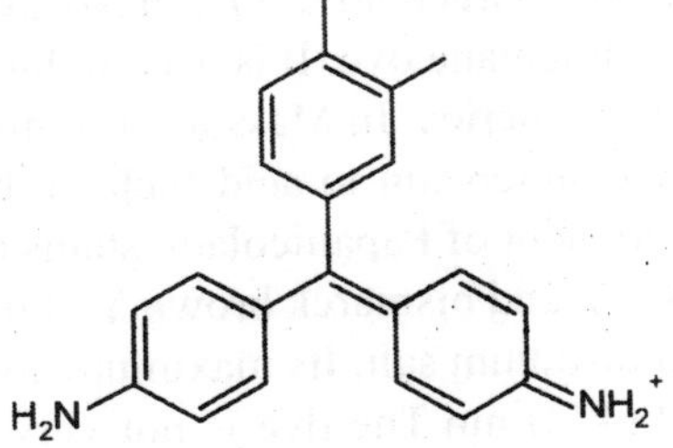

Fuchsine, fuchsin, rosanilin, or rosaniline hydrochloride is a magenta dye with chemical formula $C_{19}H_{17}N_3 \cdot HCl$. It becomes magenta when dissolved in water; as a solid, it forms dark green crystals. Its inventor, the 19th century French company Renard, named the dye after the German translation of the company's name, fuchs (fox). As well as dying textiles, fuchsine is used to stain bacteria and sometimes as a disinfectant. Fucshine when mixed with sodium bisulphite is known as Schiff's reagent. Invented by Hugo (Ugo) Schiff (1834-04-26 - 1915-09-08) was a German Chemist. Born in Frankfurt am Main, Schiff was a student of Friedrich Wöhler in Göttingen. In 1879 he founded the Chemical Institute of the University of Florence. He discovered Schiff bases and other imines, and was responsible for research into aldehydes and had the Schiff test named after him. He also worked in the field of amino acids and the Biuret reagent. Schiff died in Florence.

The Schiff test

The Schiff test named after Hugo Schiff is a chemical test for the detection of aldehydes. An unknown sample is added to the decolorized Schiff reagent and when aldehyde is present a characteristic magenta or purple colour develops. The Schiff reagent is the reaction product of Fuchsine and sodium bisulfite. Human skin also contains aldehydes and gets stained as well.

Schiff reagents are used for various staining methods, *e.g.* Feulgen stain and periodic acid-Schiff stain.

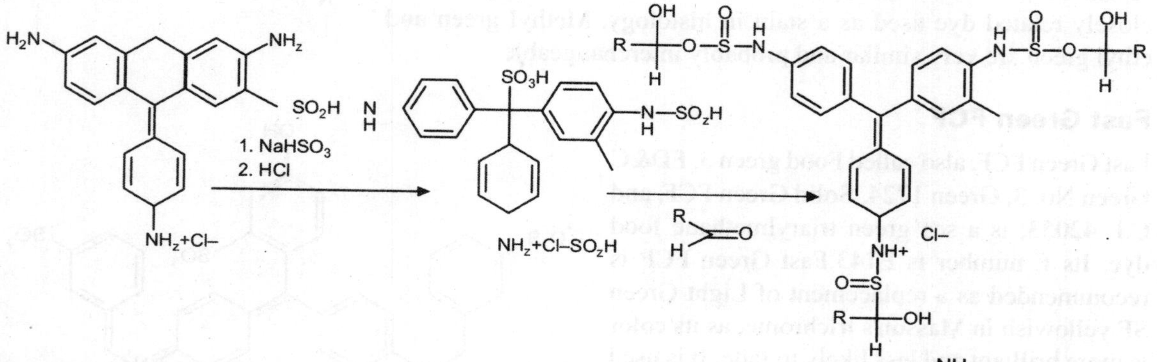

New Fuchsine

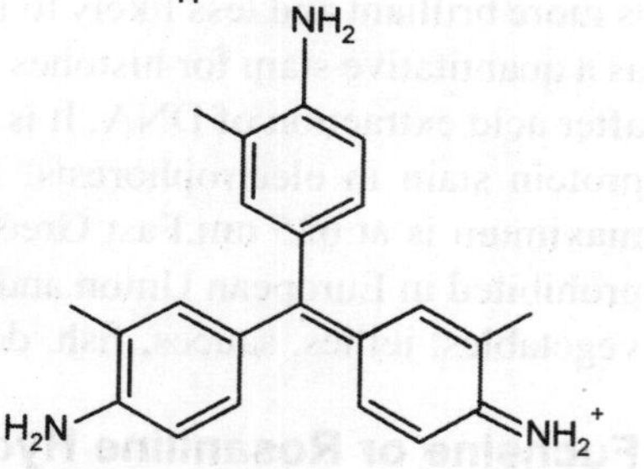

New fuchsine, new fuchsin, Magenta III, Basic Violet 2, or C.I. 42520 (from German "fuchs", fox) is a magenta dye having chemical formula $C_{22}H_{24}N_3Cl$. It is closely related to fuchsine and fuchsine acid. It can be used for staining acid fast organism, e.g. by Ziehl-Neelsen stain.

Light Green SF Yellowish

Light Green SF yellowish, or Light Green, Acid Green, Lissamine green SF, Acid Green 5', Food Green 2, FD&C Green no. 2, Green No. 205, Acid Brilliant Green 5, Pencil Green SF, or C.I. 42095, is a green triarylmethane dye. It is used in histology for staining collagen; for that purpose it is a standard dye in North America. In Masson's trichrome it is used as a counterstain to acid fuchsin. It is a critical component of Papanicolaou stains together with eosin Y and bismarck brown Y. It usually comes as a disodium salt. Its maximum absorption is at 630 (422) nm.The dye is not very durable -- it has a tendency to fade. When fading is to be avoided, it is replaced with Fast Green FCF, which also has more brilliant color. Fast Green FCF can also substitute Light Green SF yellowish in many other procedures.

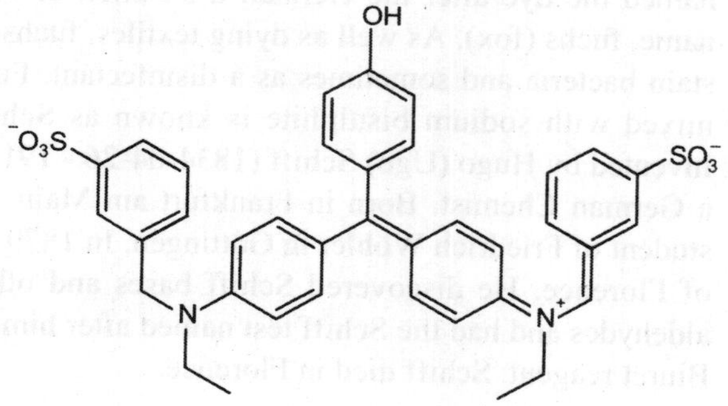

Malachite Green

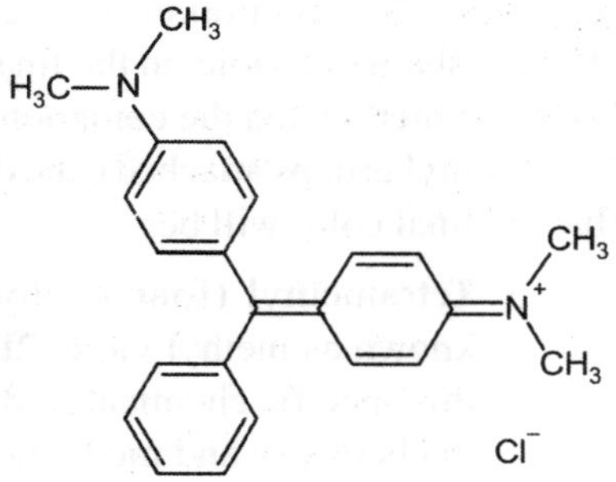

Malachite green, also called aniline green, basic green 4, diamond green B, or victoria green B, IUPAC name:4-[(4-dimethylamino phenyl)-phenyl-methyl]-N,N-dimethyl-aniline) is a toxic chemical primarily used as a dye. When diluted, it can be used as a topical antiseptic or to treat parasites, fungal infections, and bacterial infections in fish and fish eggs. It is also used as a bacteriological stain.

However in 1992 in Canada, it was determined that there is a significant health risk to humans who eat fish contaminated with malachite green. The chemical was classified a Class II Health Hazard because it was found to be toxic to human cells and might cause liver tumor formation. However, due to its ease and low cost to manufacture, it is still used in certain countries with less restrictive laws for non-aquaculture purposes. In 2005 eels and fish imported from China were found in Hong Kong with traces of this chemical.

Malachite green (first transition) (pH indicator)		
below pH *0.2*		*above pH* *1.8*
0.2	↔	**1.8**

Malachite green (second transition) (pH indicator)		
below pH *11.5*		*above pH* *13.2*
11.5	↔	**13.2**

Methyl Blue

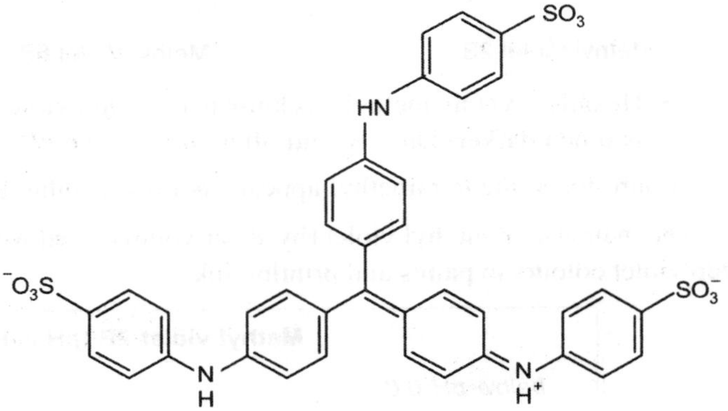

Methyl blue, also known as Cotton blue, Helvetia blue, Acid blue 93, or C.I. 42780, is a chemical compound used as a stain in histology. Methyl blue stains collagen blue in tissue sections. It is soluble in water and slightly soluble in ethanol. Methyl blue's molecular formula is $C_{37}H_{27}N_3Na_2O_9S_3$. Methyl blue should not be confused with methyl violet or methylene blue, two other stains. Methyl blue is also available in mixture with water blue, under name Aniline Blue WS, Aniline blue, China blue, or Soluble blue.It can be used in the Mallory's connective tissue stain and Gomori's one-step trichrome stain. It is used in differential staining.

Methyl Violet

Methyl violet is the name given to a group of similar chemicals used as pH indicators and dyes. Methyl violets are mixtures of tetramethyl, pentamethyl and hexamethyl pararosanilins. By blending the different versions, the dyemaker can create different shades of violet in the final dye. The more methylated the compound (the more methyl groups attached), the deeper blue the final color will be:

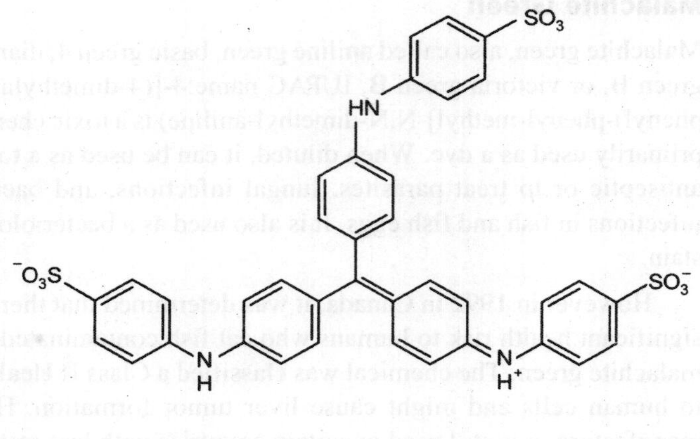

- Tetramethyl (four methyls) is known as methyl violet 2B, and this specific chemical finds uses in chemistry and medicine.
- Pentamethyl (five methyls) is known as methyl violet 6B, and is darker (in dye form) than 2B.

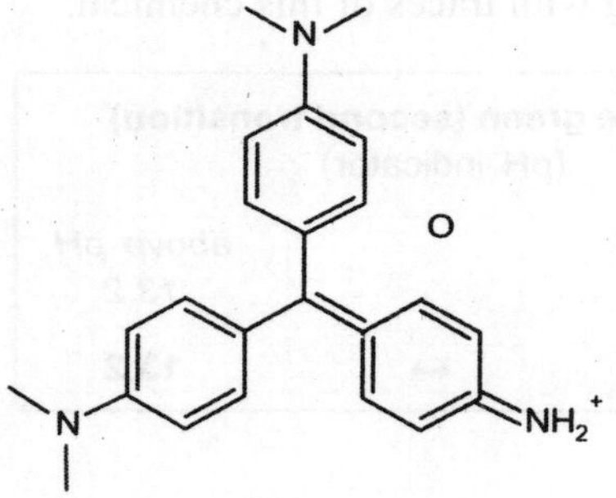

Methyl Violet 2B 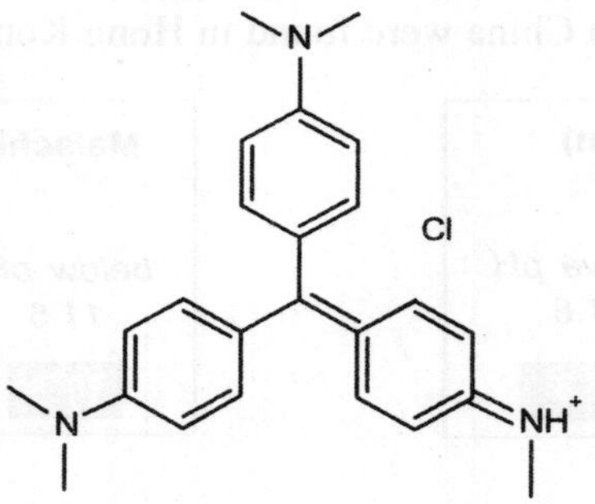 Methyl Violet 6B 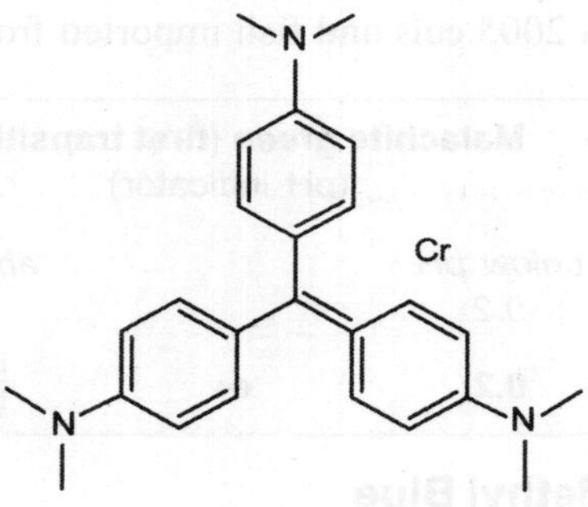 Methyl Violet 10B

- Hexamethyl (six methyls) is known as methyl violet 10B, or specifically as crystal violet. This is much darker than 2B, and often darker than 6B.

In pure form, the tetramethyl appears as lustrous blue-green crystals that melt at 137°C (279°F).

The main use of methyl violet (by sheer volume used worldwide) is to dye textiles purple and give deep violet colours in paints and printing ink.

Methyl violet 2B (pH indicator)		
below pH 0.0		*above pH 1.6*
0.0	↔	1.6

Pararosaniline

Pararosaniline, Magenta 0, Basic Red 9, or C.I. 42500 is a magenta dye having chemical formula $C_{19}H_{18}N_3Cl$. It is closely related to fuchsine, new fuchsine, and fuchsine acid. It makes the best Schiff's reagent.

Patent Blue V

Patent Blue V, also called Food Blue 5 or Sulphan Blue, is a dark bluish synthetic dye used as a food colouring. As a food additive, it has E number E131. It is a sodium or calcium salt of [4-(α-(4-diethylaminophenyl)-5-hydroxy-2,4-disulfophenyl-methylidene)-2,5-cyclohexadien-1-ylidene] diethylammonium hydroxide inner salt. It has the appearance of a violet powder.

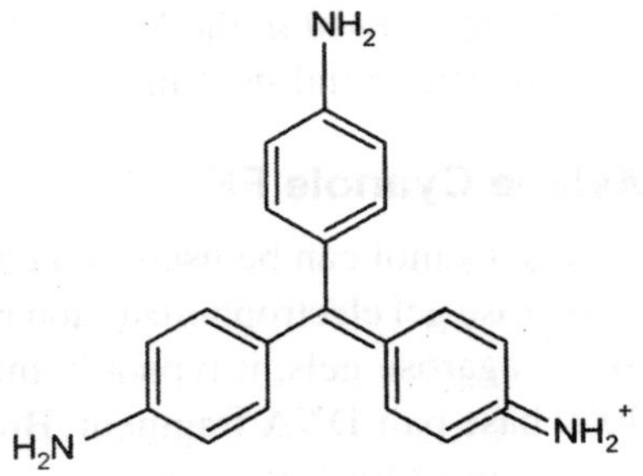

It is not widely used, but can be found in e.g. Scotch eggs. Patent Blue V is banned as a food dye in Australia, USA, and Norway.In medicine, Patent Blue V is used in lymphangiography as a dye to colour lymph vessels. It is also used in dental disclosing tablets as a stain to show dental plaque on teeth.May cause allergic reactions, with symptoms ranging from itching and nettle rash to nausea, hypotension, and in rare cases anaphylactic shock. Not recommended for children.

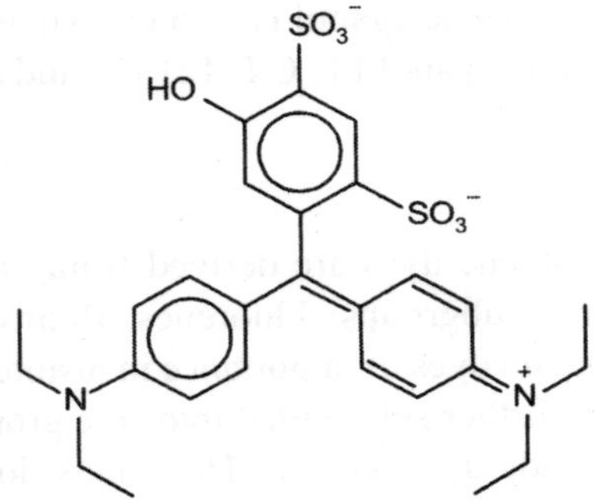

Victoria Blue BO

Victoria Blue BO, also known as C.I. Basic Blue 7 and C.I. 42595, is a chloride salt of a synthetic blue triarylmethane dye. Its chemical formula is $C_{33}H_{40}N_3Cl$. It has the appearance of a reddish blue powder. It is a photosensitizer.Victoria Blue BO base, also known as Solvent Blue 5 and C.I. 42595:1, is a hydroxide instead of chloride. Its chemical formula is $C_{33}H_{41}N_3O$.

Victoria Blue BO is used to dye anionic substrates, eg. wool, silk, nylon, and acrylics, where bright dying is required. It is also used for staining in microscopy, where it is used to stain mitochondria. As Solvent Blue 5, it is used in some pyrotechnic compositions for blue colored smoke.

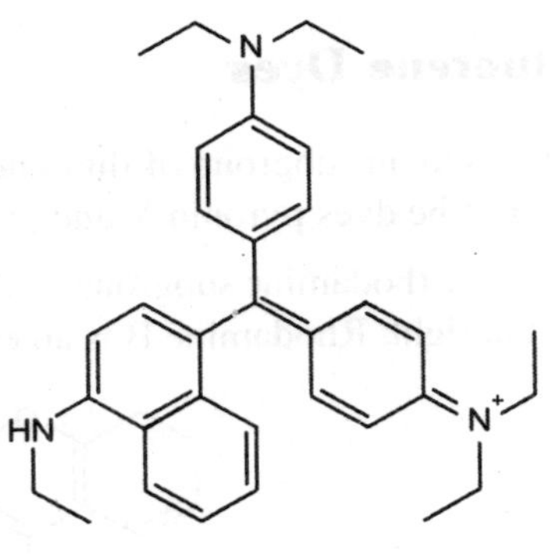

Water Blue

Water blue, also known as aniline blue, Acid blue 22, Soluble Blue 3M, Marine Blue V, or C.I. 42755, is a chemical compound used as a stain in histology. Water blue stains collagen blue in tissue sections. It is soluble in water and slightly soluble in ethanol. Water blue's molecular formula is $C_{32}H_{25}N_3O_9S_3Na_2$. Water blue is also available in mixture with methyl blue, under name Aniline Blue WS, Aniline blue, China blue, or Soluble blue.

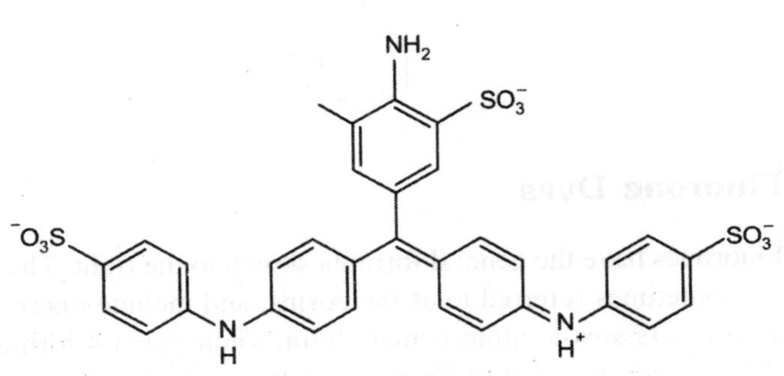

It can be used in the Mallory's connective tissue stain and Gomori's one-step trichrome stain. It is used in differential staining.

Xylene Cyanole FF

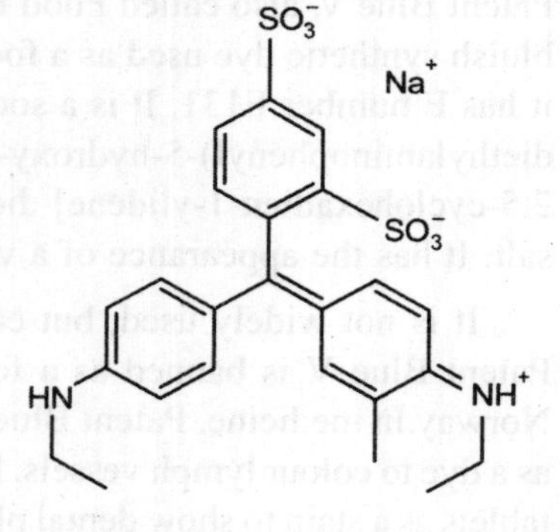

Xylene cyanol can be used as a color marker to monitor the process of agarose gel electrophoresis and polyacrylamide gel electrophoresis; in 1% agarose gels, it typically migrates at about the same rate as a 4000 base pair DNA fragment. Bromophenol blue and orange G can also be used for this purpose.

Xylene cyanol is also known as xylene cyanole, Acid Blue 147, xylene cyanol FF, C.I. 42135, ,and sometimes as xylene cyanole FF.

SO_3^- Na^+ SO_3^- HN NH^+

XANTHENE DYES

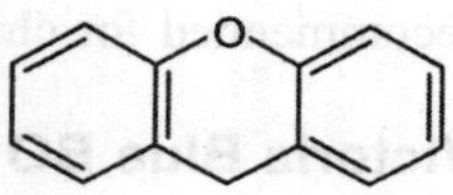

Xanthene dyes are derived from xanthene. This class of dyes is divided into three subgroups: Fluorenes, fluorones and rhodols. Fluorenes and fluorones contain dyes of importance in histotechnology, the rhodols do not. The fluorenes are further subdivided into five groups. Of these the pyronins and rhodamines include dyes we use. The others do not.

O

Fluorene Dyes

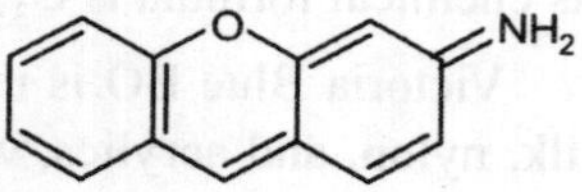

The pyronin subgroup of fluorenes have the general formula shown to the right. The dyes pyronin Y and pyronin B are examples.

O NH_2

The rhodamine subgroup of fluorenes have the general formula shown to the right. Rhodamine B is an example.

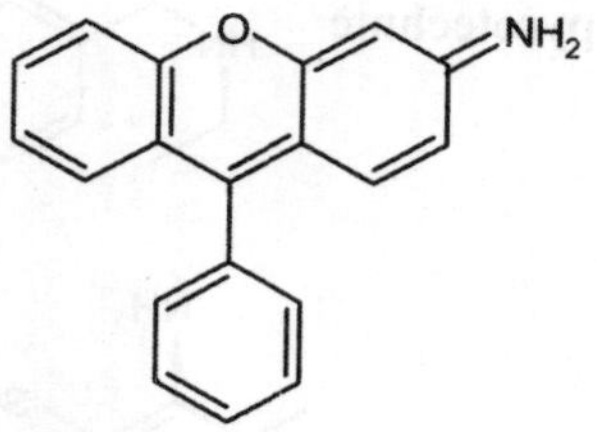

O NH_2

Fluorone Dyes

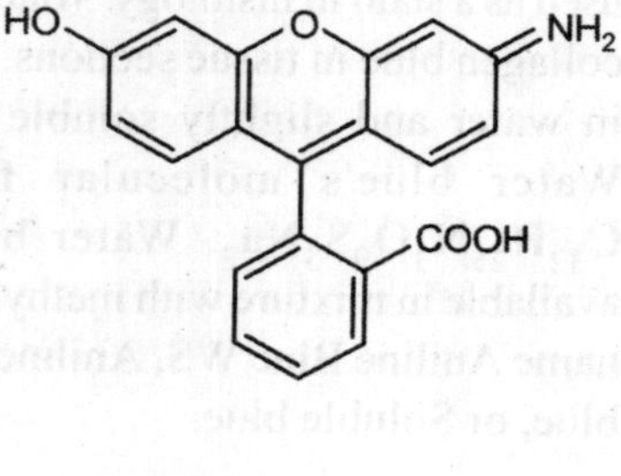

HO O NH_2 COOH

Fluorones have the general formula shown to the right. The fluorones are sometimes referred to as the eosins, and include many dyes used as counterstains to alum hematoxylin. Their general formula shown above is actually that of fluorescien, from which they are derived.

Pyronin Y

Property	Data
Common name	Pyronin Y
Suggested name	Pyronin Y
Other names	Pyronin G
C.I. number	45005
Class	Pyronin
Ionisation	Basic
Solubility aqueous	8.96%
Solubility ethanol	0.6%
Absorption maximum	552 (Conn), 545 (Gurr)
	548 (Aldrich)
Colour	Red
Empirical formula	$C_{17}H_{19}N_2OCl$
Formula weight	302.8

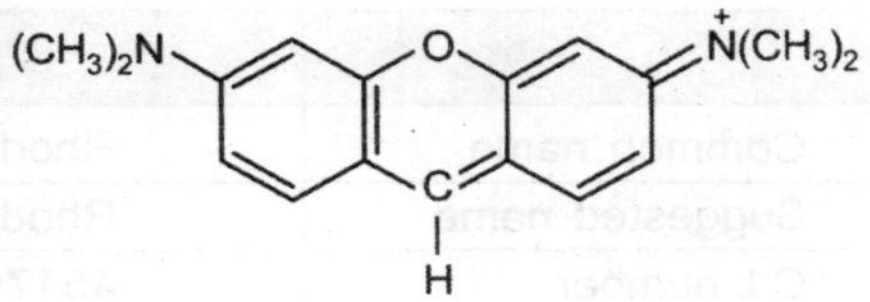

Pyronin Y is used with methyl green to selectively demonstrate RNA (red) in contrast to DNA (green) with the Unna-Pappenheim method.

Pyronin B

Pyronin B is sometimes substituted for pyronin Y in the Unna-Pappenheim stain for nucleic acids. According to Merck the dye is supplied as a complex with ferric chloride. This probably accounts for the difference in formula weights given by Conn compared to that of Aldrich.

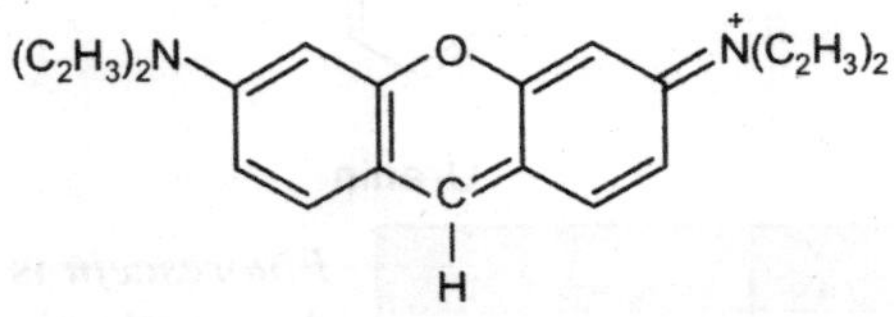

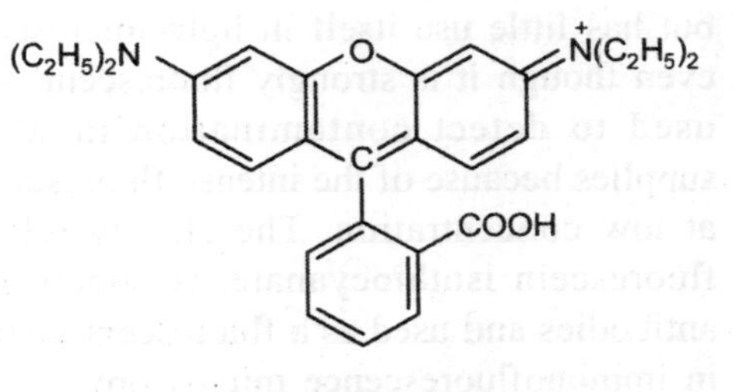

Pyronon B
according to Merck

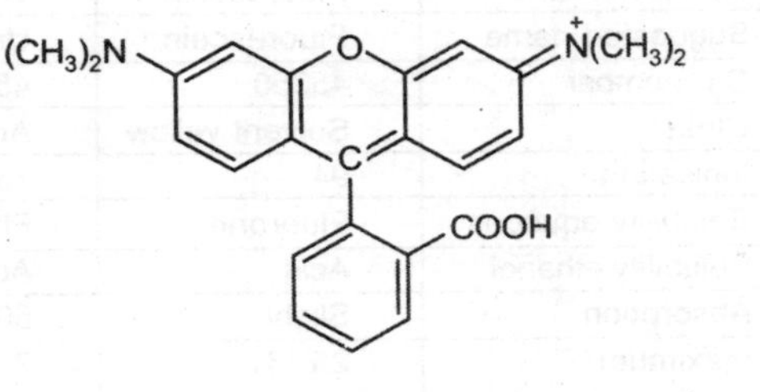

Pyronin B
according to Conn

Rhodamine B

Property	Data
Common name	Rhodamine B
Suggested name	Rhodamine B
C.I. number	45170
C.I. name	Basic violet 10
Class	Rhodamine
Ionisation	Basic
Solubility aqueous	Very
Solubility ethanol	Very
Absorption maximum	556.5 (Conn), 548 (Gurr)
	543,355 (Aldrich)
Colour	Red
Empirical formula	$C_{28}H_{31}N_2O_3Cl$
Formula weight	479

Rhodamine B is an amphoteric dye, although usually listed as basic as it has an overall positive charge. It is most commonly used as a fluorochrome, an example being in mixture with auramine O to demonstrate acid fast organisms.

The structural formula given in the Merck Index is slightly different than in Conn, having methyl groups instead of ethyl groups.

Fluorescein

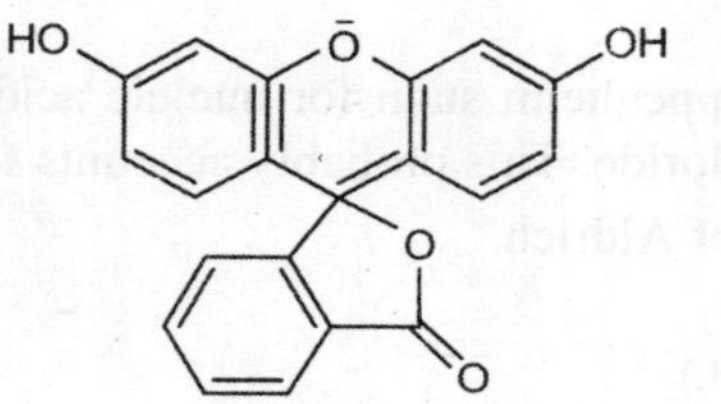

Fluorescein

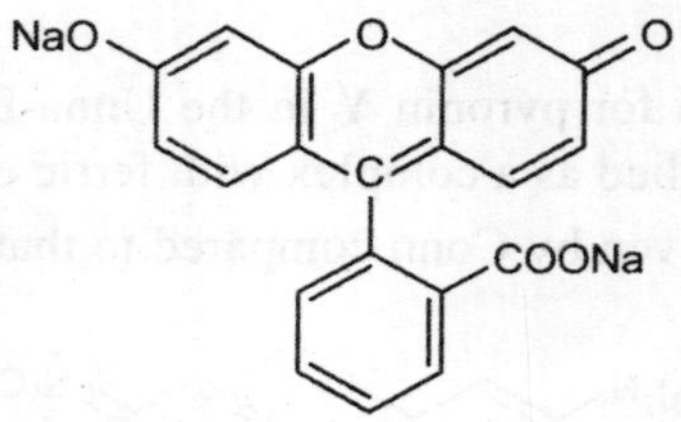

Uranin

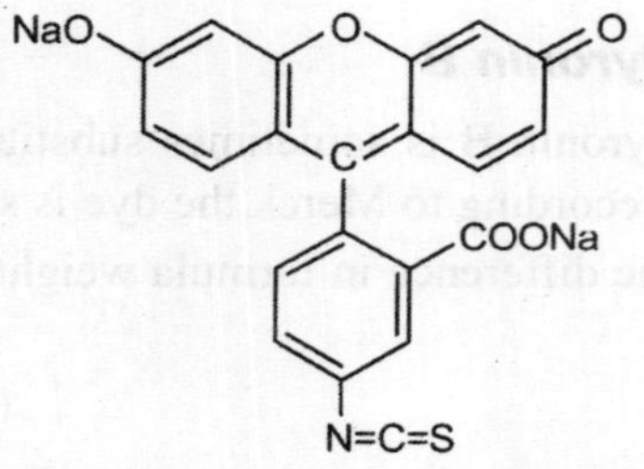

Fluorescein isothiocyanate

Property	Data	Data
Common name	Fluorescein	Uranin
Suggested name	Fluorescein	Uranin
C.I. number	45350	45350
Class	Solvent yellow	Acid yellow
Ionisation	94	73
Solubility aqueous	Fluorone	Fluorone
Solubility ethanol	Acid	Acid
Absorption	Slight	50%
maximum	2.21%	7.19%
	493.5,460	493.5 (Conn)
Colour	485 (Gurr)	485 (Gurr)
Empirical formula	Yellow	Yellow
Formula weight	$C_{20}H_{12}O_5$	$C_{20}H_{10}Na_2O_5$
	332.3	376.3

Fluorescein is the precursor of the group of dyes collectively referred to as the eosins, but has little use itself in light microscopy even though it is strongly fluorescent. It is used to detect contamination in water supplies because of the intense fluorescence at low concentration. The closely related fluorescein isothiocyanate, is attached to antibodies and used as a fluorescent marker in immunofluorescence microscopy.

Uranin is the sodium salt of fluorescein.

Flurescein isothiocyanate is prepared by heating resorsinol phthalic anhydride and oxalic acid

Phthalic anhydride + Resorcinol → (Oxalic acid, Heat) fluorescein

Erythrosin B

Property	Data
Common name	Erythrcsin B
Suggested name	Erythrosin B
C.I. number	45430
C.I. name	Acid red 51
Class	Fluorone
Ionisation	Acid
Solubility aqueous	11%
Solubility ethanol	1.8%
Absorption maximum	524 (Merck) 525 (Gurr) 524-527 (Conn)
Colour	Red
Empirical formula	$C_{20}H_6O_5I_4Na_2$
Formula weight	879.9

This dye is not commonly used in North America, although it is a perfectly satisfactory substitute for Eosin Y ws. In Europe it is used instead of Phloxine in the hemtoxylin phloxine saffron (HPS) stain, which is the same as the hematoxylin erythrosine saffron (HES) method in Europe.

Rose Bengal

Property	Data
Common name	Rose bengal
Suggested name	Rose bengal
C.I. number	45440
C.I. name	Acid red 94
Class	Fluorone
Ionisation	Acid
Solubility aqueous	36%
Solubility ethanol	7.5%
Absorption maximum	544-548 (Conn) 548-549 (Aldrich) 544 (Gurr)
Colour	Red
Empirical formula	$C_{20}H_2O_5I_4Cl_4Na_2$
Formula weight	1017.7

Rose bengal has been used to replace Eosin Y ws in the H&E stain for photomicrography as early colour film recorded eosin Y poorly. It is now rarely used.

Merbromin

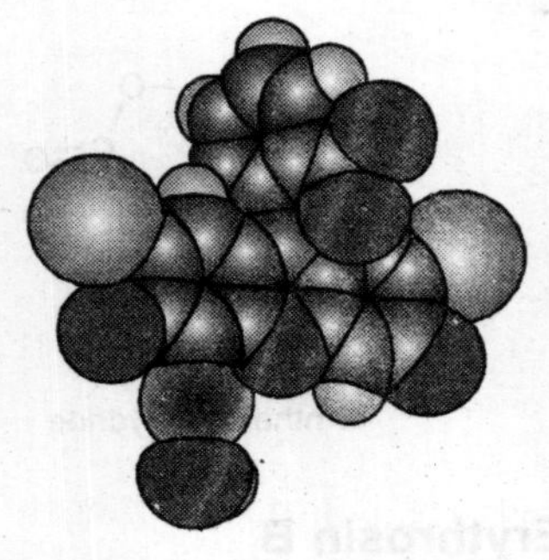

Merbromin (marketed as Mercurochrome, Merbromine, Sodium mercurescein, Asceptichrome, Supercrome and Cinfacromin) is a topical antiseptic used for minor cuts and scrapes. It is no longer sold in the USA because of its mercury content. Merbromin is an organomercuric disodium salt compound and a fluorescein.

It is prepared by the action of dibromoflurescein and mercuric acetate or mercuric oxide (red and acetic acid and sodium hydroxide).

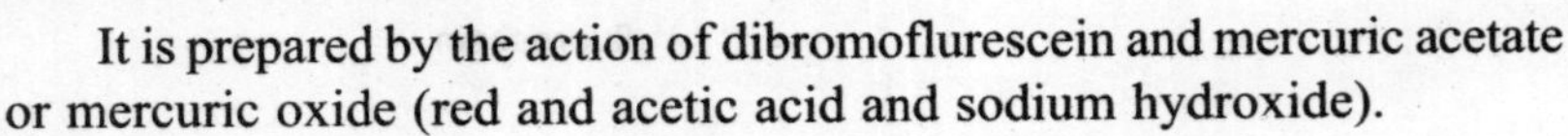

Mercuric acetate / Sodium hydroxide

Dibromoflurescein → Dibromohydroxymercurifluorescein

Mercurochrome and Tinctures

Mercurochrome is the trade name of merbromin and (usually) of merbromin tinctures made of merbromin and alcohol or water (usually 2% merbromin and 4% merbromin to 98% alcohol or water).

HETEROCYCLIC DYES

Hetercyclic dyes consists of Safranines, quinoneimine azine cyanine, etc dyes

Safranines

Molecular structure of the Safranine cation.

Safranines are the azonium compounds of symmetrical 2,8-dimethyl-3,7-diamino-phenazine. They are obtained by the joint oxidation of one molecule of a para-diamine with two molecules of a primary amine; by the condensation of para-aminoazo compounds with primary amines, and by the action of para-nitrosodialkylanilines with secondary bases such as diphenylmetaphenylenediamine. They are crystalline solids showing a characteristic green metallic luster; they are readily soluble in water and dye red or violet. They are strong bases and form stable monacid salts. Their alcoholic solution shows a yellow-red fluorescence. Phenosafranine is not very stable in the free state; its chloride forms green plates. It can be readily diazotized, and the diazonium salt when boiled with alcohol yields aposafranine or benzene induline, $C_{18}H_{12}N_3$. F. Kehrmann showed that aposafranine could be diazotized in the presence of cold concentrated sulfuric acid, and the diazonium salt on boiling with alcohol yielded phenylphenazonium salts. Aposafranone, $C_{18}H_{12}N_2O$, is formed by heating aposafranine with concentrated hydrochloric acid. These three compounds are perhaps to be represented as ortho- or as para-quinones. The "safranine" of commerce is a ortho-tolusafranine. The first aniline dye-stuff to be prepared on a manufacturing scale was mauveine, which was obtained by Sir W.H. Perkin by heating crude aniline with potassium bichromate and sulfuric acid. Mauveine was converted to parasafranine (1,8-dimethyl Safranine) by Perkin in 1879 by oxidative/reductive loss of the 7-N-para-tolyl group.

Safranin O

H3C, CH3, N, N+, H2N, NH2

Dimethyl safranin

H3C, CH3, N, N+, H2N, NH2, H3C

Trimethyl safranin

Property	Data
Common name	Safranin
Suggested name	Safranin O
C.I. number	50240
C.I. name	Basic red 2
Class	Safranin
Ionisation	Basic
Solubility aqueous	5.45%
Solubility ethanol	3.41%
Absorption maximum	530 (Conn, Aldrich) 517 (Gurr)
Colour	Red
Empirical formula	$C_{20}H_{19}N_4Cl$ (Dimethyl) $C_{21}H_{21}N_4Cl$ (Trimethyl)
Formula weight	350.8 (Dimethyl) 364.9 (Trimethyl)

Safranin O is a mixture of the two compounds shown according to Conn. Aldrich and Gurr give only the dimethyl compound.

Safranin O is most commonly used for counterstaining nuclei red. In medical histotechnology it has little other use, although it has been used as a metachromatic method for cartilage, which is stained yellow.

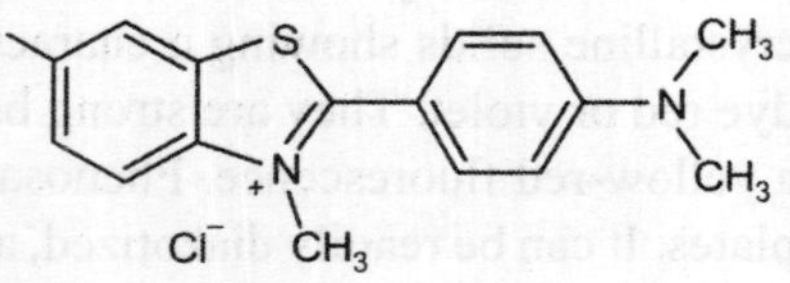

Thioflavin T (Basic Yellow 1 or CI 49005) is a benzothiazole salt obtained by the methylation of dehydrothiotoluidine with methanol in the presence of hydrochloric acid. The dye is used to visualize plaques composed of amyloid beta found in the brains of Alzheimer's disease patients. When it binds to beta sheets, such as those in amyloid oligomers, the dye undergoes a characteristic 120 nm red shift of its excitation spectrum that may be selectively excited at 450 nm, resulting in a fluorescence signal at 482 nm.

Thioflavin S is a mixture of compounds that results from the methylation of dehydrothiotoluidine with sulfonic acid. It is also used to stain Alzheimer's plaques.

Mauve

Mauve was first named in 1856. Chemist William Henry Perkin, then eighteen, was attempting to create artificial quinine. An unexpected residue caught his eye, which turned out to be the first aniline dye-specifically, mauveine, sometimes called aniline purple. Perkin was so successful in recommending his discovery to the dyestuffs industry that his biography by Simon Garfield is titled Mauve.

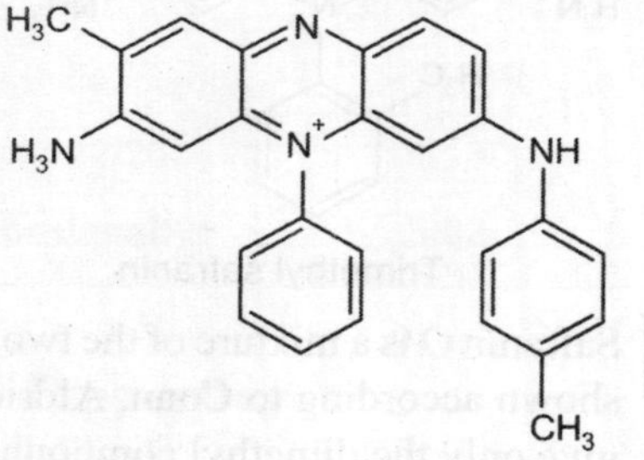

Mauvenine A

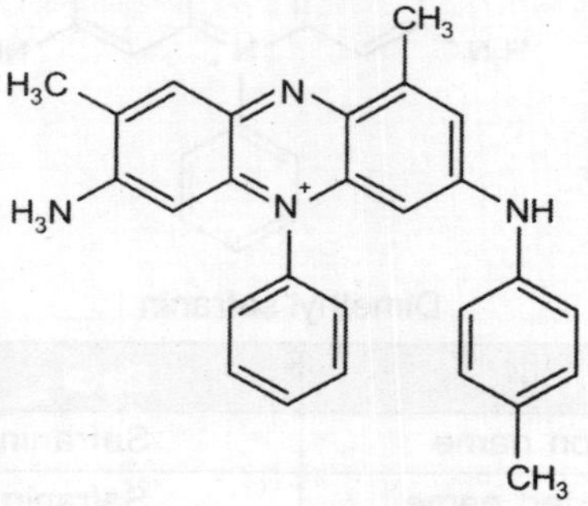

Mauvenine B

Mauveine, also known as aniline purple, was the first synthetic organic dye. The chemical name is 3-amino-2,±9-dimethyl-5-phenyl-7-(p-tolylamino) phenazinium acetate. The formula is $C_{26}H_{23}N_4^+X^-$ (mauveine A) and $C_{27}H_{25}N_4^+X^-$ (mauveine B)

Gallocyanin

Property	Data
Common name	Gallocyanin
Suggested name	Gallocyanin
CI. number	51030
C.I. name	Mordant blue 10
Class	Oxazin

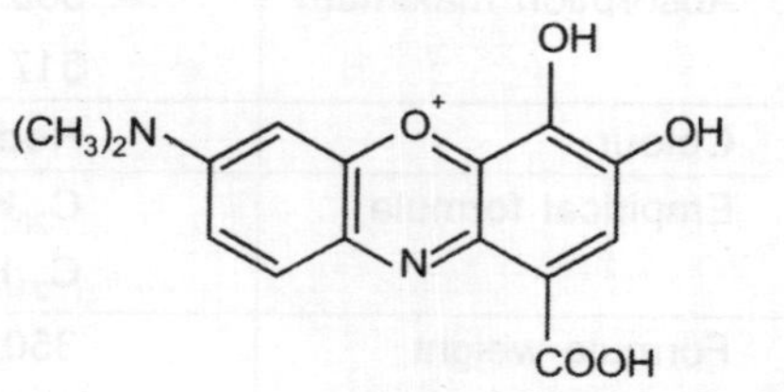

Ionisation	Basic
Solubility aqueous	Poor
Solubility ethanol	Soluble
Absorption maximum	636 (Conn), 601 (Aldrich)
Colour	Blue
Empirical formula	$C_{15}H_{13}N_2O_5Cl$
Formula weight	336.7

When mordanted with iron alum, gallocyanin makes a useful nuclear stain, suitable as an alum hematoxylin substitute in H&E. Einarson used it with chrome alum to demonstrate nucleic acids. Gallocyanine dye is prepared by the action of gallic acid on p-nitrosodimethylaniline.

Proflavine

Proflavine also called proflavin and diaminoacridine, is an acriflavine derivative, a disinfectant bacteriostatic against many gram-positive bacteria. It has been used in the form of the dihydrochloride and hemisulfate salts as a topical antiseptic, and was formerly used as a urinary antiseptic. Proflavine is also known to have a mutagenic effect on DNA by intercalating between nucleic acid base pairs.

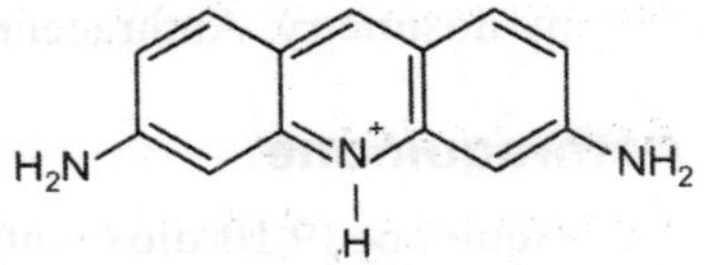

Anthracene

Anthracene is a solid polycyclic aromatic hydrocarbon consisting of three benzene rings derived from coal-tar. Anthracene is used in the artificial production of the red dye alizarin. It is also used in wood preservatives, insecticides, and coating materials. Anthracene is colorless but exhibits a blue (400-500nm peak) fluorescence under ultraviolet light.

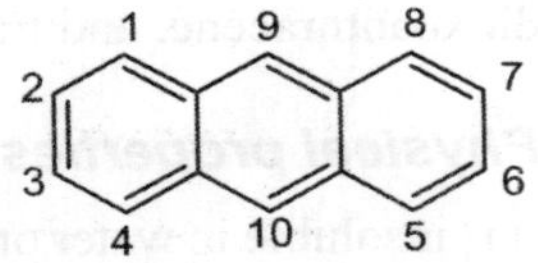

Synthesis

A classis method for the preparation of anthracene in the laboratory is by cyclodehydration of o-methyl- or o-methylene-substituted diarylketones in the so-called Elbs reaction.

Reactions

Anthracene has the ability to photodimerize with irradiation by UV light. This results in considerable changes in the physical properties of the material.

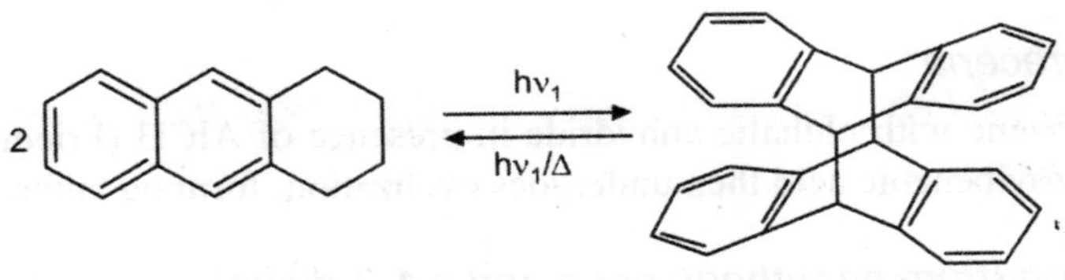

The dimer is connected by two covalent bonds resulting from the [4+4] cycloaddition. The dimer reverts to anthracene thermally or with UV irradiation below 300 nm. The reversible bonding and photochromic properties of anthracenes is the basis of many potential applications using poly and monosubstituted anthracene derivatives. The reaction is sensitive to oxygen.In most other reactions of anthracene, the central ring is also targeted, as it is the most highly reactive. Electrophilic substitution occurs at the "9" and "10" positions of the center ring, and oxidation of anthracene occurs readily, giving anthraquinone, $C_{14}H_8O_2$

Uses

Anthracene can also have a hydroxyl group to form 1-hydroxyanthracene and 2-hydroxyanthracene, homologous to phenol and napthol, and hydroxyanthracene is also called anthrol, and anthracenol. Hydroxyanthracene derivatives are pharmacologically active, and are contained in aloe for example.Anthracene is an organic semiconductor.Anthracene is used as a scintillator for detectors of high energy photons, electrons and alpha particles. Plastics such as polyvinyltolulene can be doped with Anthracene to produce a plastic scintillator that is aproximately water equivalent for use in radiation therapy dosimetry. Anthracenes emission spectrum peaks at between 400nm and 440nm.

Anthraquinone

Anthraquinone (9,10-dioxoanthracene) is an aromatic organic compound. It is a derivative of anthracene. It has the appearance of yellow or light gray to gray-green solid crystalline powder. Its other names are 9,10-anthracenedione, anthradione, 9,10-anthrachinon, anthracene-9,10-quinone, 9,10-dihydro-9,10-dioxoanthracene, and trade names Hoelite, Morkit, Corbit, and others.

O

O

Physical properties

It is insoluble in water or alcohol, but dissolves in nitrobenzene and aniline. It is chemically fairly stable under normal conditions.

Natural Occurrences

Anthraquinone naturally occurs in some plants (eg. aloe, senna, rhubarb, and Cascara buckthorn), fungi, lichens, and insects, where it serves as a basic skeleton for their pigments. Natural anthraquinone derivatives tend to have laxative effects.

Chemistry

There are several ways to obtain anthraquinone:

Oxidation of Anthracene

Condensation of benzene with phthalic anhydride in presence of AlCl3 (Friedel-Crafts substitution). The resulting o-benzoylbenzoic acid then undergoes cyclization, forming anthraquinone.

Diels-Alder reaction (from naphthoquinone and a 1,3-diene)

In a classic organic reaction called the Bally-Scholl Synthesis (1905), anthraquinone condenses with glycerol forming Benzanthrone. In this reaction the quinone is first reduced with copper metal in sulfuric acid (converting one ketone group into a methylene group) after which the glycerol is added.

Industrial applications

Anthraquinone is used in production of dyes, such as alizarin. Many natural pigments are derivatives of anthraquinone. Anthraquinone is also used as a catalyst in production of wood pulp in pulp and paper industry. Another use is as a bird repellant on seeds.A derivative of anthraquinone (2-ethylanthraquinone) is used to produce hydrogen peroxide commercially.

Acridine

Acridine, $C_{13}H_9N$, is an organic compound and a nitrogen heterocycle. Acridine is also used to describe compounds containing the $C_{13}N$ tricycle.

Acridine is structurally related to anthracene with one of the central CH groups is replaced by nitrogen. Acridine, a colourless solid, was first isolated from coal tar. It is a raw material used for the production of dyes and some valuable drugs. Many acridines also have antiseptic properties such as Proflavine. Acridine and related derivatives bind to DNA and RNA due to their abilities to intercalate. Acridine Orange (3,6-dimethylaminoacridine) is a nucleic acid-selective metachromatic stain useful for cell cycle determination.

Acridine Orange

The chemical structure of Acridine orange is shown at the right. The absorption and fluorescence emission spectra are shown below (Fig. 28.5).

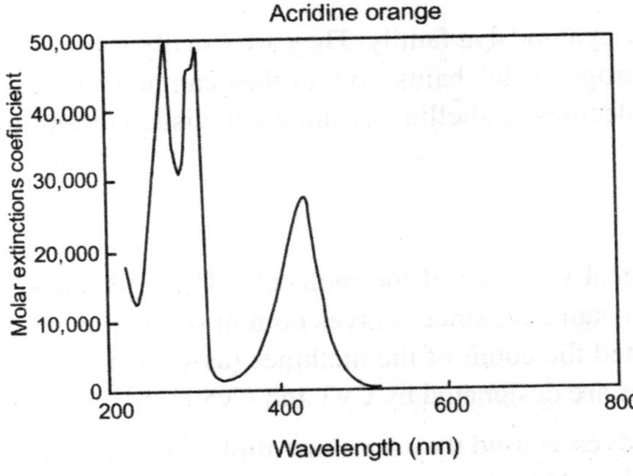

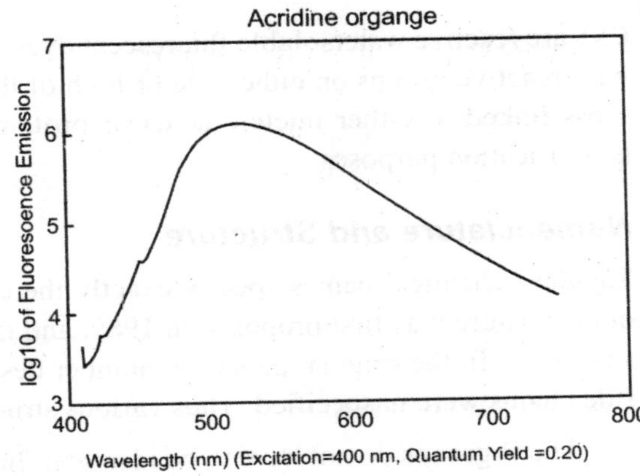

Fig. 28.5 The Absorption spectrum of Acridine orange

CYANINE DYES

Synthetic dyes with the general formula

$R_2N[CH{=}CH]_nCH{=}N{+}R_2 5 \ll R_2N{+}{=}CH[CH{=}CH]_nNR_2$ (*n* is a small number) in which:

- The nitrogen and part of the conjugated chain usually form part of a heterocyclic system, such as imidazole, pyridine, pyrrole, quinoline and thiazole,
- Cyanine is a non-systematic name of a synthetic dye family with the common molecular formula:

  $ArN{+}{=}CH[CH{=}CH]_n{=}NAr$

  where two quaternized nitrogens are joined by a poly-methine chain. Both nitrogens are each independently part of a heteroaromatic moiety, such as imidazole, pyridine, pyrrole, quinoline, thiazole, etc.
- Cyanines were first synthesized over a century ago, and there are a large number reported in the literature.

Cy3 and Cy5

Cy3 and Cy5 are dyes used in comparative genomic hybridization, and in gene chips which are used in transcriptomics. They are also used to label proteins for various studies including proteomics. Cy3 and

Cy3

Cy5

Cy5 are reactive watersoluble fluorescent dyes of the cyanine dye family. They are usually synthesized with reactive groups on either one or both of the nitrogen side chains, so that they can be chemically cross-linked to either nucleic acids or protein molecules. Labelling is done for visualization and quantification purposes.

Nomenclature and Structure

Standard chemical names specify exactly the chemical structure of the molecule. The Cy3 and Cy5 nomenclature was first proposed in 1989, and is non-standard, since it gives no hint of their chemical strucutres. In the original paper the number designated the count of the methines (as shown), and the side chains were unspecified. Thus various structures are designated by Cy3 and Cy5 in the literature.

The R groups do not have to be identical. In the dyes as used they are short aliphatic chains one or both of which ends in a highly reactive moieties such as N-hydroxysuccinimide or maleimide.

Spectral characteristics

Cy3 is excited maximally at 550 nm and emits maximally at 570 nm, in the green part of the spectrum; quantum yield is 0.15; FW=766.

Cy5 is excited maximally at 649 nm and emits maximally at 670 nm, in the red part of the spectrum; quantum yield is 0.28. FW=792.

The scanners actually use different laser emission wavelengths (typically 532 nm and 635 nm) and filter wavelengths (550-600 nm and 655-695 nm) to avoid background contamination. They are thus able to easily distinguish between two samples when one sample has been labeled with Cy3 and the other labeled with Cy5. They are also able to quantitate the amount of labeling in either sample.

Application Nucleic acid labelling

In microarray experiments DNA or RNA is labeled with either Cy3 or Cy5 that has been synthesized to carry the reactive group N-hydroxysuccinimide (NHS). Since, N-hydroxysuccinimide reacts readily only with aliphatic amine groups, which nucleic acids lack, nucleotides have to be modified with aminoallyl groups. This is done through incorporating aminoallyl-modified nucleotides during polymerase reactions. A good ratio is a label every 60 bases such that the labels are not too close to each other, thus resulting in quenching effects.

Protein Labelling

For protein labeling, Cy3 and Cy5 dyes usually bear maleimide reactive groups instead. The maleimide group allows conjugation of the fluorescent dye to the sulfhydryl group of cysteine residues. Cysteins can be added and removed from the protein domain of interest via PCR mutagenesis.

Cy5, is sensitive to the electronic environment it resides in. Changes in the conformation of the protein it is attached to will produce an enhancement or quenching of the emission. The rate of this change can be measured to determine enzyme kinetic parameters. The dyes can be used for similar purposes in FRET experiments.

Cy3 and Cy5 are used in proteomics experiments so that samples from two sources can be mixed and run together thorough the separation process. This eliminates variations due to differing experimental conditions that are inevitable if the samples were run separately. These variations make it extremely difficult, if not impossible, to use computers to automate the acquisition of the data after the separation is complete. Using these dyes make makes the automation trivial.

Curcumin

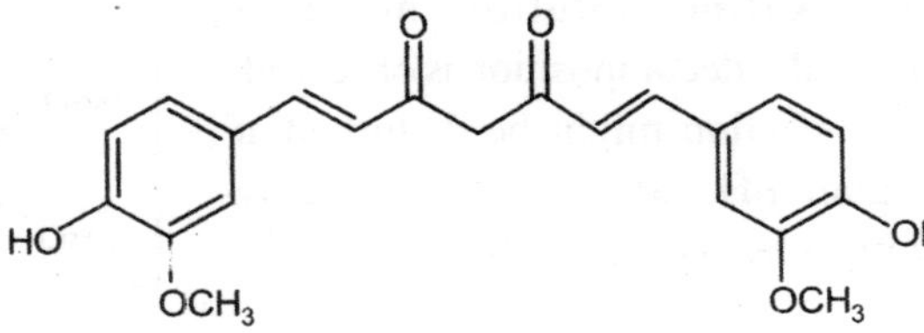

Curcumin is the principal curcuminoid of the Indian curry spice turmeric. The curcuminoids are polyphenols and are responsible for the yellow color of turmeric. Curcumin can exist in at least two tautomeric forms, keto and enol. The enol form is more energetically stable in the solid phase and in solution Curcumin can be used for boron quantification in the so-called curcumin method. It reacts with boric acid forming a red colored compound, known as rosocyanine. Since curcumin is brightly colored, it may be used as a food colouring. As a food additive, its E number is E100.

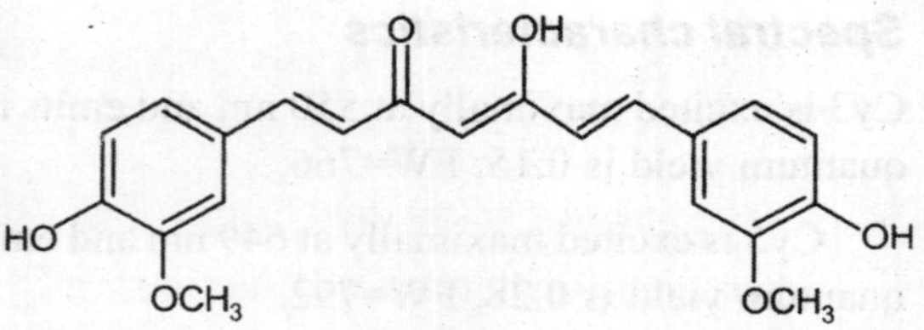

Rosocyanine

Rosocyanine and Rubrocurcumin are two red colored materials, which are formed by the reaction between curcumin and borates.

Application

The color reaction between borates and curcumin is used within the spectrophotometrical determination and quantification of boron present in food or materials. Curcumin is a yellow coloring natural pigment, found in the root stocks of some Curcuma species, especially in Curcuma longa (cf. turmeric) in concentrations up to 3%. In the so called curcumin method for boron quantification it serves as reaction partner for boric acid. The reaction is very sensitive and also smallest quantities of boron can be detected. The maximum absorbance at 540 nm for rosocyanine is used in this colorimetric method. The formation of rosocyanine depends on the reaction conditions. The reaction is carried out preferentially in acidic solutions containing hydrochloric or sulfuric acid. The color reaction also takes place under different conditions, but in alkaline solution however gradually decomposition is observed. The reaction might be disturbed at higher pH values, interferencing with other compounds.

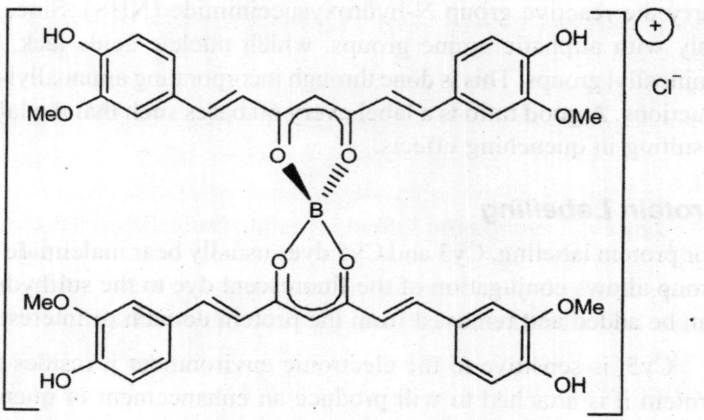

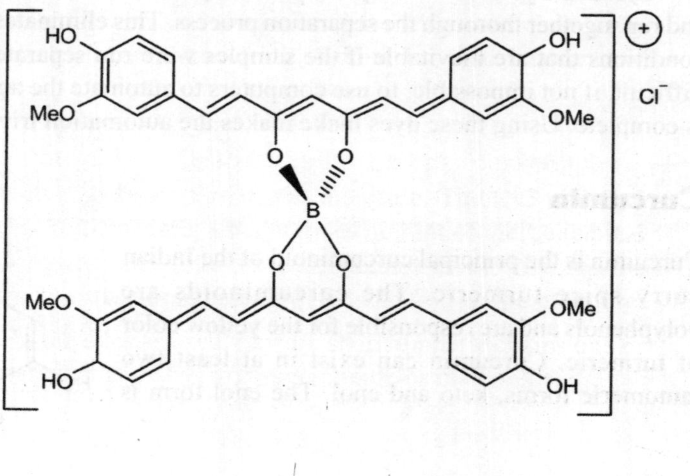

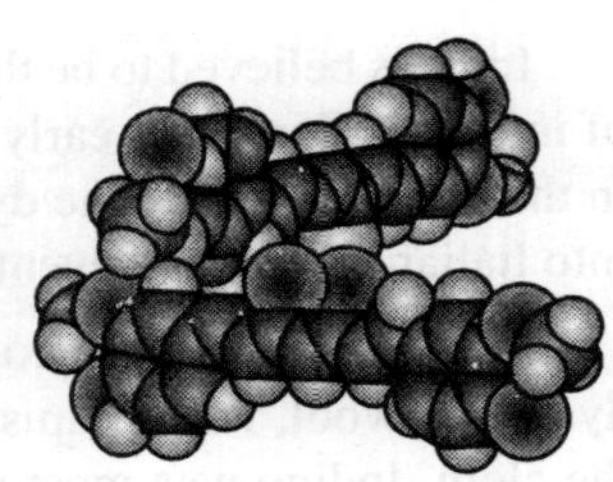

Rosocyanine is formed as 2 : 1-complex from curcumin and boric acid in acidic solutions. Curcumin possesses a 1,3-diketone structure and can therefore considered as a chelating agent. The formed boron complexes are called dioxaborines (here a 1,3,2-dioxaborine). The structural formula 1 given for rosocyanine (a cationic dicurcuminatoboron complex, here written with chloride as counter ion) is idealized. Investigations on the structure show that the positive charge is distributed all over the molecule, and therefore a structure as indicated in formula 2 rather describes the molecule condition. In rosocyanin the two curcumin molecules halves are not plain within the same level, but are twisted against each other. The same applies to rubrocurcumin.

In order to exclude the disturbing presence of other materials during the boron quantification using the curcumin method, a variant was developed. 2,2-Dimethyl-1,3-hexanediol or 2-ethyl-1,3-hexanediol are added, in addition to curcumin, to a neutral solution of the boron-containing solution. The formed complex between boron and the 1,3-hexanediol derivate, is removed from the aqueous solution by extraction in an organic solvent. After acidifying the organic phase, rubrocyanine is formed, which could be measured by colorimetric methods. The reaction of curcumin with borates in presence of oxalic acid therefore produces the coloring material rubrocurcumin.

Characteristics

Rosocyanine - [sum formula $[B(C_{21}H_{19}O_6)_2]Cl$, here given as chloride] - is as dark-green colored solid with glossy-metallic shine, forming red colored solutions. In water and some organic solvents it is almost insoluble, in ethanol it is very slightly soluble (up to 0.01%), but in pyridine, sulfuric acid and acetic acid a clearly better solubility is observed (approx. 1%). An alcoholic solution of rosocyanine temporarily turns deeply blue on treatment with alkali. In rubrocurcumin one molecule curcumin is replaced with oxalic acid. Rubrocurcumin produces a similar red colored solution.Rosocyanine is build from ions, while rubrocurcumin is a neutrally charged composition.

INDIGOID DYES

This group of dyes include the thio indigo and indigo dyes.

Indigo

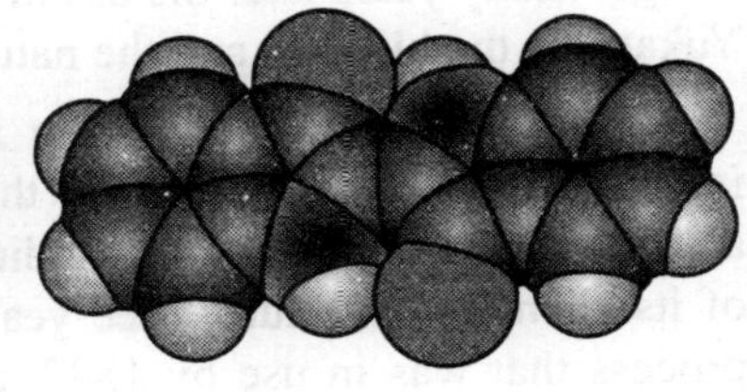

Indigo (earlier indico, from Lat. indicum, the Indian substance or dye; the Sans. name was *nit*', from *nila,* dark blue, and this through Arab. *al-nil, annil,* gives "*aniline*") one of the most important and valuable of all dyestuffs.

History

Indigo is among the oldest dyes to be used for textile dyeing and printing. Many Asian countries, such as India, China, and Japan, have used indigo as a dye for centuries. The dye was also known to ancient civilizations in Mesopotamia, Egypt, Greece, Rome, Britain, Peru, Iran, and Africa.

India is believed to be the oldest center of indigo dyeing in the Old World. It was a primary supplier of indigo to Europe as early as the Greco-Roman era. The association of India with indigo is reflected in the Greek word for the dye, which was indikon. The Romans used the term indicum, which passed into Italian dialect and eventually into English as the word indigo.

In Mesopotamia, a Neo-Babylonian cuneiform tablet of the 7th century BC gives a recipe for the dyeing of wool, where lapis-coloured wool (uqnatu) is produced by repeated immersion and airing of the cloth. Indigo was most probably imported from India.

The Romans used indigo as a pigment for painting and for medicinal and cosmetic purposes. It was a luxury item imported to the Mediterranean from India by Arab merchants.

Indigo remained a rare commodity in Europe throughout the Middle Ages. Woad, a dye derived from the plant Isatis tinctoria (Brassicaceae), was used instead.

In the late fifteenth century, the Portuguese explorer Vasco da Gama discovered a sea route to India. This led to the establishment of direct trade with India, the Spice Islands, China, and Japan. Importers could now avoid the heavy duties imposed by Persian, Levantine, and Greek middlemen and the lengthy and dangerous land routes which had previously been used. Consequently, the importation and use of indigo in Europe rose significantly. Much European indigo from Asia arrived through ports in Portugal, the Netherlands, and England. Spain imported the dye from its colonies in South America. Many indigo plantations were established by European powers in tropical climates; it was a major crop in Jamaica and South Carolina, with much or all of the labor performed by enslaved Africans and African-Americans. Indigo plantations also thrived in the Virgin Islands. However, France and Germany outlawed imported indigo in the 1500s to protect the local woad dye industry.

Indigo was the foundation of centuries-old textile traditions throughout West Africa. The use of indigo here pre-dated synthetics. From the Tuareg nomads of the Sahara to Cameroon, clothes dyed with indigo signified wealth. Women dyed the cloth in most areas, with the Yoruba of Nigeria and the Manding of Mali particularly well known for their expertise. Among the Hausa male dyers working at communal dye pits were the basis of the wealth of the ancient city of Kano, and can still be seen plying their trade today at the same pits.

In Japan, indigo became especially important in the Edo period when it was forbidden to use silk, so the Japanese began to import and plant cotton. It was difficult to dye the cotton fiber except with indigo. Many years later the use of indigo is very much appreciated as a color for the summer Kimono Yukata, as the blue sea and the nature are recalled on this traditional clothing.

In 1865 the German chemist Johann Friedrich Wilhelm Adolf von Baeyer began working with indigo. His work culminated in the first synthesis of indigo in 1880 from o-nitrobenzaldehyde and acetone upon addition of dilute sodium hydroxide, barium hydroxide, or ammonia and the announcement of its chemical structure three years later. BASF developed a commercially feasible manufacturing process that was in use by 1897, and by 1913 natural indigo had been almost entirely replaced by synthetic indigo. In 2002, 17,000 tons of synthetic indigo were produced worldwide.

In the nineteenth century, the British obtained much indigo from India. With the coming of the synthetic substitute, the demand for natural indigo dropped, and for many indigo farmers the raising of indigo became unprofitable.

A variety of plants, including woad, have provided indigo throughout history, but most natural indigo is obtained from those in the genus Indigofera, which are native to the tropics. In temperate climates indigo can also be obtained from woad (Isatis tinctoria) and dyer's knotweed (*Polygonum tinctorum*), although the Indigofera species yield more dye. The primary commercial indigo species in Asia was true indigo (*Indigofera tinctoria*, also known as *Indigofera sumatrana*). In Central and South America the two species *Indigofera suffruticosa* (Anil) and *Indigofera arrecta* (Natal indigo) were the most important.

Natural indigo was the only source of the dye until about 1900. Within a short time, however, synthetic indigo had almost completely superseded natural indigo, and today nearly all indigo produced is synthetic. In the United States, the primary use for indigo is as a dye for cotton work clothes and blue jeans. Over one billion pairs of jeans around the world are dyed blue with indigo. For many years indigo was used to produce deep navy blue colors on wool.

Indigo does not bond strongly to the fiber, and wear and repeated washing may slowly remove the dye.Indigo is also used as a food coloring, and is listed in the USA as FD&C Blue No. 2. The specification for FD&C Blue No. 2 includes three substances, of which the major one is the sodium salt of Indigotindisulphonate.

Indigotinesulphonate is also used as a dye in renal function testing, as a reagent for the detection of nitrates and chlorates and in the testing of milk.

Synthesis of Indigo

Indigo may be synthetically manufactured in a number of different ways. The original method, first used to synthesise indigo by Heumann in 1897, involves heating phenylglycine-o-carboxylic acid to 200°C in an inert atmosphere with sodium hydroxide. This produces indoxyl-2-carboxylic acid, a material that readily decarboxylates and oxidises in air to form indigo.

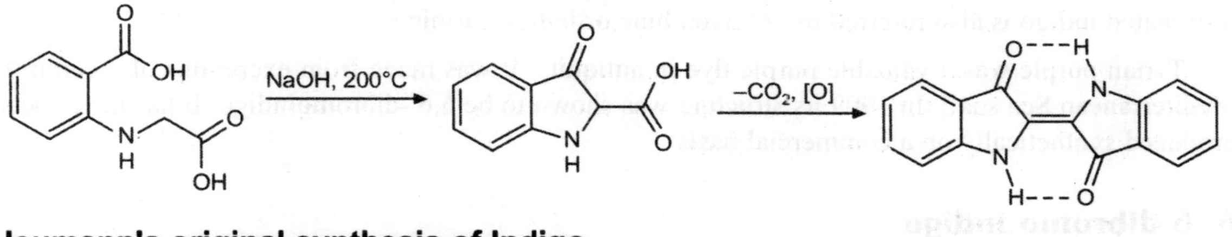

NaOH, 200°C → $-CO_2$, [O] →

Heumann's original synthesis of Indigo

The modern synthesis of indigo is slighty different from that route originally used and its discovery is credited to Pfleger in 1901. In this process, phenylglycine is treated with an alkaline melt of sodium and potassium hydroxidees containing sodamide. This produces indoxyl, which is subsequently oxidised in air to form indigo.

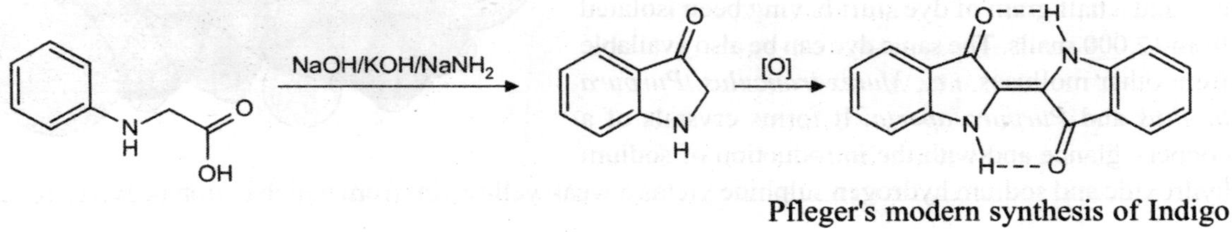

$NaOH/KOH/NaNH_2$ → [O] →

Pfleger's modern synthesis of Indigo

Chemical Properties of Indigo

Indigo is a dark blue crystalline powder that melts at 390°-392°C. It is insoluble in water, alcohol, or ether but soluble in chloroform, nitrobenzene, or concentrated sulfuric acid. The chemical structure of indigo corresponds to the formula $C_{16}H_{10}N_2O_2$.

The naturally occurring substance is indican, which is colorless and soluble in water. Indican can easily be hydrolyzed to glucose and indoxyl. Mild oxidation, such as by exposure to air, converts indoxyl to indigo.

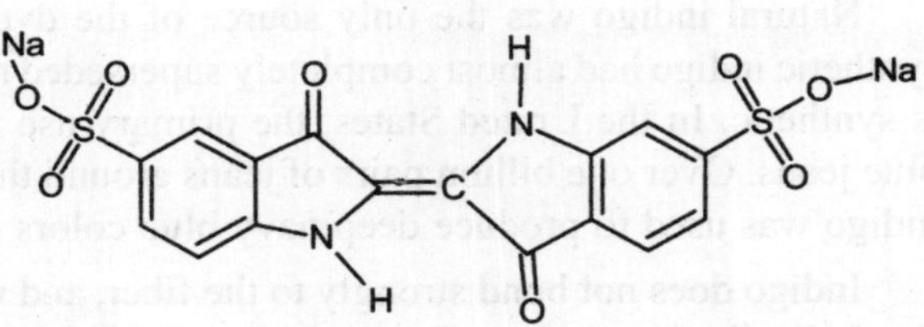

The manufacturing process developed in the late 1800s is still in use throughout the world. In this process, indoxyl is synthesized by the fusion of sodium phenylglycinate in a mixture of sodium hydroxide and sodamide.

Several simpler compounds can be produced by decomposing indigo; these compounds include aniline and picric acid. The only chemical reaction of practical importance is its reduction by urea to indigo white. The indigo white is reoxidized to indigo after it has been applied to the fabric.

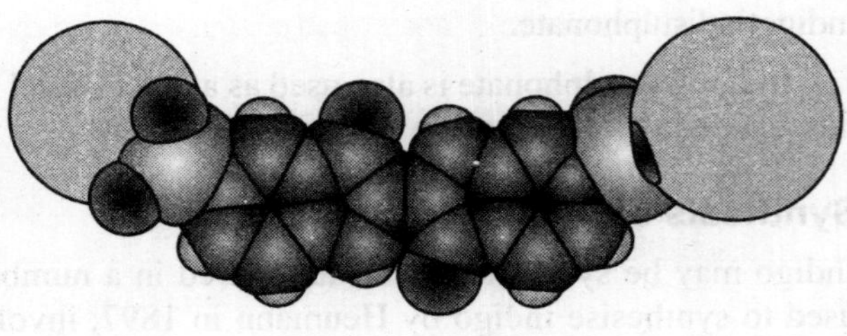

Indigo treated with sulfuric acid produces a blue-green color. It became available in the mid-1700s. Sulfonated indigo is also referred to as Saxon blue or indigo carmine.

Tyrian purple was a valuable purple dye in antiquity. It was made from excretions of a common Mediterranean Sea snail. In 1909 its structure was shown to be 6,6′-dibromoindigo. It has never been produced synthetically on a commercial basis

6, 6 dibromo indigo

It is a substance very similar to the purple tyiran. This has been proved by direct comparison of the synthetic product with that prepared from the secretion of the purple snail Murex Brandaris. About one and a half gram of dye stuff having been isolated from 12,000 snails. The same dye can be also available from other molluses, *i.e., Murex tranculus, Purpura lapillus* and *Purpura aperta.* It forms crystals of a coppery glance and with the introduction of sodium hydroxide and sodium hydrogen sulphide yields a weak yellow tint from which cotton is dyed a reddish

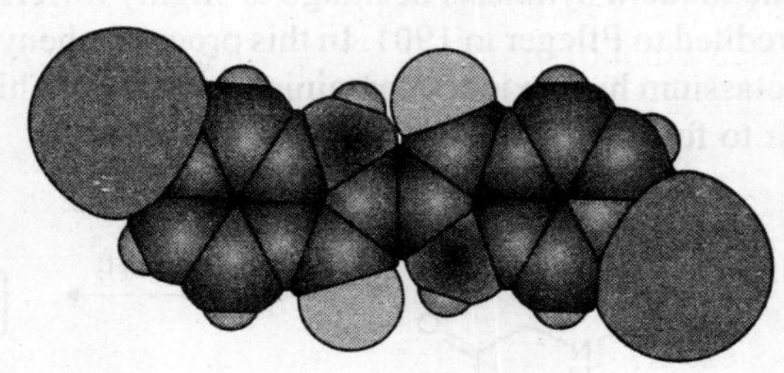

- violet shade. This remarkable displacement of the colour of indigo results not only from the introduction of bromine, but also of chlorine and methoxyl group in the 6,6' positions. Hence substitution in the *p*-position to the CO - group exerts a quite specific influence. In the same way the effect *p*-substitution may be traced in the derivatives of thio-indigo.

Indigosols - In order to obviate the shrinking of wooden and silk goods when used with indigo in alkaline vats, solubilised indigo was introduced by Bader in 1924. Indigo is reduced to indigo white and latter when treated with pyridine - sulphur trioxide, when a sulphuric ester of leuco - indigotin is formed. The sodium salt of this compound is indigosol O. it is stable in neutral solutions, from which it is directly absorbed by wool, silk and cotton. The absorbed compound is then treated on the fabric with and acid oxidizing agent. When the sulphuric ester groups are hydrolysed and the regenerated indigo white converted into indotin. Other indigosols may be prepared from derivatives of indigo in a similar manner.

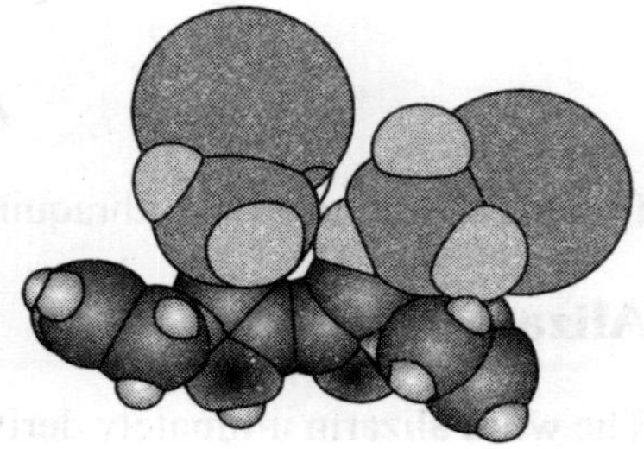

THIOINDIGO DYES

Dye stuffs of this group where discovered by **Friedlander** in 1906 as a result of experiments directed towards replacing the imino groups of indigo by sulphur atoms. The simplest compound is the thioindigo (thioindigo red B, or ciba red) is prepared from the anthrnilic acid in the following stages. By diazotization and treatment with sodium disulphide it is converted into di thio-salicylic acid, which on reduction with iron and alkali yields thio-salicylic acid. The latter combines with chloroacetic acid to form.

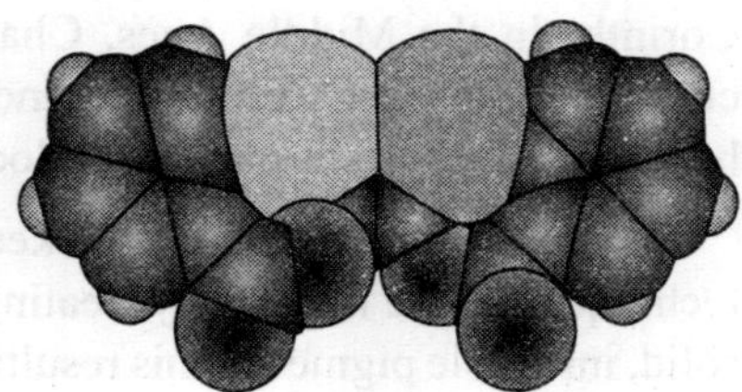

Thio indigo red B

Phenyl-thioglycollic-o-carboxylic acid and this on fusion with alkali or sodamide gives thio-indioxyl. On oxidation with air or potassium ferricyanide, thio-indoxyl is converted into thio-indigo, a bright bluish red dye which resembles indigo in its fastness to light and oxidation. Other dyes of this group are made by corresponding thio-indigoxyl with various ketonic compounds. Thus ciba scarlet G an excellent if some-what expensive red dye is obtained by the use of acenapthenequinone. It is of a special value in the printing.

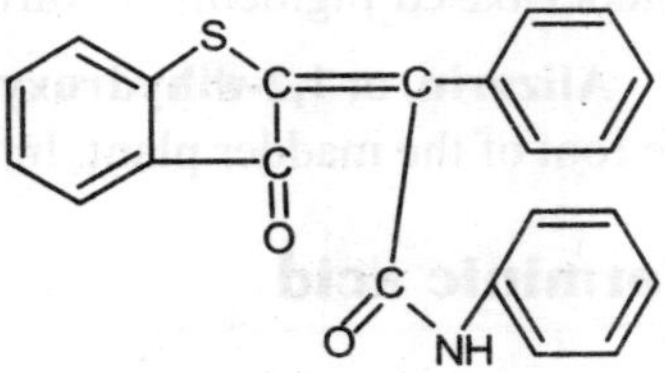

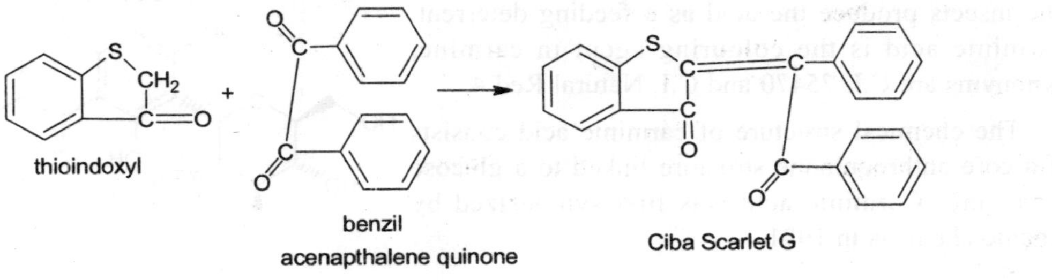

thioindoxyl + benzil (acenapthalene quinone) ⟶ Ciba Scarlet G

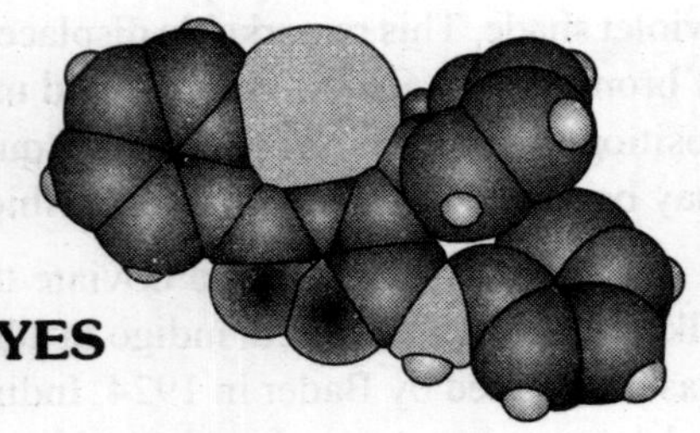

Thio-indigo R : It is made in the similar manner by the reaction of thio-indoxyl and isatin. In this compound the bluish shade of the thio-indigo is being replaced by a brilliant scarlet.

ANTHRAQUINONE DYES

Detaital description of Anthraquinone dyes are given below:

Alizarin

The word alizarin ultimately derives from the Arabic al-usara, juice.

History

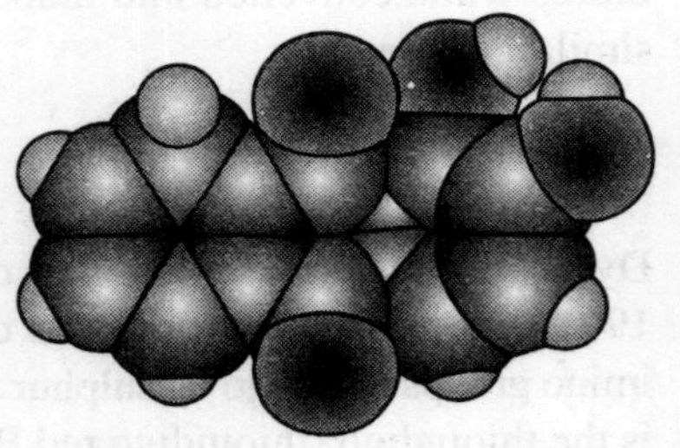

Madder has been cultivated as a dyestuff since antiquity in central Asia and Egypt, where it was grown as early as 1500 BC. Cloth dyed with madder root pigment was found in the tomb of the Pharaoh Tutankhamun and in the ruins of Pompeii and ancient Corinth. In the Middle Ages, Charlemagne encouraged madder cultivation. It grew well in the sandy soils of the Netherlands and became an important part of the local economy.

By 1804, the English dye maker George Field had developed a technique to lake madder by treating it with alum. This turned the water-soluble madder extract into a solid, insoluble pigment. This resulting madder lake had a longer-lasting colour, and could be used more versatilely, for example by blending it into a paint. Over the following years, it was found that other metal salts, including those containing iron, tin, and chromium, could be use in place of alum to give madder-based pigments of various other colours.

Alizarin, or **1,2-dihydroxyanthraquinone or mordant red**, is the red dye originally derived from the root of the madder plant. In 1869, it became the first natural pigment to be duplicated synthetically.

Carminic acid

Carminic acid ($C_{22}H_{20}O_{13}$) is a red glucosidal hydroxyanthapurin that occurs naturally in some scale insects, such as the cochineal and the Polish cochineal. The insects produce the acid as a feeding deterrent. Carminic acid is the colouring agent in carmine. Synonyms are C.I. 75470 and C.I. Natural Red 4.

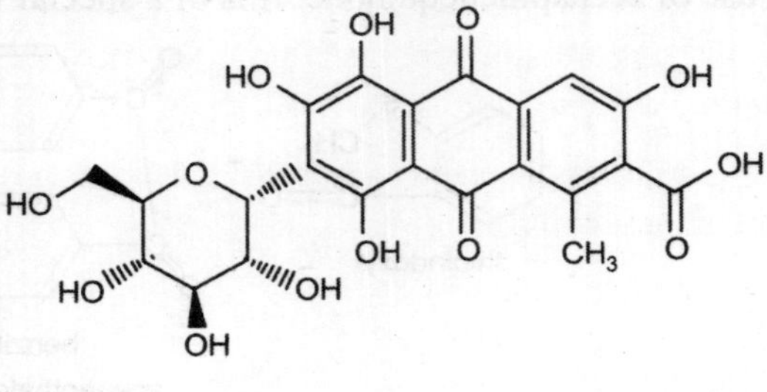

The chemical structure of carminic acid consists of a core anthroquinone structure linked to a glucose sugar unit. Carminic acid was first synthesized by organic chemists in 1991.

Disperse Red

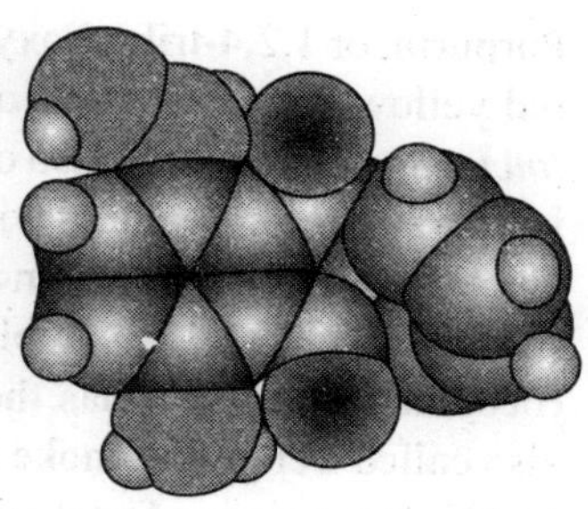

Disperse Red 9, also called C.I. 60505 and 1-methylamino anthraquinone, is a red dye derived from anthraquinone. It has the appearance of a red powder with melting point of 170-172 °C.

Disperse Red 9 is used in some older red and violet-red coloured smoke formulations. It is used in the M18 colored smoke grenade and also often used for dye packs. Its properties can be improved by coating the dye particles with an inert material, eg. an epoxy resin.

Disperse Red 11

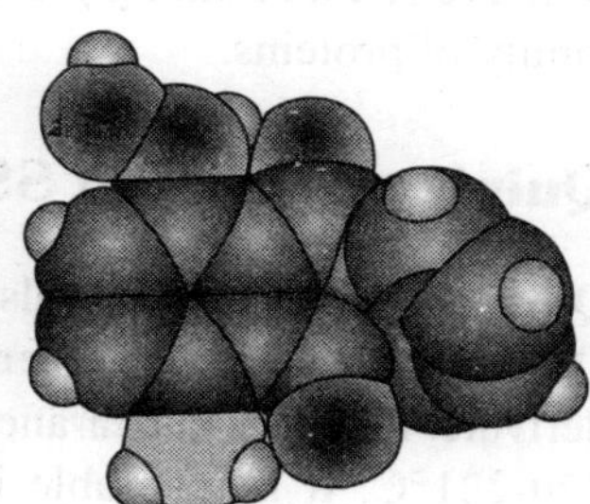

Disperse Red 11, also called C.I. 62015 and 1,4-diamino-2-methoxy anthraquinone, is a red dye derived from anthraquinone.Disperse Red 11 can be used in plastics and textiles industry to dye polyvinylchloride, acetate, polyester, polyamide, polyacryl and polyurethane materials and synthetic fibers, and foam materials (eg. polyurethane foams). It is also used in cosmetics.Disperse Red 11 is used in some red and violet-red colored smoke formulations.

Oil Blue 35

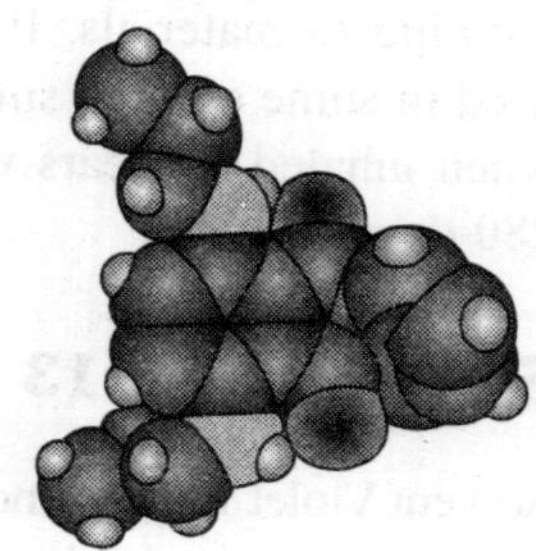

Oil Blue 35, also called Solvent Blue 35, Blue 2N, Blue B, Oil Blue B, 1,4-bis(butylamino) anthraquinone and CI 61554, is a blue dye derived from anthraquinone. It has the appearance of a dark blueish-black powder, soluble in benzene and toluene and insoluble in water, with melting point 104-105 °C. When exposed to 5% hydrochloric acid solution, it becomes dirty green.

Oil Blue A

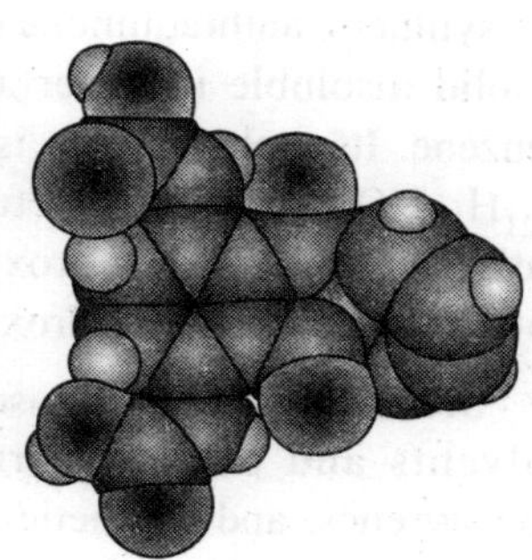

Oil Blue A, also called Solvent Blue 36, Solvent Blue A, Oil Blue G, Blue AP, 1,4-bis(isopropylamino) anthraquinone and CI 61551, is a blue dye derived from anthraquinone. It has the appearance of a dark bluish-violet powder, soluble in acetone, benzene and toluene and insoluble in water, with melting point 133-135 °C.

Oil Blue A is used as a dye for polystyrene and acrylic resins, and in petroleum and inks. It has good resistance to light.

A chemically similar dye is Solvent Blue 35.

Purpurin

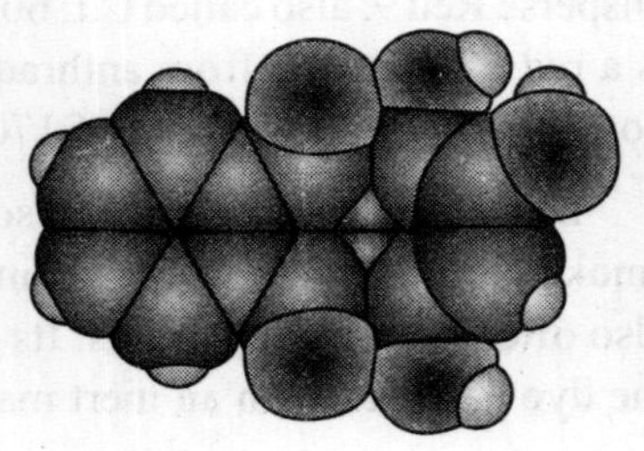

Purpurin, or 1,2,4-trihydroxyanthraquinone, is a naturally occurring red/yellow dye in the roots of the plant madder (or known as *Rubia tinctorum* L). Purpurin also occurs in madder with alizarin. Purpurin is a crystalline compound with the formula of $C_{14}H_5O_2(OH)_3$ that is colourless until dissolved in alkalic solutions. It is soluble in ethanol (becomes red) and is soluble in water at boiling and alkalis water (becomes yellow). It has the appearance of dark red needles. It is also called Verantin, Smoke Brown G, Hydroxylizaric acid, and C.I. 58205. Purpurin is a fast dye for cotton printing and forms complexes with various metal ions. As relating to biochemistry, purpurin is a glycosaminoglycan binding protein as well as a retinol binding protein. Purpurin's pattern of sequences has shown it to belong to the lipocalin family of proteins.

Quinizarine Green SS

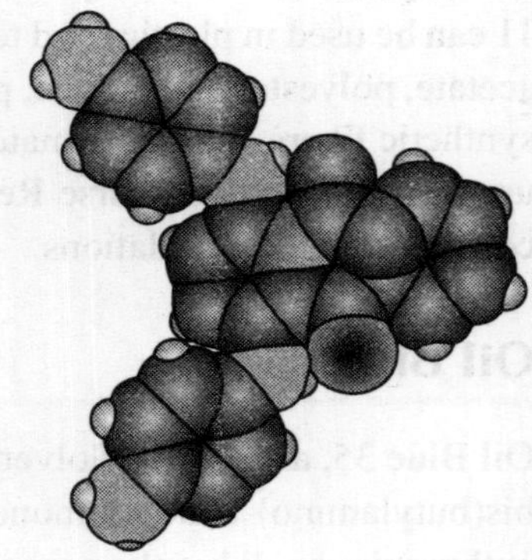

Quinizarine Green SS, also called **Solvent Green 3, C.I. 61565, Oil Green G, D&C Green #6**, is a green dye, an anthraquinone derivate. It has the appearance of a black powder with melting point of 220-221°C. It is insoluble in water. It is used for adding greenish coloring to materials. It is used in cosmetics and medications. It is used in some colored smoke formulations, in both old and new ones; when inhaled, it clears very slowly from the lung, with half time of 280 days

Solvent Violet 13

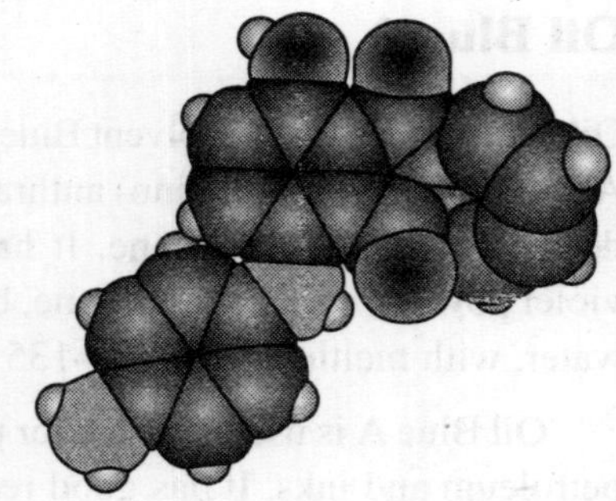

Solvent Violet 13, also known as D&C Violet No.2, oil violet, Solvent Blue 90, Alizarine Violet 3B, Alizurol Purple, Duranol Brilliant Violet TG, Ahcoquinone Blue IR base, Disperse Blue 72, and C.I. 60725, is a synthetic anthraquinone dye with bright bluish violet hue. It is a solid insoluble in water and soluble in acetone, toluene, and benzene. Its melting point is 142-143°C. Its chemical formula is $C_{21}H_{15}NO_3$, and its structure is 1-hydroxy-4-(p-tolylamino)-anthraquinone, or 1-hydroxy-4-[(4-methylphenyl) amino]-9,10-anthracenedione or 1-hydroxy-4-(4-methylanilino) anthraquinone.

Solvent Violet 13 is used to dye hydrocarbon products like solvents and petrol, thermoplastics, synthetic resins, e.g. polystyrenes, and synthetic fibre. It is also used in cosmetics, eg. in hair and skin care products. In pyrotechnics, it is used in some violet colored smoke compositions.

Vat Yellow 4

Vat Yellow 4, also known as Golden Yellow GK, dibenzochrysenedione, dibenzpyrenequinone, Tyrian Yellow I-GOK, and C.I. 59100, is a yellow synthetic anthraquinone vat dye. Chemically it is dibenzo[b,def]chrysene-7,14-dione, or 3,4:8,9-dibenzopyrene-5,10-dione, or $C_{24}H_{12}O_2$. It has the appearance of a viscous orange liquid.

Vat Yellow 4 is used mostly as a dye for textiles and paper.

Together with benzanthrone, it is used in some older pyrotechnic compositions for green and yellow colored smokes.

Vat Yellow 4 is a Group 3 carcinogen according to the IARC, as the evidence of its carcinogenity to humans is inadequate

Sulphur Dyes

These dyes are formed when nitrophenols are reduced by sodium sulphide. These dyes are generally insoluble due to reduction and becomes soluble in water after reduction. These dyes are generally used for dyeing cotton fibres and are usually popular because of their deep shades, they come in different colours such as blue brown green etc. of reasonable fastness to light amd ordinary washing asd a effective cost. These dyesa are second to the azo group in quantity produced.

Sulphur Black I

This dye is prepared by the reaction of m-dinitrop[henol and sodium polysulphide

2,3-dinitrophenol + Na_2S_4 ⟶ Sodium 3-hydroxy-4,5-dinitrobenzenesulfonate + Sulphuric acid

Sulphur black I

The both reactants are fused and then dissolved in water and air is blown into the mixture to separate the dyes. Then it is washed filtered and rewashed, the colour of these dyes depend upon the reaction with the intermediate. For example; when m - toluenediamine reacts with sulphur, dye is produced.

2-methylbenzene-1,4-diamine + S ⟶ 3,4-diamino-2-methylbenzenethiol + H_2S

Red shades are obtained by the fusion of azine derivative and sulphur.

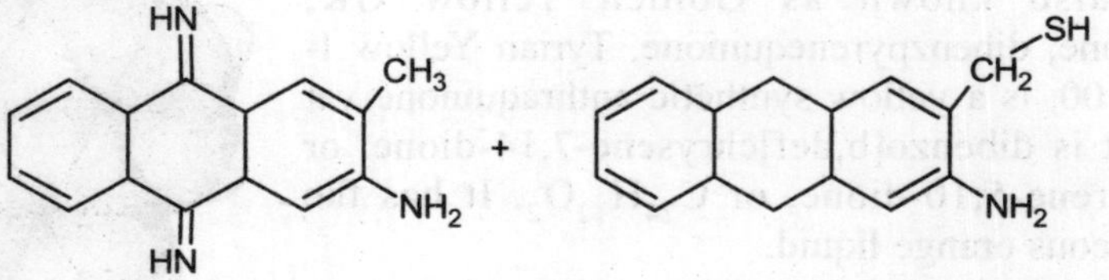

Blue sulphur dyes are prepared by reacting diphenylamine derivative with sodium polysulphide

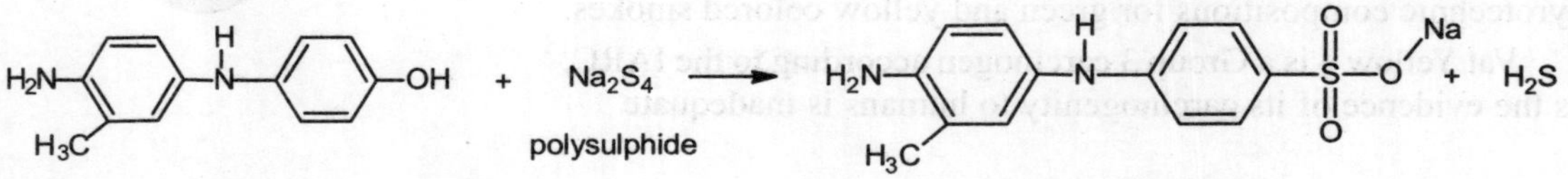

polysulphide

sulphur blue 7

REACTIVE DYES

In a reactive dye a chromophore contains a substituent that is activated and allowed to directly react to the surface of the substrate.Reactive dyes first appeared commercially in 1956, after their invention in 1954 by Rattee and Stephens at the ICI Dyestuffs Division site in Blackley, Manchester, UK. They are used to dye cellulosic fibres. The dyes contain a reactive group, either a haloheterocycle or an activated double bond, that, when applied to a fibre in an alkaline dye bath, forms a chemical bond with an hydroxyl group on the cellulosic fibre. Reactive dyeing is now the most important method for the coloration of cellulosic fibres. Reactive dyes can also be applied on wool and nylon, in the latter case they are applied under weakly acidic conditions.

These dyes contain dichlrotriazinyl group group. These reactive dyes are prepared by the reaction of cyanuric chlroide and phenylene blue.

Procion Red I

It is obtained by coupling diazotized aniline with hydroxy-5-nitronaphthalene-2,7-disulfonic acid in the presence of sodium carbonate and 2,4,6-trichloro-1,3,5-triazine at low temperature.

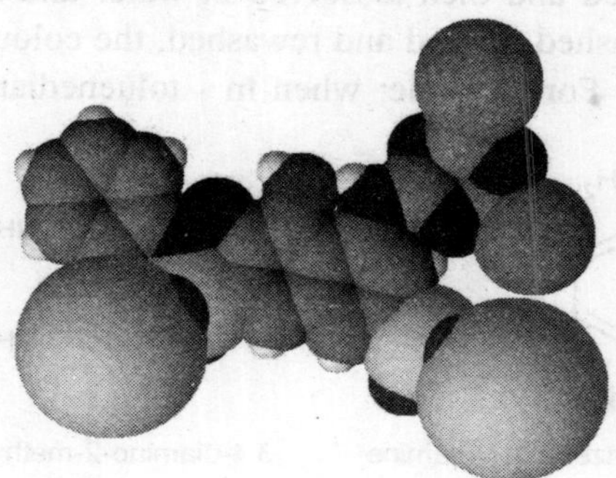

4-hydroxy-5-nitronaphthalene-2,7-disulfonic acid + benzene diazomium chloride → 4-amino-5-hydroxy-6-(phenyldiazenyl)naphthalene-2-sulfonic acid

2,4,6-trichloro-1,3,5-triazine + Na_2CO_3

procion red I

Procion Blue

This dye is obtained by condensing 1-amino-4-bromo-3-anthraquinone sulphonci acid with

1. 2-sulpho-p-phenylenesdiamine
2. The resultant is then condensed with 2,4,6-trichloro-1,3,5-triazine in presence of sodium carbonate at a low temperature. It is then condensed with sulphanilic acid and ultimately converted into its alkaline salt for preservation.

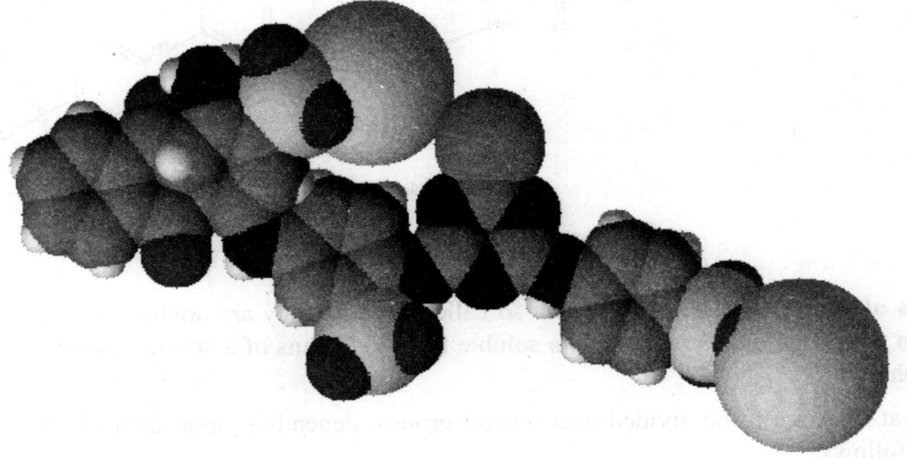

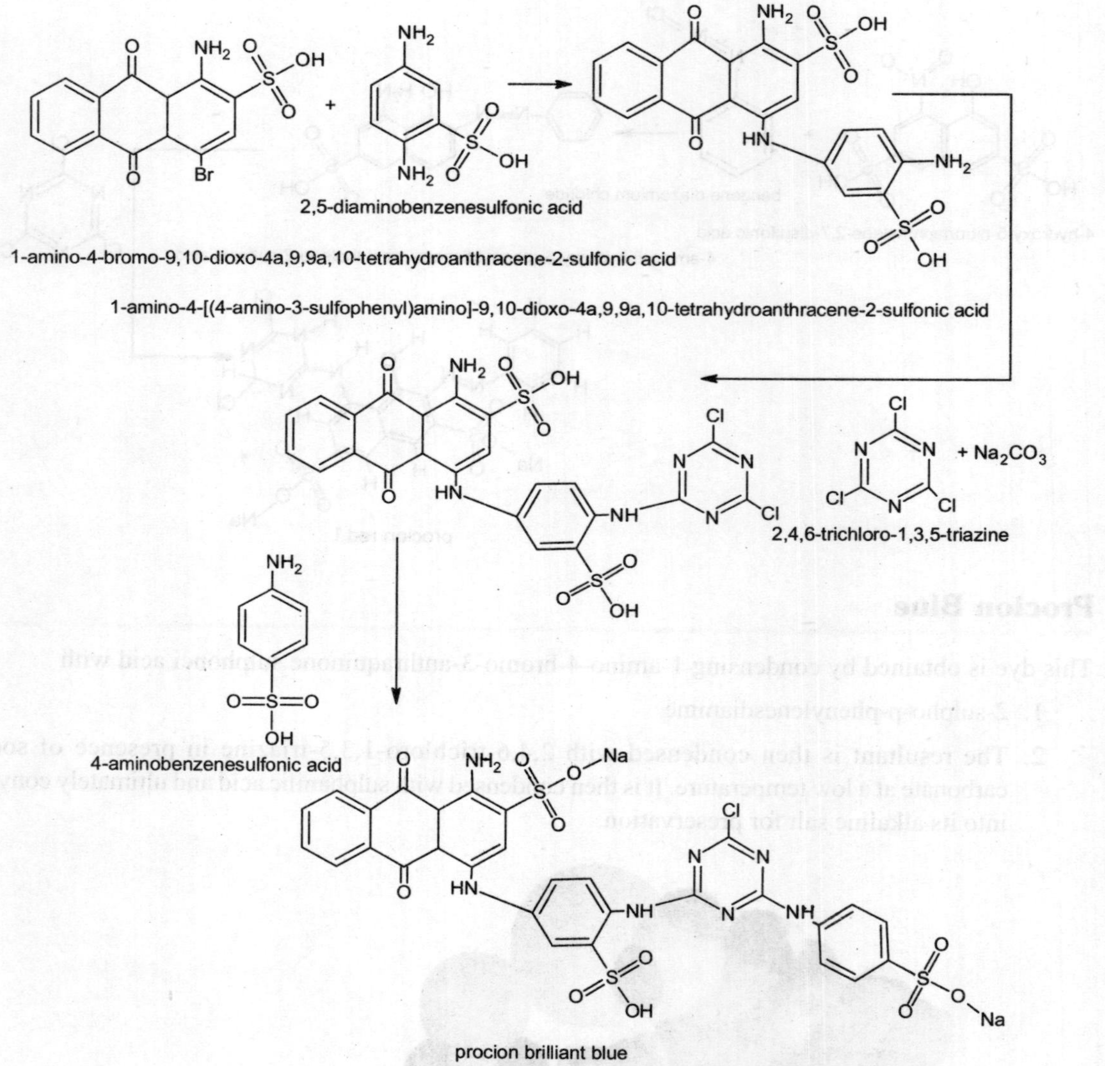

2,5-diaminobenzenesulfonic acid

1-amino-4-bromo-9,10-dioxo-4a,9,9a,10-tetrahydroanthracene-2-sulfonic acid

1-amino-4-[(4-amino-3-sulfophenyl)amino]-9,10-dioxo-4a,9,9a,10-tetrahydroanthracene-2-sulfonic acid

+ Na_2CO_3

2,4,6-trichloro-1,3,5-triazine

4-aminobenzenesulfonic acid

procion brilliant blue

THE VAT DYES

1. Classes of Vat Dyes. The vat dyes are so called because they are applied in a special kind of a dyebath in which the dye is reduced to a soluble form by means of a strong reducing agent, such as hydrosulphite.

The vat dyes are to be divided into several groups, depending upon their chemical nature and origin, as follows:

(a) Indigo, including both natural and Synthetic

(b) Thio-indigo dyes, containing sulphur.

(c) Indigo derivatives, such as the brom-indigos; usually not derived directly from indigo itself, but built up synthetically.

(d) Anthraquinone derivatives, including the various Indanthrene, Marione, Algol dyes, some Helindone, and others.

(e) Carbazol derivatives, of which Hydron Blue is the chief representative

In a broad sense, all of the vat dyes at present known appear to be members of three distinct chemical groups:

(a) Indigoids including Indigo and its various derivatives such as the Thio-indigos, Helindone and Ciba dyes, and some of the red Algol dyes. The chemical constitution of tills class is similar to that of Indigo. They may be applied in a neutral or slightly alkaline bath, and hence may be used for both wool and cotton dyeing. Their reduction products are pale yellow or almost colorless (similar to indigo), which exposure to the air re-oxidizes to the original color. As distinguished from the next class they are sublimed as coloured vapours from the fiber when heated.

(b) Anthraquinone dyes, [Of the different classes of vat dyes in a general way it may be said that the anthraquinone dyes are the fastest, the carbazol dyes are next and the indigoid dyes are the least fast. On the other hand, however, the indigold dyes are easier to dye. Indigo is faster on wool than it is on cotton.] including the Indanthrene, the Marione, and most of the Algol dyes. They are complex derivatives of anthraquinone and require a strongly alkaline vat in dyeing, consequently they are only useful for the dyeing of cotton. In common with all the dyes of the anthracene class their reduced compounds are not colorless but have about the same color as the original dye. When the dyed fiber is heated these dyes do not sublime or form colored vapors as with the indigoids.

(c) Carbazol Dyes, or Hydron Blue. With the exception of Indigo the vat dyes are all of comparatively recent introduction. The dyes themselves are highly insoluble in water, but readily yield products on reduction which are soluble in alkaline liquids. The dye bath, therefore, consists of a mixture of the dyestuff, a strong reducing agent and an alkali; and such a mixture is termed a "vat". The vat dyes at present include quite a wide range of colors; Indigo is a blue dyestuff, the thio-indigo dyes and their derivatives are mostly reds and scarlets, the anthraquinone dyes include blue, yellow, brown, green, violet, gray, orange, etc. The carbazol dyes are blue. The vat dyes are characterized in general by great fastness to light, washing, acids, alkalies, and in many cases to bleaching with hypochlorites. [Though the vat colors in general are fast to bleaching with hypochlorite liquors (known as fast to chlorine), they do not as a rule withstand kier-boiling with caustic soda (an operation which usually precedes bleaching). Their fastness in this respect has been found to be improved by boiling with sodium perborate. It has also been found that by introducing a small quantity of potassiumbromate (1 oz. per gallon) into the kier liquor, the bleeding of the color may be largely prevented. The use of a small amount of anthraquinone is also employed for the same purpose.] This makes them very valuable products, especially for cotton goods, and they have been coming more and more into use, and show every indication of ranking in kier-boiling cotton pieces containing yarns dyed with the vat colors it will sometimes be found that the goods at the bottom of the kier will be stained by the marking off of the color. This is supposed to be due to the fact that the alkali in the kier liquor together with the impurities removed from the cloth form a sort of local vat which reduces and dissolves the color, thus allowing it to run. Certain products, such as Ludigol (\meta-nitro-benzene-sulphonic acid) have been recommended as additions to the kier to prevent the staining of the goods, but it is a question as to why they are used ? As the principal cotton dyes of the future. As they are rather difficult to manufacture and require complicated processes in their preparation, the vat dyes are quite expensive, and on this account are used chiefly for the dyeing of raw cotton or yarns to be used for colored stripes in otherwise white fabrics, so that only a relatively small amount of dyed yarn is used in. the total fabric. Though the vat dyes[1] may be applied to all fibers they are more suited to the dyeing of cotton, as most of them require a rather strongly alkaline dyevat. The material to be dyed is simply immersed in the "vat" or dye bath until the goods are thoroughly impregnated with the solution. The material is then squeezed and exposed to the air, which causes the oxidation of the reduced "leuco" compound and the formation of the color. The temperature of the vat is usually lukewarm for the purpose of facilitating the impregnation of the fiber with the solution. In some cases the dipping and oxidation have to be repeated several times in order to build up a heavy color. The vat dyes have come to be very essential dyes for cotton, as it is only by the use of these dyes that laundry-fast colors in cotton wash fabrics can be obtained. They are necessary dyes for the production of colours in shirtings, blouse material, cotton skirtings, and hosiery and such fabrics or garments that require to be frequently laundered. No other class of colors will stand the bleaching effect of the hypochlorite liquors used in whitening cotton goods in the modern laundry. There are but few other dyes with this property (Chloramine Yellow). It was on the great fastness of the vat dyes that the reputation of the excellent quality and fastness of the German-made dyes became so strongly fixed in the mind of the public. Many of the vat dyes are faster than Indigo, though of course this latter dye is itself to be considered but a member of the general class of vat dyes.

[1]The different classes of vat dyes vary considerably in their ease of application; the anthracene dyes are the most difficult to apply, while the indigoid and carbazol colors are much easier to dye, especially with regard to penetration and even colors. The halogenated anthracene dyes are much better in this respect than the non-halogenated dyes The hydron series is the only one which includes a navy blue at a reasonable price, the other classes of colors have always been very high in price. In the dyeing of compound shades with the vat dyes there is frequently considerable difficulty experienced, especially in obtaining even colors. It may also be remarked that while the self shades of two colors may be fast, when dyed in mixture the compound shade may not be as fast, but on the other band the opposite may also be true; Anthraflavone, for example, is not very fast as a self shade, but when dyed in combination with 1ndanthrene Blue the compound shade is very fast. It must also be remembered that some of the vat dyes are applied best at one temperature and some at another and in dyeing mixtures a proper balance of temperature must, be maintained, as the correct temperature in the dyeing of vat dyes is of the utmost importance if the best results are to be obtained. The anthracene vat dyes require the use of more alkali in tile vat than the dyes of the indigoid or hydron series (about four times as much), but on the other hand they require much less hydrosulphite. The amount of caustic soda to be used, strange to say, is to be calculated on the volume of the liquor employed and not on the amount of dyestuff used, so that in dyeing a light shade just as much caustic soda is required as for a heavy shade. In using a mixture of vat dyes for the production of compound shades it is always best to first reduce each dyestuff separately and then mix them in the dyebath.

The vat dyes in some cases have been proposed for use with wool but owing to the fact that the vat is strongly alkaline with caustic soda it is difficult to apply properly the color to wool. Most of the vat dyes show no affinity for wool below a temperature of 160°F., and glue or sulphonated oil soap must be used in the bath to protect the fiber from the action of the alkali; furthermore in order to obtain fast colors it is necessary to boil the dyed goods in sulphuric acid to destroy the hydrosulphite, and this is a great disadvantage.

In the practical use of the vat dyes it is usually the custom to first prepare a stock vat or solution of the reduced dyestuff, and this is used in such quantities as may be necessary for there plenishing of the dye vat. In the preparation of this stock solution the following is a typical method:-

100 lbs. of dyestuff (which is generally in the form of a paste containing 20 per cent of dry dye);

20 gallons of water at 71°C F.;2-6 gallons of caustic soda solution of 76' Tw.;10-40 lbs. of hydrosulphite powder (anhydrous sodium hydrosulphite).

The exact amounts of hydrosulphite and caustic soda will depend on the particular dyestuff employed. Very frequently some Turkey-red oil (or similar sulphonated soluble oil) is also added.

The vat is best made up in a wooden barrel or tank fitted with a steam pipe so that the contents may be maintained at a temperature of 71°C. until the reduction is completed, which usually requires about one hour or somewhat less. The solution may then be made up to 40 or 50 gallons with water and is ready for use. It is important that the water employed for both the stock solution and the dyevat should first have the dissolved air corrected by the addition of a little hydrosulphite and caustic soda; 100 gallons of water will usually require about 4 ozs. of caustic soda and 3 ozs. of hydrosulphite. As far as possible soft water should be used. The Whittaker gives the following typical vats for the three classes of vat dyes:

10 per cent Chloranthrene Blue BD (10 per cent paste) Anthracene Series

30 per cent caustic soda (76' Tw.)

21 per cent hydrosulphite powder cone.

2 per cent Ciba Blue 2R powder Indigoid Series

.7 per cent caustic soda (76' Tw.)

7 per cent hydrosulphite powder cone.

6 per cent Hydron Blue G paste Carbazol Series

6.6 per cent caustic soda (76' Tw.)

6 per cent hydrosulphite powder cone.

The colour should first be stirred up to a smooth paste with the caustic soda, hot water added, and then the hydrosulphite. In some cases it may be found necessary to heat the liquor even to the boil in order to obtain a complete solution purpose of the correction is to remove the dissolved oxygen and the hardness in the water so as to avoid precipitation of the dyestuff. The stock solution must be preserved from undue exposure to the air, otherwise dyestuff Will be precipitated and cause bad shades. This same precaution also applies to the dye vat. The latter is prepared by heating the necessary volume of water to about 100°F., adding the necessary amount of caustic soda and hydrosulphite required to counteract the dissolved oxygen, and then adding the required amount of the stock dye solution. The vat is then gently stirred and allowed to rest awhile before use for dyeing. [In the preparation of the dyebath with the vat dyes it is sometimes a question as to when the proper degree of reduction is obtained. With some dyes the reduced or leuco-compound is of a different color than the dye, and in this case it is easy to determine if the vat is completely reduced; for instance, Chloranthrene Yellow gives a reduced vat which is blue in color. On the other hand, some of the dyes give vats of the same general color as that of the dye, and it is difficult to tell just when the color is reduced by the appearance of the vat. Chloranthrene Blue, for example, gives a reduced vat which is also blue in color. To determine if reduction in such a case is complete Whittaker recommends drawing out some of the vat liquor in a pipette arid allowing it to run slowly down the side of a clean test tube held against the light; if the dye is reduced the liquor will show clear, but if it is not completely reduced undissolved particles of color will be detected in the liquid. With the blues of the indigold class, like Ciba Blue, the reduced vats are bright golden yellow in color and consequently there is little difficulty in ascertaining when reduction is complete.] When yarn is dyed it should first be well boiled-out, and if open dye vats are used the yarn should be entirely submerged beneath the liquor by being hung on bent iron rods similar to those recommended for use in the dyeing of sulphur colors. In dyeing yarn it is also important that it should be evenly wrung out after steeping in the vat, and more even results are always obtained if several dips are given. After dyeing and wringing the yarn is then exposed to the air for about thirty minutes to oxidize completely the leuco-compound to the dyestuff. It is then boiled in a bath containing about 2 lbs. of soap per 100 gallons of liquor in. [Owing to the large amount of caustic alkali used in the bath with the vat dyes, after dyeing the cotton the goods must be soured in a weak acid bath to neutralize the alkali, then washed, soaped well and washed again before drying. The soaping is a very necessary operation, as it both develops and brightens the shade, besides softening the goods and ensuring neutralization of all acid.] This is for the purpose of completely developing the color and removing all unfixed dyestuff, which would otherwise dull the shade and cause crocking. On account of the difficulty of obtaining even shades and good penetration of color, it is more satisfactory to dye vat colors in the loose stock rather than on yarn or piece-goods.

Indigo-brown, indigo-gluten and some mineral matters

In the plant itself the coloring matter is supposed to exist in the form of a glucoside called indican. The process of extraction of the dyestuff from the plant is both interesting and complicated. The indigo plant, which is a shrub growing 3 to 4 ft. in height, is cut, in summer. The cut plants are tied up in bundles and packed into long vats, which are then filled with water; in a short time fermentation sets in which is allowed to continue for ten to fifteen hours. This process converts the indican of the plant into soluble matters which are extracted from the plant by the water and pass into solution. The liquor is then run into another vat, placed at a lower level, and here it is churned and beaten up either by hand or mechanically for the purpose of exposing it to the oxidizing action of the air, whereby the insoluble indigotine is formed and precipitated out. This collects at the bottom of the vats in the form of a, paste or mud which is washed, filtered, pressed into cakes, and dried. This constitutes the raw indigo of trade, and comes in the form of large cubical blocks. The best natural Indigo comes from Java and Bengal, and contains from 60 to 75 per cent of coloring matter. Madras Indigo is usually somewhat inferior, while that from Guatemala, China, Africa, and Egypt is very variable.

The raw indigo of trade is a dark blue, earthy-looking substance. When scratched with the finger nail good qualities will exhibit a coppery streak. At, the present time a great, deal of the crude Indigo undergoes a refining process, for the purpose of eliminating the many impurities liable to occur in the raw product; it also comes into trade ready ground either as a powder or a paste in order to facilitate its use by the dyer.

Two crops are usually gathered from the same plants each year.

The Indigo: is extracted chiefly from the leaf of the plant; this contains, oil the average, about 0.5 per cent of coloring matter. Before Indigo can be used by the dyer for purposes of reduction it must be ground to a very fine impalpable powder. In former times where the natural Indigo was bought in the form of blocks or lumps it had to be ground in special indigo mills for a long time. This accounts for the fact that it is now so much used in the form of a 20 per cent paste (with water or glycerin). Synthetic Indigo is practically altogether marketed in the form of such a paste, as it is then ready for direct use in preparing the stock solution of indigo-white. The principle of indigo dyeing has always differed entirely from that of other classes of dyestuffs, and has constituted an art by itself. Indigo is perfectly insoluble in water, and hence cannot be applied in dyeing in this form. By the action of various reducing agents, however, it may be converted into a substance known as indigo-white, which is soluble in alkalies; in this form, it is applied to the fiber, and by subsequent oxidation by simple exposure to the insoluble blue indigotine, which thus remains permanently fixed in the fiber. Indigo may also be converted into a soluble blue coloring matter by treatment with strong sulphuric acid. This body, known as Indigo Carmine, or indigo sulphonate, maybe classed as an ordinary acid dyestuff) being applied in the usual form of acid dyebath; but it does not possess the great fastness and other valuable properties of Indigo itself.

Methods of Dyeing' Indigo. Indigo is extensively used for both wool and cotton dyeing, though it is being used proportionately less for the dyeing of wool since the introduction of the fast alizarine and anthracene blue dyes. It is not much employed for the dyeing of silk. In calico printing it has an extensive application, principally for discharge styles. The vats used for cotton dyeing are generally more strongly alkaline than those for wool, while the proportion of Indigo used in them is also higher. In cotton dyeing, too, the vats are usually worked cold. Indigo dyeing is known as " vat " dyeing because it is carried out in a specially prepared vat. According to the character of the reducing agent employed, these vats are classified as follows:

Zinc vat, Fermentation vat, Hydrosulphite vat, Copperas vat.

The fermentation vat : Is the oldest form of indigo dyeing, and is still used to a considerable extent for wool dyeing. Its action depends on the chemical activity of certain ferments which reduce the indigotine to the soluble indigo-white. This vat is used warm, while the other vats are usually worked cold.

The copperas vat: Was the earliest form of chemical vat; it was exclusively adopted for cotton. At the present time, however, it is almost obsolete. The reducing agent employed was copperas, or ferrous sulphate.

The zinc vat : Was formerly the favorite one employed for cotton, and even at the present time it is quite largely used. The reducing agent employed is an alkaline solution of zinc dust.

Indigo is apparently fixed on the fiber mechanically; that, is to say, the coloring matter is deposited in the fiber in a fine state of division; If the color is deposited too rapidly it will lack fastness, especially to rubbing. Hence it is not advisable to use concentrated vats for dyeing heavy shades of Indigo, but to build up the color by means of several successive dips in weaker vats. Wool reacts somewhat differently with the reduced Indigo in the vat than cotton. The former, on account, perhaps, of its somewhat alkaline character, has considerable affinity for the acid indigo-white and consequently fairly deep shades of good fastness can be obtained with one dip on wool. Cotton has much less attraction for the reduced indigo and takes up a much smaller quantity of the dye. It is customary, therefore, to dye wool in the warm (fermentation) vat, while cotton is dyed in a cold vat as its affinity for the dye decreases as the temperature rises. The hydrosulphite 'vat is the one of latest origin; it is employed very largely at the present time for all classes of indigo dyeing both on wool and cotton. The reducing agent employed in this vat is sodium hydrosulphite, $NaHSO_2$, prepared by the action of zinc dust on sodium bisulphite. It is gradually replacing the other forms of vats as it is the most simple and scientific and the most easily regulated.

The alkali used for dissolving the indigo-white is the same for all forms of vats; it maybe either lime or caustic soda, or a mixture of the two, depending upon whether the vat is to be employed for wool or cotton dyeing. Ammonia does not appear to dissolve indigo-white very readily, and the alkaline carbonates are still less suitable. Indigo-white behaves like a very weak acid, and it requires an excess of rather strong caustic alkali to bring it into solution, and it is readily precipitated again by the addition of any acid. On this account the vat must always be kept alkaline.

Before Indigo is introduced into the vat (of whatever variety) it must be in a very finely divided state, otherwise the reduction NN-111 always be incomplete. The grinding of Indigo is a rather important consideration; it is usually first ground in the dry state, and then ground a second time with a little water (to which a small amount of alkali may be added) to the form of a paste, Indigo paste of this character may be purchased ill the market by the dyer, and may be added to the vat directly; it should contain 20 per cent of indigotine. During the reduction of Indigo in the vat, the process is usually accompanied with secondary chemical reactions varying ill their nature and degree with the character of the vat. This results in the conversion of smaller or larger amounts of the dyestuff into substances other than indigo-white and a resultant loss of coloring matter. This is especially large in the copperas vat, it also amounts to considerable in the fermentation and zinc vats; in the hydrosulphite vat it is reduced to a minimum of about 2 per cent.Indigo vats when used for dyeing should not have a concentration of more than 3parts of indigotine per 1000 parts of liquor. More than this tends to the production of shades which are liable to crock and also lose in washing. The vat must also possess an excess of reducing agent. This may act in several ways. It prevents the premature oxidation of the indigo-white arising from the vat liquor coming in contact with the air or water in the pores of the material being dyed, and so prevent or retard the penetration of the indigo-white. Again, the more thoroughly the indigo white in the vat is reduced the more completely will it work its way into the material and the faster will be the color. A certain excess of reducing agent is also of advantage in the subsequent oxidation of the indigo-white to indigo-blue, as then the action of the oxygen is slower and more uniform, giving better penetrated colors and also causing the dyestuff to be precipitated in a finer state of division, which results in faster and better colors. If there is not sufficient excess of reducing agent in the vat, on washing after dyeing and oxidizing a large part of the color will be removed, whereas if more reducing agent were present, the loss on washing should be little.

The time required for the reduction of the Indigo in the various vats is about as follows:

The fastness of Indigo is said to be improved by an after-treatment with bluestone and acetic acid. The use of glue in the vat also has the same effect.** Indigo is frequently bottomed by first dyeing with certain substantive or sulphur dyes; and indigo blue on cotton may be topped by dyeing with basic colors; the goods after dyeing in the vat being mordanted with tannin, fixed with tartar emetic and dyed.Redder shades may be obtained with Indigo on cotton by steaming after dyeing in the vat, but this somewhat decreases the fastness to washing. Heavier shades may be obtained by first mercerizing the cottonBy passing the cloth before dyeing through a solution of Turkey-red oil and alumina the fastness to alkali and chlorine is much increased. By treatment with a strong solution of caustic soda.* The mercerized fiber shows a greater attraction for the Indigo than the untreated cotton. In order to save Indigo it has been suggested to mercerize only one side of the cloth to be dyed, and when this is run through the vat the mercerized side will dye up much darker than the other. With improvements in mechanical devices it has become possible to dye Indigo on cotton in the form of cops, tubes, cheeses, beamed warps, etc. In such machines the material remains stationary and the indigo vat liquor is forced through the fiber. Only the hydrosulphite vat can be used for this purpose as there must be no sediment or undissolved particles, because the cotton material in this case acts as a filter to the liquid. Therefore great care must be taken in preparing the vat for this method of dyeing. Special apparatus must be used for dyeing Indigo (and the other vat dyes as well) differing from that employed for the ordinary dyestuffs, as provision must be made to draw air through the dyed material in order to oxidize the color. In piece dyeing two forms of indigo vats are used: (a) immersion vat, and (b) continuous vat. In the first form of vat sinking frames are used on which the goods are spirally attached by means of hooks. These frames are immersed in the vat for the required time, then lifted out, and exposed to the air for oxidation, when another dip is given until the required depth of color is obtained. Usually between each dip the frame is turned bottom up so as to get even dyeings. Immersion vats are chiefly used for heavy goods that do not readily dye through, for the frame may be left in the liquor for any length of time necessary, whereas in continuous dyeing machines this is not possible. Heavy linens, moleskins, and such fabrics are often left in the vat overnight, or even for several days in order to obtain proper penetration. Immersion vats are also used for goods to be dyed on one side of the piece only; in such a case two pieces are fixed back to back on the frame. When the dyed pieces are exposed to the air only the outer sides are oxidized and the Indigo is chiefly developed there, the other side being dyed a considerably lighter shade. For dyeing on immersion frames the zinc vat is more suitable than the hydrosulphite vat, as the latter contains hydrosulphite and caustic soda, which cannot be squeezed out in this case after dyeing, and consequently uneven colors are liable to result.

*Copperas vat ....two to three hours Zinc-lime vat ....four to five hours Hydrosulphite vat ....one-half to one hour

**Treatment, with bluestone causes the shade to become somewhat greener. In using the glue treatment it is recommended to pad the cloth previous to dyeing with a solution of glue (11 to 21 ozs. per gallon). This causes the shade to be brighter and redder and increases the fastness to rubbing.

To produce full shades of blue the Hoechst Co. recommend passing the goods, before dyeing in the vat, through a solution containing 1 to 11 lbs. of starch to 100 gallons of water.

For producing very heavy shades of Indigo cotton is sometimes first dyed with a weak Aniline Black, as follows: For 100 lbs. of cotton yarn, work for, one hour at 100°F. in a bath containing 31. lbs. aniline salt, 312 lbs. sodium bichromate, and 7 lbs. of hydrochloric acid; wring out well and treat in a fresh bath with 141 lbs. of soda ash for one-half hour at 100°F. Then rinse twice and hydro-extract and dye in the indigo vat. In this way very coppery shades of blue maybe obtained with very little Indigo. Instead of using Aniline Black a light shade of manganese bronze may also be employed as a bottom.

In continuous machines the pieces are run through successive vats and exposed to the air for oxidation between the dips. Usually four to six vats are employed in one range soas to obtain heavy shades. The hydrosulphite vat liquor is most generally employed for this form of continuous dyeing, as there is no sediment and the vat is easily regulated. The depth of color may be regulated by varying the speed of running and the number of immersions. It is always preferable to enter the goods in the wet, state. By drying the goods first after dyeing and then souring and washing heavier shades of blue are obtained.

Fermentation Vat. The essential ingredients of the fermentation vat are: Indigo, lime, woad, bran, and madder. The woad furnishes the proper kind of ferment for the reduction of the Indigo, the bran and madder serve as nourishment for the growth of the ferment, while the lime serves to neutralize the acids liberated during the fermentation and also furnishes the alkali necessary for the solution of the reduced Indigo. In dyeing carbonized wool or shoddy in the indigo vat care must be had to have the goods thoroughly neutralized with soda before entering the vat, as any acid in the wool may cause, disturbances in the vat by neutralizing the alkali. There are, however, a large variety of substances used in the preparation of the fermentation vat, and almost every indigo dyer has his own special formula, but the essential ingredients are those given. According to the make-up of its constituents, the fermentation vats are classified as follows:

Potash vat, in which potash is used as the chief alkali.

Soda vat, also known as the German vat, in which soda is the chief alkali used.

In Eastern countries all manner of substances are added to the indigo vat for purposes of aiding the fermentation or supplying nourishment to the ferment; among some of these substances may be enumerated dates, raisins, honey, plant seeds, glucose, etc.

(a) Saxon vat. This is one of the earliest forms of Indigo dyeing in Europe, and is still practiced in the same primitive manner by the peasants of Saxony, where the celebrated Saxon blue is dyed. The following experiment will illustrate this method: Take10 grams Indigo paste (20%) and mix with 10 grams potash dissolved in 50 cc. water; place 50 grams raw unscoured wool in a wooden or earthenware vessel, and pour over it .the above solution. sufficiently diluted to just cover the wool. Set aside in a warm place for a week or ten days. A moderate fermentation sets in which causes the reduction of the Indigo, which is absorbed by the wool, and thus the dyeing is accomplished. When sufficiently colored, remove the wool, squeeze, allow to oxidize in the air, and finally wash in a soap solution The shades obtained in this manner are especially beautiful, and they are highly prized on account of their fastness to rubbing.

(b) *Woad vat.* It is very difficult to obtain any very satisfactory results on a small experimental scale with the fermentation vat, but the following will illustrate the method of setting this vat: Place 6 liters of water in a wooden or stoneware vessel and heat to about160°F.; add 50 grams of woad previously broken up and soaked for several hours in a little warm. water; next stir in 20 grams bran, 8 grams soda ash, 3 grams lime, 20 grains madder, and 12 grams Indigo paste (20%). Stir well, and then cover with a cloth and allow to stand in a warm place for twenty-four hours. During this time the fermentation has become quite active; the liquor should be yellowish in colour and be covered with a light blue froth. [Cotton dyed in the fermentation vat acquires a peculiar "indigo smell" which is insisted upon by buyers in some countries.] It should now be stirred up well, and if any large quantity of gas is given off, a little lime should be added; after which it is again covered and left to ferment for a few hours more. When the reduction of the Indigo is complete, the liquor of the vat will be yellow in color with the surface covered with a dark blue layer, which if skimmed off should be granular in appearance. During dyeing the vat should be maintained at a temperature of 120' F. If the vat does not have a satisfactory appearance, a little more lime should be added, the liquor stirred up, and then covered and allowed to stand for a couple of hours.

When the vat has been brought to a proper condition, steep a handful of well-scoured wool in the liquor for a few minutes. On being taken out the wool should be of a greenish yellow color; squeeze and expose to the air until the blue color is completely developed, then wash in a warm soap bath. The latter treatment should cause the wool to lose but a small amount of color. In dyeing, care should be taken not to disturb the sediment in the vat, otherwise streaked and uneven colors will result. The vat may now be used for dyeing a variety of woolen material (loose wool, tops, yarn, and 1170Vell pieces). Heavier shades may be produced by giving several dips in the vat, squeezing and oxidizing in the air after each dip. Care should be taken not to agitate the liquor too much, as otherwise it will rapidly oxidize and turn blue, and no longer be fit for dyeing. The vat should be contained in a tall-shaped vessel, as about one-third of the vat is made up of the sediment which it is not desirable to disturb while dyeing.

The vat may be maintained continuously for a long period of time. After being worked for some time it becomes partially exhausted and oxidized; then a little glucose (syrup), bran, madder, and lime may be added together with more Indigo paste. It is well stirred up, covered over, and allowed to stand for several hours or overnight, when it is again ready for dyeing. For good results the amount of Indigo in the vat should not rise above three parts per 1000.

If solid Indigo is used only about 2 to 4 lbs. should be used; but in this case the dvestuff should be very carefully ground in a ball or roller mill prepared from stale mine, salt, madder, and Indigo, the alkali being supplied by the ammonium carbonate. Properly to prepare and maintain a fermentation vat requires considerable skill and experience, especially with regard to the proper amounts of and the proper times for adding the lime. The fermentation must be regulated in such a manner as to reduce the Indigo sufficiently by the generation of the proper amount of hydrogen, and yet kept sufficiently under control as to prevent the danger of putrid fermentation setting in, which will result in the rapid destruction of the Indigo. W lien putrid fermentation starts, the vat is said to have "gone sick," and lime must be added and the vat well stirred up. If the secondary fermentation, however, has gone too far and cannot be stopped in this manner, the vat must be boiled up in order to prevent a total loss of the Indigo therein. After this, of course, the vat must be set all over again. The addition of lime always tends to reduce the fermentation, if too much is added the fermentation may be lessened beyond that point necessary for the complete reduction of the Indigo. If the fermentation is proceeding too slowly it may be increased by the addition of bran. If too little lime is present, the acids liberated by the fermentation will throw the Indigo out of solution, hence the vat will become weak, and bluish in colour. In dyeing heavy shades with Indigo it is best to build up the color with several successive dips in weaker vats, rather than to dye it to the full shade by a single dip in a very strong vat. In this manner the pigment is More thoroughly absorbed by the fiber and will not be so liable to crock off as otherwise. In using synthetic Indigo the following fermentation vat is recommended. Use 25 lbs. of Indigo paste (20%), 12 lbs. of bran, 12 lbs. of soda ash, and 8 lbs. of madder. The dye will be reduced in about twenty-four to thirty-six hours. The liquor at first has a muddy appearance, this gradually becomes greenish, and after the addition of lime shows a golden yellow color. The fresh vat has a sickly smell, but this gradually disappears, giving place to a pungent odor. In order to keep up the fermentation in the vat after use an addition of 5 ozs. of molasses to each pound of indigo used is made.

The Copperas Vat. This form of indigo vat is not much used at the present time, as it is not very suitable for continuous dyeing on account of the large amount of sediment it contains. The essential ingredients of this vat are ferrous sulphate and slaked lime; these react in the following manner:

$$FeSO_4 + Ca(OH)_2 = Fe\ (OH)_2 + CaSO_4$$

The copperas vat was chiefly used for dyeing skein yarn. Its chief advantage was that it was easily set and kept in condition. A considerable amount of Indigo is always lost in the copperas vat, due to over-reduction and combination of the dye with tile hydrate of iron.

The ferrous hydrate thus formed acts as a reducing agent in the presence of water:

$$2Fe(OH)_2 + 2H_2O = Fe_2(OH)_6 + H_2.$$

The indigo-white formed by the reduction dissolves in the excess of lime present.

The vat employed should be narrow and deep to accommodate the large amount of sediment formed. The temperature of the vat should be kept at about 70 to 75°F.

To prepare the copperas vat proceed as follows: 36 grams of quicklime are slaked to a thin paste with water; while warm stir in 30 grams Indigo paste (20%). Then add 30 grams ferrous sulphate (copperas) dissolved in about 100 cc. water at 140° F. Then dilute with water to 500 cc.

Have this solution in a covered flask; allow to stand for four to six hours with occasional stirring, in which Wine the liquid should have become yellow in colour with a coppery-looking bead. Before adding the stock vat to the dyevat, 1 lb. of ferrous sulphate and 11 to 2 lbs. of quicklime should be added per 100 gallons of water.

The copperas vat is also known as the vitriol vat. As a rule it is not replenished, but is worked three times a day, being well stirred after each dyeing. In about ten days the vat should be exhausted. It is mostly used for yarn dyeing and "resist" dyeing.

When cotton is dyed in a vat containing lime and which has considerable sediment, the material must always be washed with acid (1 to 2 % of sulphuric or hydrochloric acid is used) after dyeing in order to remove all particles from the fibre, which would otherwise tender the cotton on drying. After the acid treatment the cotton must be thoroughly washed.

Darker shades are obtained in this vat if the yarn is dried before acidifying, and redder shades can be produced by drying at a high temperature. Also by steaming the shade is made more violet and bloomy. These remarks hold true for cotton dyed in any form. of vat. This coppery appearance, however, is changed by washing towards black.

It is probable that Indigo forms a chemical compound with ferrous sulphate and lime, and this entails a considerable loss of dyestuff, for under the most favorable conditions only 75 to 80 per cent of the Indigo placed in the vat can be found again. A part of the Indigo remains in the sediment probably combined with ferrous hydrate.

In setting the copperas vat it is customary to put the Indigo and copperas into the bath first, and then to add the milk-of-lime. To save time, however, and to obtain a better reduction of the Indigo it is advisable to prepare a stock vat. This may be prepared conveniently by mixing 25 lbs. of Indigo paste (20 %) with 20 lbs. of copperas previously dissolved in hot, water, and then add 25 lbs. of lime in the form of a thin cream. Have the temperature of the Vat at about 120° F., stir up well and allow to stand until fully reduced, which will require about three hours. In practice the vat is usually prepared in the evening and allowed to stand overnight.

The dyevat is usually a stone or wooden circular vat 6 to 9 ft. deep and 21 to 5 ft. in diameter, and generally sunk into the floor of the dyehouse so as to make it convenient for working. In starting a new vat the necessary amount of water is run in, and then for each 100 gallons 1 lb. of copperas and 2 lbs. of lime are added in order to counteract the effect of the oxygen in the water. The necessary amount of the stock vat is then added, the liquid stirred up and left for two to three hours. The liquor should then be clear and of a brownish amber color, and on gently stirring it, dark blue streaks should appear with a coppery scum or flurry float on the surface. Before entering the goods to be dyed this flurry should be skimmed off and added to the stock vat.

If the liquor is greenish it indicates that part of the Indigo is not reduced, and more copperas has to be added. If it has a darkish appearance more alkali is needed and additional lime is added. An excess of either copperas or lime, however, should be avoided. After a day's working the vat should be well raked up and if necessary replenished by additions from the stock vat. The sediment in the copperas vat contains a considerable amount of Indigo, hence this should be saved and the Indigo recovered by treatment with hydrochloric acid.

According to the Badische Co. bright reddish shades may be obtained by previously treating the cotton goods with bone glue. For this purpose the goods are run in a solution containing 2 to 5 parts of glue per 1000 parts of water, squeezed and dyed. Better results are said to be obtained if the goods are dried before dyeing. This method of treatment is especially recommended for dyeing in the hydrosulphite vat.

Golden Yellow GK

Vat Yellow 4, also known as Golden Yellow GK, dibenzochrysenedione, dibenzpyrenequinone, Tyrian

Yellow I-GOK, and C.I. 59100, is a yellow synthetic anthraquinone vat dye. Chemically it is dibenzo[b,def]chrysene-7,14-dione, or 3,4:8,9-dibenzopyrene-5,10-dione, or $C_{24}H_{12}O_2$. It has the appearance of a viscous orange liquid.

Vat Yellow 4 is used mostly as a dye for textiles and paper.

Together with benzanthrone, it is used in some older pyrotechnic compositions for green and yellow colored smokes. Vat Yellow 4 is a Group 3 carcinogen according to the IARC, as the evidence of its carcinogenity to humans is inadequate

PIGMENTS

A **pigment** is a material that changes the colour of light it reflects as the result of selective colour absorption. This physical process differs from fluorescence, phosphorescence, and other forms of luminescence, in which the material itself emits light.

Many materials selectively absorb certain wavelengths of light. Materials that humans have chosen and developed for use as pigments usually have special properties that make them ideal for coloring other materials. A pigment must have a high tinting strength relative to the materials it colors. It must be stable in solid form at ambient temperatures.

For industrial applications, as well as in the arts, permanence and stabililty are desirable properties. Pigments that are not permanent are called fugitive. Fugitive pigments fade over time, or with exposure to light, while some eventually blacken.

Pigments are used for coloring paint, ink, plastic, fabric, cosmetics, food and other materials. Most pigments used in manufacturing and the visual arts are dry colorants, usually ground into a fine powder. This powder is added to a vehicle (or matrix), a relatively neutral or colorless material that acts as a binder.

A distinction is usually made between a pigment, which is insoluble in the vehicle, and a dye, which is either a liquid, or is soluble in its vehicle. A colorant can be both a pigment and a dye depending on the vehicle it is used in. In some cases, a pigment can be manufactured from a dye by precipitating a soluble dye with a metallic salt. The resulting pigment is called a lake pigment.

History

Naturally occurring pigments such as ochres and iron oxides have been used as colorants since prehistoric times. Archaeologists have uncovered evidence that early humans used paint for aesthetic purposes such as body decoration. Pigments and paint grinding equipment believed to be between 350,000 and 400,000 years old have been reported in a cave at Twin Rivers, near Lusaka, Zambia. Before the Industrial Revolution, the range of color available for art and decorative uses was technically limited. Most of the pigments in use were earth and mineral pigments, or pigments of biological origin. Pigments from unusual sources such as botanical materials, animal waste, insects, and mollusks were harvested and traded over long distances. Some colors were costly or impossible to mix with the range of pigments that were available. Blue and purple came to be associated with royalty because of their expense.Biological pigments were often difficult to acquire, and the details of their production were kept secret by the manufacturers. Tyrian Purple is a pigment made from the mucus of one of several species of Murex snail. Production of Tyrian Purple for use as a fabric dye began as early as 1200 BCE by the Phoenicians, and was continued by the Greeks and Romans until 1453 CE, with the fall of Constantinople. The pigment was expensive and complex to produce, and items colored with it became associated with power and wealth. Greek historian Theopompus, writing in the 4th century BCE, reported that "purple for dyes fetched its weight in silver at Colophon [in Asia Minor]." Mineral pigments were also traded over long distances. The only way to achieve a deep rich blue was by using a semi-precious stone, lapis lazuli, and the best sources of lapis were remote. Flemish painter Jan Van Eyck, working in the 15th century, did not ordinarily include blue in his paintings. To have one's portrait commissioned and painted with blue was considered a great luxury. If a patron wanted blue, they were forced to pay extra. When Van Eyck used lapis, he never blended it with other colors. Instead he applied it in pure form, almost as a decorative glaze. Spain's conquest of a New World empire in the 16th century introduced new pigments and colors to peoples on both sides of the Atlantic. Carmine, a dye and pigment derived from a parasitic insect found in Central and South America, attained great status and value in Europe. Produced from harvested, dried, and crushed cochineal insects, carmine could be used in fabric dye, body paint, or in its solid lake form, almost any kind of paint or cosmetic.

Natives of Peru had been producing cochineal dyes for textiles since at least 700 CE, but Europeans had never seen the color before. When the Spanish invaded the Aztec empire in what is now Mexico, they were quick to exploit the color for new trade opportunities. Carmine became the region's second most valuable export next to silver. Pigments produced from the cochineal insect gave the Catholic cardinals their vibrant robes and the English "Redcoats" their distinctive uniforms. The true source of the pigment, an insect, was kept secret until the 18th century, when biologists discovered the source.

Pigment Groups

- *Biological origins:* Chlorophyll and Melanin Alizarin, Alizarin Crimson, Gamboge, Indigo, Indian Yellow, Cochineal Red, Tyrian Purple, Rose madder
- *Arsenic pigments:* Paris Green
- *Carbon pigments:* Carbon Black, Ivory Black, Vine Black, Lamp Black
- *Cadmium pigments:* cadmium pigments, Cadmium Green, Cadmium Red, Cadmium Yellow, Cadmium Orange
- *Iron oxide pigments:* Caput Mortuum, oxide red, Red Ochre, Sanguine, Venetian Red, Mars Black
- *Prussian blue*
- *Chromium pigments:* Chrome Green, Chrome Yellow
- *Cobalt pigments:* Cobalt Blue, Cerulean Blue, Cobalt Violet, Aureolin
- *Lead pigments:* lead white, Naples yellow, Cremnitz White, red lead
- *Copper pigments:* Paris Green, Verdigris, Viridian, Egyptian Blue, Han Purple
- *Titanium pigments:* Titanium White, Titanium Beige, Titanium yellow, Titanium Black

- *Ultramarine pigments:* Ultramarine, Ultramarine Green Shade, French Ultramarine
- *Mercury pigments:* Vermilion
- *Zinc pigments:* Zinc White
- *Organic:* Pigment Red 170, Phthalo Green, Phthalo Blue, Quinacridone Magenta.
- *Lapis lazuli.*

Biological Pigments

In biology, a pigment is any material resulting in color of plant or animal cells. Many biological structures, such as skin, eyes, fur and hair contain pigments (such as melanin) in specialized cells called chromatophores. Many conditions affect the levels or nature of pigments in plant and animal cells. For instance, Albinism is a disorder affecting the level of melanin production in animals.Pigment color differs from structural colour in that it is the same for all viewing angles, whereas structural color is the result of selective reflection or iridescence, usually because of multilayer structures. For example, butterfly wings typically contain structural color, although many butterflies have cells that contain pigment as well.

Chlorophyll

Chlorophyll is a green photosynthetic pigment found in most plants, algae, and cyanobacteria. Its name is derived from ancient Greek: chloros = green and phyllon = leaf. Chlorophyll absorbs most strongly in the blue and red but poorly in the green portions of the electromagnetic spectrum, hence the green color of chlorophyll-containing tissues like plant leaves.

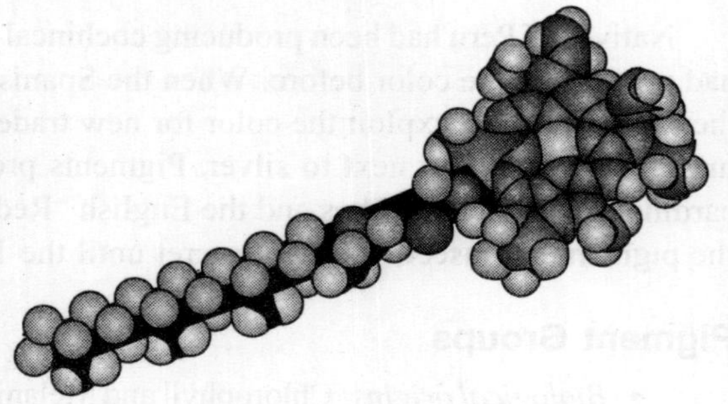

Chemical structure

	Chlorophyll *a*	Chlorophyll *b*	Chlorophyll *c1*	Chlorophyll *c2*	Chlorophyll *d*
Molecular formula	$C_{55}H_{72}O_5N_4Mg$	$C_{55}H_{70}O_6N_4Mg$	$C_{35}H_{30}O_5N_4Mg$	$C_{35}H_{28}O_5N_4Mg$	$C_{54}H_{70}O_6N_4Mg$
C3 group	$-CH=CH_2$	$-CH=CH_2$	$-CH=CH_2$	$-CH=CH_2$	-CHO
C7 group	$-CH_3$	-CHO	$-CH_3$	$-CH_3$	$-CH_3$
C8 group	$-CH_2CH_3$	$-CH_2CH_3$	$-CH_2CH_3$	$-CH=CH_2$	$-CH_2CH_3$
C17 group	$-CH_2CH_2COO-$ Phytyl	$-CH_2CH_2COO-$ Phytyl	-CH=CHCOOH	-CH=CHCOOH	$-CH_2CH_2COO-$ Phytyl
C17-C18 bond	Single	Single	Double	Double	Single
Occurrence	Universal	Mostly plants	Various algae	Various algae	Cyanobacteria

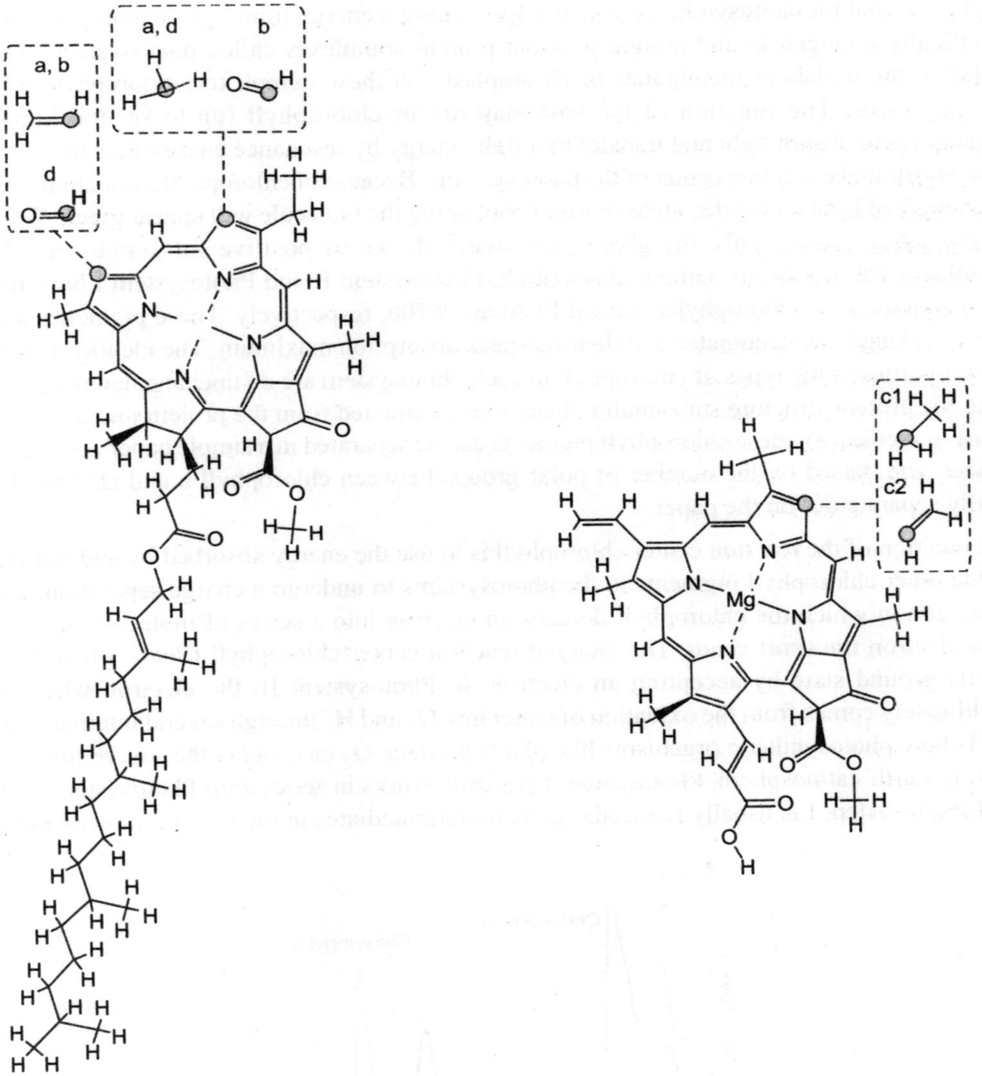

Fig. 28.6. The Structure of Chlorophyl

Chlorophyll is a chlorin pigment, which is structurally similar to and produced through the same metabolic pathway as other porphyrin pigments such as heme. At the center of the porphyrin ring is a magnesium ion. The chlorin ring can have several different side chains, usually including a long phytol chain. There are a few different forms that occur naturally:

Chlorophyll and photosynthesis

Chlorophyll is vital for photosynthesis, which helps plants get energy from light. Chlorophyll molecules are specifically arranged in and around pigment protein complexes called photosystems, which are embedded in the thylakoid membranes of chloroplasts. In these complexes, chlorophyll serves two primary functions. The function of the vast majority of chlorophyll (up to several hundred per photosystem) is to absorb light and transfer that light energy by resonance energy transfer to a specific chlorophyll pair in the reaction center of the photosystems. Because of chlorophyll's selectivity regarding the wavelength of light it absorbs, areas of a leaf containing the molecule will appear green. When a leaf was tested using iodine, only the green areas were shown as positive for starch, meaning that photosynthesis will not occur without chlorophyll. Photosystem II and Photosystem I have their own distinct reaction center chlorophylls, named P680 and P700, respectively. These pigments are named after the wavelength (in nanometers) of their red-peak absorption maximum. The identity, function and spectral properties of the types of chlorophyll in each photosystem are distinct and determined by each other and the protein structure surrounding them. Once extracted from the protein into a solvent (such as acetone or methanol), these chlorophyll pigments can be separated in a simple paper chromatography experiment, and, based on the number of polar groups between chlorophyll a and chlorphyll b, will chemically separate out on the paper.

The function of the reaction center chlorophyll is to use the energy absorbed by and transferred to it from the other chlorophyll pigments in the photosystems to undergo a charge separation, a specific redox reaction in which the chlorophyll donates an electron into a series of molecular intermediates called an electron transport chain. The charged reaction center chlorophyll (P680+) is then reduced back to its ground state by accepting an electron. In Photosystem II, the electron which reduces P680+ ultimately comes from the oxidation of water into O_2 and H^+ through several intermediates. This reaction is how photosynthetic organisms like plants produce O_2 gas, and is the source for practically all the O_2 in Earth's atmosphere. Photosystem I typically works in series with Photosystem II, thus the $P700^+$ of Photosystem I is usually reduced, via many intermediates in the thylakoid membrane,

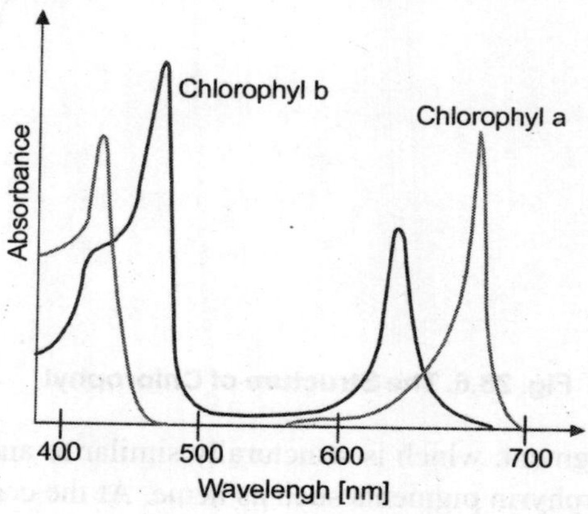

Fig. 28.7. The Absorbence curve of Chlorophyll a green

by electrons ultimately from Photosystem II. Electron transfer reactions in the thylakoid membranes are complex, however, and the source of electrons used to reduce $P700^+$ can vary. The electron flow produced by the reaction center chlorophyll pigments is used to shuttle H^+ ions across the thylakoid membrane, setting up a chemiosmotic potential mainly used to produce ATP chemical energy, and those electrons ultimately reduce $NADP^+$ to NADPH a universal reductant used to reduce CO_2 into sugars as well as for other biosynthetic reductions.

Absorbance spectra of free chlorophyll a (green) and b (red) in a solvent (Fig. 28.7). The spectra of chlorophyll molecules are slightly modified in vivo depending on specific pigment-protein interactions.

Reaction center chlorophyll-protein complexes are capable of directly absorbing light and performing charge separation events without other chlorophyll pigments, but the absorption cross section (the likelihood of absorbing a photon under a given light intensity) is small. Thus, the remaining chlorophylls in the photosystem and antenna pigment protein complexes associated with the photosystems all cooperatively absorb and funnel light energy to the reaction center. Besides chlorophyll a, there are other pigments, called accessory pigments, which occur in these pigment-protein antenna complexes. They include other forms of chlorophyll, such as chlorophyll b in green algal and higher plant antennae, while other algae may contain chlorophyll c or d. In addition, there are many non-chlorophyll accessory pigments, such as carotenoids or phycobiliproteins which also absorb light and transfer that light energy to the photosystem chlorophylls. Some of these accessory pigments, particularly the carotenoids, also serve to absorb and dissipate excess light energy, or work as antioxidants. The large, physically associated group of chlorophylls and other accessory pigments is sometimes referred to as a pigment bed, though this term is losing prominence with the advent of detailed knowledge of the structural organization of the photosystem and antenna complexes.

The different chlorophyll and non-chlorophyll pigments associated with the photosystems all have different spectra, either because the spectra of the different chlorophyll pigments are modified by their local protein environment, or because the accessory pigments have intrinsically different absorption spectra from chlorophyll. The net result is that, in vivo the total absorption spectrum is broadened and flattened such that a wider range of red, orange, yellow and blue light can be absorbed by plants and algae. Most photosynthetic organisms do not have pigments which absorb green light well, thus most remaining light under leaf canopies in forests or under water with abundant plankton is green, a spectral effect called the "green window". Some organisms, such as cyanobacteria and red algae, contain accessory phycobiliproteins that can absorb green light relatively well and thus they can exploit the little remaining green light in these habitats.

An easy experiment can show how chlorophyll is needed for photosynthesis. Some plants have variegated leaves with white and green areas. Take such a plant and put it in the dark for a few days. Take a leaf and make a "mask" to cover it out of aluminium foil. Cut two holes (one round, one square)in the foil such that one hole exposes a green part of the leaf and the other a white part of the leaf. Place the leaf in the light for an hour or so. If the leaf is exposed to a solution of iodine the green part of the leaf exposed to the light will show up black and the white part will not stain black. Furthermore, the black patch will have the same shape as the hole that you had cut in the foil. The iodine-stained starch only piles up in areas of the leaf that were green, showing that only those areas are photosynthetic. This proves that photosynthesis doesn't occur in the areas where there was no chlorophyll.

Melanin

Broadly, melanin is any of the polyacetylene, polyaniline, and polypyrrole "blacks" and "browns" or their mixed copolymers. The most common form of biological melanin is a polymer of either or both of two monomer molecules: indolequinone, and dihydroxyindole carboxylic acid. Melanin exists in the plant, animal and protista kingdoms, where it serves as a pigment.

Rose madder

Rose madder, sometimes referred to as **Rose Madder Genuine** is the crushed root of the Common Madder plant (*Rubia Tinctorium*). The ancient Egyptians used rose madder to create pinkish rose-coloured textile dyes. The Colour Index name used by paint and textile chemists for Rose Madder is Natural Red 9 (NR9)

It had limited use as an oil paint during the Renaissance era because it was considered a weak color. In the 19th century, chemists were able to manufacture a pigment that made rose madder a stronger and more durable oil paint. However, during the latter part of the 19th century, alizarin crimson was created and was considered at the time to be a superior replacement to rose madder. It is the synthetic form of rose madder, and its Colour Index name is Pigment Red 83 (PR83). Alizarin Crimson was soon discovered by artists to be a perfect color on the palette, making beautiful violets when mixed with blue, and perfect blacks and neutrals when mixed with a dark green like Viridian or Pthalo Green.

Ironically, while both alizarin crimson and rose madder are now considered "fugitive pigments or colors," genuine rose madder is now believed to be the more permanent of the two.

The term "genuine" is appended to the name "rose madder" if the paint has been manufactured by traditional means (using the root of the madder plant). The English manufacturer of oil paints Winsor & Newton sells "Rose Madder Genuine". It is still used today by some artists, however it is considered by many to be "too fugitive". Quinacridone rose is often used a substitute for rose madder. It is, however a far stronger color with a more pure hue than the genuine Rose Madder. Some artists find this disconcerting.

In an attempt to produce a colour that closely approximates the color of rose madder or alizarin crimson with pigments that are more permanent, many manufacturers produce colors made of various mixtures under names such as Permanent Alizarin. Not all of these mixtures are more permanent than the original however. For example Grumbacher manufactures a color called "rose madder hue" which is a mixture of Arylide Yellow (PY3) and Dihydroxy Anthraquinone Lake (PR83), which is simply an organic yellow mixed with Alizarin Crimson.

The relative weakness of rose madder limited its industrial uses mainly to textiles or artist's colors, but alizarin enjoyed widespread industrial use until the introduction of superior organic reds in this shade range started in the 1950s. By the beginning of the new millennium usage of rose madder and alizarin was in decline.

Rose Madder Genuine: Transparency Rating = 4 (high transparency); Stainability Rating = 0 (low); non-toxic.

External source: Red Oil Paints

SYNTHETIC INORGANIC PIGMENTS

Paris Green is a common name for copper(II)-acetoarsenite, or C.I. Pigment Green 21, an extremely toxic blue green chemical with four main uses: pigment, animal poison (mostly rodenticide), insecticide, and blue colorant for fireworks.

Other names for the chemical are Emerald Green, Parrot Green, Schweinfurth Green, Imperial Green, Vienna Green, and Mitis Green. It is almost never called Paris Green when referencing its use as a pigment. Since the use of Emerald Green as a pigment has been abandoned (around 1960), if one comes across the chemical today it is usually referred to as Paris Green. It was once a popular pigment, C.I. Pigment Green 21, used in artists' paints. When used as a pigment it was almost always given a color-based name, usually Emerald Green. The brilliance of this pigment has not been matched by modern pigment chemistry. The modern, significantly less vivid, substitution is a mixture of phthalocyanine green (blue shade), an organic lemon yellow, and white. Modern imitations either call themselves "Emerald Green" or "Permanent Green". The closest match to true Emerald Green in watercolors, Winsor & Newton's "Emerald Green", is impermanent and was recently discontinued.

It was once used to kill rats in Parisian sewers, hence the common name Paris Green. It was also used in America and elsewhere as an insecticide for produce, such as apples, around 1900, where it was blended with lead arsenate. This outrageously toxic mixture is said to have burned the trees and the grass around the trees. Old pianos may contain this mixture, or either of its components. It may still be found in limited use as an insecticide, primarily in the developing world, and as a fireworks colorant. It is reportedly very difficult to obtain a good blue in fireworks with any other chemical. Paris Green was once a popular pigment for painting ships, because its toxicity prevented the accumulation of barnacles.

Scheele's green is a chemically simpler, less brilliant, and less permanent, synthetic copper-arsenic pigment used for a rather short time before Emerald Green was first synthesized, which was approximately 1814. It was popular as a wallpaper pigment, and would degrade, with moisture and moulds, to arsine gas. Emerald green may have also been used in wallpaper to some extent and may have also degraded similarly. Both pigments were once used in printing ink formulations.

The color of Emerald Green is said to range from a pale, but vivid, blue green when very finely ground, to a deeper true green when coarsely ground. The molecule's vividness comes from hydrogen bonds. Similar natural compounds are the minerals Chalcophyllite $Cu_{18}Al_2(AsO_4)_3(SO_4)_3(OH)_{27}\cdot 36(H_2O)$, Conichalcite $CaCu(AsO_4)(OH)$, Cornubite $Cu_5(AsO_4)_2(OH)_4\cdot(H_2O)$, Cornwallite $Cu_5(AsO_4)_2(OH)_4\cdot(H_2O)$, and Liroconite $Cu_2Al(AsO4)(OH)_4\cdot 4(H_2O)$. These vivid minerals range from greenish blue to slightly yellowish green. The ancient Romans used one of them, possibly conichalcite, as a green pigment. The Emerald Green paint used by the Impressionists is said to have been composed of relatively coarse particles. Later, the chemical was produced with increasingly small grinds and without carefully removing impurities; its permanence suffered. It is likely that it was ground more finely for use in watercolors and inks, too.

The exterior of the home of Ulysses S. Grant was painted with this pigment, making it extremely striking. It was also the pigment used to paint window shutters, signs, etc. It blackens when exposed due to the instability of the molecule, instability that appears to be increased when the chemical is produced with a very fine particle size and isn't washed to remove impurities. Much of the blackening may be due to its reactivity with sulfur, as sulfur-containing compounds are common in painting (such as cadmium sulfide) and in air due to air pollution. In oils, artists often isolated Emerald Green with varnish to reduce its tendency to darken, a practice that is suggested for all arsenic pigments, such as orpiment. It has become the norm in parts of America to paint shutters a dark green color because people mistakenly believe the tradition was to paint them such a dark hue. In fact, shutters were commonly painted with Emerald Green.

Arsenic based green also is a very old dye for cloth, but its use was eventually abandoned because those who wore clothes so dyed tended to come to early ends, without the substance's toxicity being formally recognized. To this day, French theater costumes traditionally eschew the colour green.

An artist who produced his own oil paint with Paris Green obtained from a fireworks supplier suffered arsenic poisoning from vapours which emanated from the finished paint. Either impurities were the cause, or the molecule itself spontaneously degrades, creating a highly toxic arsine gas. This was Cezanne's favorite pigment, and it dominates many of his paintings. In his watercolours, thin washes have turned brown but thicker applications have remained bright green. The pigment was also used heavily by other artists of his era, such as Van Gogh. Cezanne developed severe diabetes, which is a symptom of chronic arsenic poisoning. Monet's blindness and Van Gogh's neurological disorders are likely directly related to their use of Emerald Green, as well as lead pigments, mercury-based Vermilion, and solvents such as turpentine.

The permanent purple pigment "Cobalt Violet", also used by the Impressionists, was once formulated with arsenic. It was reformulated successfully without arsenic in two varieties, "light" with ammonium or lithium and "deep" without. It would be useful to artists and others in need of a brilliant green pigment for a compound similar to Emerald Green to be synthesized without arsenic.

Carbon Black

Carbon black is a material, today usually produced by the incomplete combustion of petroleum products. Carbon black is a form of amorphous carbon that has an extremely high surface area to volume ratio, and as such it is one of the first nanomaterials to find common use. It is similar to soot but with a much higher surface area to volume ratio. Carbon black is often used as a pigment and reinforcement in rubber and plastic products. The current IARC evaluation is that, "Carbon black is possibly carcinogenic to humans (Group 2B)". Short-term exposure to high concentrations of the carbon black dust may produce discomfort to the upper respiratory tract, through mechanical irritation.

Common Uses

The most common usage [70%] of carbon black is as a pigment and reinforcing phase in automobile tires. Carbon black also helps conduct heat away from the tread and belt area of the tire, reducing thermal damage and increasing tire life. Carbon black particles are also employed in some radar absorbent materials and in printer toner.

Total production in 2006 is expected to be about 78,932,789 kilotonnes. About 20% of world production goes into belts, hoses, and other rubber goods. The balance is used in inks and as a pigment for products other than tires.

Reinforcing Carbon Blacks

The highest volume use of carbon black is as a reinforcing filler in rubber products, especially tires. While a pure gum vulcanizate of SBR has a tensile strength of no more than 2.5 MPa, and almost nonexistent abrasion resistance, compounding it with 50% of its weight of carbon black improves its tensile strength and wear resistance as shown in the table below.

Types of Carbon Black Used in Tires

Name	Abbrev.	ASTM Desig.	Particle Size nm	Tensile Strength MPa	Relative Laboratory Abrasion	Relative Roadwear Abrasion
Super Abrasion Furnace	SAF	N110	20-25	25.2	1.35	1.25
Intermediate SAF	ISAF	N220	24-33	23.1	1.25	1.15
High Abrasion Furnace	HAF	N330	28-36	22.4	1.00	1.00
Easy Processing Channel	EPC	N300	30-35	21.7	0.80	0.90
Fast Extruding Furnace	FEF	N550	39-55	18.2	0.64	0.72
High Modulus Furnace	HMF	N683	49-73	16.1	0.56	0.66
Semi-Reinforcing Furnace	SRF	N770	70-96	14.7	0.48	0.60
Fine Thermal	FT	N880	180-200	12.6	0.22	--
Medium Thermal	MT	N990	250-350	9.8	0.18	--

Practically all rubber products where tensile and abrasion wear properties are crucial use carbon black, so they are black in color. Where physical properties are important but colors other than black are desired, such as white tennis shoes, precipitated or fumed silica is a decent competitor to carbon black in reinforcing ability. Silica based fillers are also gaining market share in automotive tires because they provide better fuel efficiency due to a lower rolling loss compared to carbon black filled tires. Traditionally silica fillers had worse abrasion wear properties, but the technology has gradually improved to where they can match carbon black abrasion performance.

Pigment

Carbon black (Colour Index International, PBL-7) is the name of a common black pigment, traditionally produced from charring organic materials such as wood or bone. It consists of pure elemental carbon, and it appears black because it reflects almost no light in the visible part of the spectrum. It is known by a variety of names, each of which reflects a traditional method for producing carbon black:

- Ivory black was traditionally produced by charring ivory or animal bones.
- Vine black was traditionally produced by charring desiccated grape vines and stems.
- Lamp black was traditionally produced by collecting soot, also known as lampblack, from oil lamps.

Newer methods of producing carbon black have superseded these traditional sources, although some materials are still produced using traditional methods, for artisanal purposes.

Surface Chemistry of Carbon Black

All carbon blacks have chemisorbed oxygen complexes (i.e., carboxylic, quinonic, lactonic, phenolic groups and others) on their surfaces to varying degrees depending on the conditions of manufacture. These surface oxygen groups are collectively referred to as volatile content. It is also known to be a non-conductive material due to its volatile content.

The coatings and inks industries prefer grades of carbon black that are acid oxidized. Acid is sprayed in high temperature dryers during the manufacturing process to change the inherent surface chemistry of the black. The amount of chemically-bonded oxygen on the surface area of the black is increased to enhance performance characteristics.

Cadmium pigments are a class of pigments that have cadmium as one of the chemical components. Most of cadmium produced worldwide is used in the production of Ni-Cd Batteries, but about half the remaining consumption, which is about 2,000 tons annually, is used to produce coloured cadmium pigments. The principal pigments are a family of yellow/orange/red cadmium sulfides and sulfoselenides. Cadmium yellow is cadmium sulfide (CdS), cadmium red is cadmium selenide (CdSe) and cadmium orange is an intermediate cadmium sulfoselenide. Cadmium yellow is sometimes mixed with viridian to give a bright, pale green mixture called cadmium green.

Brilliantly coloured, with good permanence and tinting power, Cadmium Yellow, Cadmium Orange, and Cadmium Red are familiar artist colors, but of little use in architectural paints. Their greatest use is in the coloring of plastics and specialty paints which must resist processing or service temperatures up to 300°C. The color-fastness or permanence of cadmium requires protection from a tendency to slowly form carbonate salts with exposure to air. Most paint vehicles accomplish this, but cadmium colors will fade in fresco or mural painting. Cadmium pigments can also colour glass and ceramic glazes, not by solution, but colloidal dispersion within the glass. The lenses of red stoplights use this technique.

Cadmium sulfide and a mixture of cadmium sulfide with cadmium selenide are commonly used as pigments in artist's paints. They have an excellent reputation for colour permanence although this is partially based on two reasons which are not necessarily directly related to their properties:

1. when introduced, there were hardly any stable pigments in the yellow to red range, especially orange and bright red was very troublesome, when the cadmium pigments replaced e.g. mercury sulfide (the original vermilion), the light-fastness was greatly improved,
2. companies sell the cadmium-containing paints at premium price. Although the pigments are certainly more expensive, the premium price is often not fully justifiable, with reasons more in the marketing area than in the actual raw material cost.

Nowadays, the cadmium pigments have been partially replaced by azo pigments. These are similar in lightfastness to the cadmium colors and have the advantage of both being cheaper and non-toxic. With respect to lightfastness the lemon yellow cadmium pigment is an exception: the azo-variety is highly superior in light-fastness. In some countries, such as Australia, consumer activists such as Michael Vernon were successful in banning the use of cadmium pigments in plastics that could be used for toy manufacture, owing to the toxicity of cadmium.

Caput Mortuum is a Latin term meaning 'death's head'. In alchemy, it signified a useless substance left over from a chemical operation such as sublimation. Alchemists represented this residue with a stylized human skull, a literal death's head. In its current limited usage, the caput mortuum represents decline and entropy.Caput mortuum (also spelled *caput mortum* or *caput mortem*) is the name given to a purple variety of iron oxide pigment, an "earth color". It is used in oil paints and paper dyes. The name for this pigment may have come from the alchemical usage, since iron oxide (rust) is the useless residue of oxidization.It is the name of a brownish paint that was originally made from the wrappings of mummies. It was most popular in the 1600s. It was suddenly discontinued in the early 19th century when its composition became generally known to artists. A London colorman claimed that he could satisfy the demands of his customers for twenty years from one Egyptian mummy. In the past years, it has been made with iron sulphate and impurities obtained from the residues of the distillation of scisti piritosi in the fabrication of sulphuric acid. The paint color is also known as Colcothar, Mummy Brown, Mummy, Egyptian Brown, and by combinations like Caput Mortem Violet.

Red ochre and **yellow ochre** (from the Greek ochros, yellow) are pigments made from naturally tinted clay. It has been used worldwide since prehistoric times. Chemically, it is hydrated iron (III) oxide.

Ochres are non-toxic, and can be used to make an oil paint that dries quickly and covers surfaces thoroughly. Many people believe that the best ochre comes from the area of Roussillon, France.

To manufacture ground ochre, ochre clay is first mined from the ground. It is then washed in order to separate sand from ochre, which can be done by hand. The remaining ochre is then dried in the sun and sometimes burned to enhance the natural colour.

Venetian red is a light and warm somewhat (unsaturated) pigment that is a darker shade of scarlet, derived from nearly pure ferric oxide (Fe_2O_3) of the hematite type. Modern versions are frequently made with synthetic red iron oxide.

Prussian blue (German: Preußisch Blau, Berliner Blau) is a dark blue pigment used in paints and formerly in blueprints. Prussian blue was discovered by accident by the painter Diesbach in Berlin and therefore it is also called Berlin blue. He originally was attempting to create a paint with a red hue. It has several different chemical names, these being iron(III) ferricyanide, ferric ferrocyanide, iron(III) hexacyaniferrate(II), and ferric hexacyaniferrate. Commonly and conveniently it is simply called PB

Composition

Despite being one of the oldest known synthetic compounds, the composition of PB was uncertain until recently. The precise identification of PB was complicated by three factors: (i) PB is extremely insoluble but also tends to form colloids, (ii) traditional syntheses tend to afford impure compositions, and (iii) even pure PB is structurally complex, defying routine crystallographic analysis. So first one must decide on a definition of PB. As summarized in the classic review by Dunbar and Heintz, the chemical formula of PB is $Fe_7(CN)_{18}(H_2O)_x$ where $14 < X < 16$. The assignment of the structure and the formula resulted from decades of study using IR spectroscopy, Moessbauer spectroscopy and X-ray and neutron crystallography. Parallel studies were conducted on related materials such as $Mn_3[Co(CN)_6]_2$ and $Co_3[Co(CN)_6]_2$ [i.e., $Co_5(CN)_{12}$]. Since X-ray diffraction cannot distinguish C from N, the locations of these lighter elements is deduced by spectroscopic means as well as distances from the Fe centers. By growing crystals slowly from 10M HCl, Ludi obtained crystals wherein the defects were ordered. These workers concluded that the framework consists of Fe(II)-CN-Fe(III) linkages, with Fe(II)-C distances of 1.92 Å and Fe(III)-N distances of 2.03 Å. The Fe(II) centers, which are low spin, are surrounded by six carbon ligands. The Fe(III) centers, which are high spin, are surrounded on average by 4.5 N centres and 1.5 O centres, the latter from water. Again, the composition is notoriously variable due to the presence of lattice defects, allowing it to be hydrated to various degrees as water molecules are incorporated into the structure to occupy four cation vacancies. The variability of PB's composition is attributable to its low solubility, which leads to its rapid precipitation vs. growth of a single phase.

Turnbull's Blue

The story of "Turnbull's Blue" (TB) illustrates the complications and pitfalls associated with the characterization of a composition obtained by rapid precipitation. One obtains PB by the addition of Fe(III) salts to a solution of $[Fe(CN)_6]^{4-}$. TB supposedly arises by the related reaction where the valences are switched on the iron precursors, i.e. the addition of a Fe(II) salt to a solution of $[Fe(CN)_6]^{3}$. One obtains an intensely blue coloured material, whose hue was claimed to differ from that of PB. It is now appreciated that TB and PB are the same because of the rapidity of electron exchange through a Fe-CN-Fe linkage. The differences in the colors for TB and PB reflect subtle differences in the method of precipitation, which strongly affects particle size and impurity content.

"Soluble" Prussian Blue

PB is insoluble, but it tends to form such small crystallites that colloids are common. These colloids act like solutions, for example they pass through fine filters. According to Dunbar and Heintz, these "soluble" forms tend toward compositions with the approximate formula $KFe_2Fe(CN)_6$.

The Colour of PB

PB is strongly coloured, and tends towards black and dark purple when mixed with other oil paints. The exact hue depends on the method of preparation, which dictates the particle size. The intense blue color of Prussian blue is associated with the energy of the transfer of electrons from Fe(II) to Fe(III). Many such mixed valence compounds absorb visible light. Red light at 680 nm is absorbed, and the transmitted light appears blue as a result.

Other Properties

Prussian Blue has been extensively studied by inorganic chemists and solid-state physicists because of its unusual properties.

- It undergoes intervalence charge transfer. Although intervalence charge transfer is well-understood today PB was the subject of intense study when the phenomenon was discovered.
- It is electrochromic - changing color from blue to white in response to a voltage. This is caused by partial oxidation of the Fe(II) to Fe(III) eliminating the intervalence charge transfer that causes PB's blue color.
- It undergoes spin-crossover behavior. Upon exposure to visible light the Fe atoms transition from low spin to high spin. This spin transition also changes the magnetic coupling between the Fe atoms, making PB one of the few known classes of material that has a magnetic response to light.

Despite the presence of the cyanide ion, PB is not especially toxic because the cyanide groups are tightly bound. Other cyanometalates are similarly stable with low toxicity. However, treatment with acids can liberate hydrogen cyanide which is extremely toxic as discussed in the article on cyanide.

Production

PB is prepared by adding a solution containing iron(III) chloride to a solution of potassium ferrocyanide. During the course of the addition, the solution thickens visibly and the colour changes immediately to the characteristic hue of PB.

This reaction is the first step to obtain ink. Dissolving it with water, and then filtering it, results in a blue residuum. The addition of a dilution of oxalic acid previously heated results in a useful blue ink.

Uses

- Colloids derived from PB are the basis for laundry bluing.
- PB is the coloring agent used in Engineer's blue.
- The formation of PB is a "wet" chemical test for cyanide. This test was a key component of the Errol Morris film Mr. Death: The Rise and Fall of Fred A. Leuchter, Jr..
- PB is the pigment formed on cyanotypes, giving them their name "blueprint".
- PB's ability to incorporate +1 cations makes it useful as a sequestering agent for certain heavy metals ions. In particular, pharmaceutical-grade PB is used for patients who have ingested radioactive caesium or thallium (also non-radioactive thallium). According to the IAEA an adult male can eat 10 grams of Prussian Blue per day without serious harm. It is also occasionally used in cosmetic products. The US FDA has determined that the "500 mg Prussian blue capsules, when manufactured under the conditions of an approved New Drug Application (NDA), can be found safe and effective for the treatment of known or suspected internal contamination with radioactive cæsium, radioactive thallium, or non-radioactive thallium."

Chromium(III) oxide, also known as chromium sesquioxide or chromia, is one of four oxides of chromium, chemical formula Cr_2O_3. It is commonly called chrome green when used as a pigment; however it was referred to as viridian when it was first discovered. The Parisians Pannetier and Binet first prepared Cr_2O_3 in 1838 via a secret process. It is manufactured from the mineral chromite, $Fe(CrO_2)_2$, which is mined in southern Africa, Asia, Turkey and Cuba. The conversion of chromite to chromia entails air oxidation to $Na_2Cr_2O_7$, which is subsequently reduced with carbon or sulfur.

Cr_2O_3 is an amphoteric oxide, dissolving in acids to give chromium(III) salts and in molten alkali to give chromites. It is also a good absorbent in process of adsorption.

Chromium oxide can be made into elemental chromium metal through the thermite reaction, though this metal oxide is much less commonly used than Fe_2O_3 and Fe_3O_4. Unlike iron oxide thermites, chromium oxide thermite creates little or no sparks, smoke or sound, but it glows with a blinding light. Because of the very high melting point of chromium, chromium thermite casting is impractical.

Precautions

Though Cr_2O_3 is not a serious health hazard, it can cause irritation of the skin and eyes, and can cause nausea and other problems if ingested. It also can cause respiratory problems when dust is inhaled. It is not a fire hazard, and does not readily react with other materials.

Chrome Yellow is a natural yellow pigment made of lead(II) chromate ($PbCrO_4$). It was first extracted from the mineral crocoite by the French chemist Louis Vauquelin in 1809. Because the pigment tends to oxidize and darken on exposure to air over time, and it contains lead, a toxic, heavy metal, it has been largely replaced by another pigment, Cadmium Yellow. It is commonly produced by mixing solutions of lead nitrate and potassium chromate and filtering off the lead chromate precipitate.

Cobalt is a cool, slightly desaturated blue colour, historically made using cobalt salts. It was discovered by Louis-Jacques Thenard in 1802. The world leading manufacturer of cobalt blue in the 19th century was Blaafarveværket in Norway, led by Benjamin Wegner. It is extraordinarily stable. Chemically it is a cobalt(II) oxide-aluminium oxide, or cobalt aluminate. Commercial production began in France in 1807.

The blue seen on many glassware pieces is cobalt blue, and it is used widely by artists in many other fields. John Varley suggested it as a good substitution for ultramarine blue for painting skies. Maxfield Parrish, famous partly for the intensity of his skyscapes, used cobalt blue, and cobalt blue is sometimes called Parrish blue as a result.

"Cobalt Blue" is also a filter used in ophthalmoscopes, and is used to illuminate the cornea of the eye following application of fluorescein dye which is used to detect corneal ulcers and scratches.

Cerulean blue Discovered in 1805 by Andreas Höpfner, the pigment was first marketed in 1860 as "coeruleum" by George Rowney of the United Kingdom. The primary chemical constituent is cobalt(II) stannate. Cerulean blue is a cerulean (light blue or azure) pigment used in artistic painting. It is particularly valuable for painting atmospheric shades because of the purity of the blue (specifically the lack of greenish hues). The pigment is regarded as permanent: in oil, no other blue pigment retains color as well.

Aureolin (sometimes called Cobalt Yellow) is a pigment used in oil and watercolor painting. Its pigment number is PY40 (40th entry on list of yellow pigments). It was first made in 1851 and its chemical composition is potassium cobaltinitrite. Aureolin is rated as an impermanent, transparent, lightly staining, light valued, intense medium yellow pigment. It is a rather expensive pigment

Naples yellow, also called antimony yellow, is a somewhat muted reddish yellow pigment containing lead(II)-antimonate. Its chemical composition is $Pb(SbO_3)_2/Pb_3(Sb_3O_4)_2$. It is also known as **jaune d'antimoine**. It is one of the oldest synthetic pigments, dating from around 1620. The related mineral pigment, bindheimite, dates from the 16th century BC. Naples yellow was used extensively by the Old Masters and well into the 20th century. The genuine pigment is toxic, and its use today is becoming increasingly rare. Most paints labeled "Naples yellow" are instead made with a mix of modern, less toxic pigments. The colors of these paints vary considerably from one manufacturer to another.

Red lead, also called minium, lead tetroxide or triplumbic tetroxide, is a bright red or orange crystalline or amorphous pigment. Its Latin name minium originates from the Minius River in northwest Spain where it was first mined.The melting point of lead tetroxide is 500°C, at which it decomposes to lead(II) oxide and oxygen.Chemically red lead is lead tetroxide, Pb_3O_4, or $2PbO.PbO_2$. It is used in the manufacture of batteries, lead glass and rust-proof paint. Red lead is virtually insoluble in water. However, it is soluble in hydrochloric acid present in stomach, therefore it is toxic when ingested. It is also insoluble in alcohol. It dissolves in hydrochloric acid, glacial acetic acid, and diluted mixture of nitric acid and hydrogen peroxide.

Verdigris is the common name for the chemical $Cu(CH_3COO)_2$, or copper(II) acetate. It commonly occurs by the action of acetic acid when copper, brass or bronze is weathered and exposed to air or seawater over a period of time. Its name comes from the Middle English *vertegrez*, from the Old French *verte grez*, an alteration of *vert-de-Grice - verd* (green), *de* (of), and *Grice* (Greece)- "green of Greece". It is mentioned in the musical "Wicked" in the song "The Wizard and I" to describe the Wicked Witch Elphaba's colour.

The vivid green color of verdigris makes it a very common pigment. Until the 19th century, verdigris was the most vibrant green pigment available and frequently used in painting. Verdigris is lightfast in oil paint, as numerous examples of 15th century paintings show. However, its lightfastness and air resistance is very low in other media. Copper resinate, made from verdigris, is not lightfast, even in oil paint. In the presence of light and air, green copper resinate becomes stable brown copper oxide. This degradation is to blame for the brown or bronze color of grass or foliage in many old paintings, although not typically those of the "Flemish primitive" painters such as Jan van Eyck, who often used normal verdigris. In addition, verdigris is a fickle pigment requiring special preparation of paint, careful layered application and immediate sealing with varnish to avoid rapid discoloration (but not in the case of oil paint). Verdigris has the curious property in oil painting that it is initially bluish-green, but turns a rich foliage green over the course of about a month. This green is stable. Vergidris fell out of use by artists as more stable green pigments became available.Verdigris is poisonous and has also been used in medicine and as a fungicide. Copper(II) acetate is soluble in alcohol and water and slightly soluble in ether and glycerol. It melts at 115°C and decomposes at 240°C. It can be prepared by reacting copper(II) oxide, CuO, or copper(II) carbonate, $CuCO_3$, with acetic acid, CH_3COOH. It is used industrially as a fungicide, a catalyst for organic reactions, and in dyeing

Viridian is a blue-green pigment, a hydrated chromium(III) oxide, of medium saturation and relatively dark in value. It is composed more of green than blue. Viridian takes its name from the Latin viridis, meaning "green".

Egyptian Blue ($CaCuSi_4O_{10}$ or $CaO.CuO.4SiO_2$) is a pigment used by Egyptians for thousands of years. It is considered to be the first synthetic pigment. The pigment was known to the Romans by the name caeruleum. Vitruvius describes in his work 'de architectura' how it was produced by grinding sand, copper and natron and heating the mixture, shaped into small balls, in a furnace. Lime is necessary for the production as well, but probably lime-rich sand was used. After the Roman era Egyptian Blue was not used anymore.

Han Purple ($BaCuSi_2O_6$) is a pigment that has been used in China for over 2,000 years.The terracotta warriors in Xian were dyed with it.

Chinese alchemists first synthesized *Han Purple* from barium copper silicates. It was used in large imperial projects and pottery.

Titanium dioxide, also known as **titanium(IV) oxide or titania**, is the naturally occurring oxide of titanium, chemical formula TiO_2. When used as a pigment, it is called titanium white, Pigment White 6, or **CI 77891**.

Natural Occurrence

Titanium dioxide occurs in four forms:

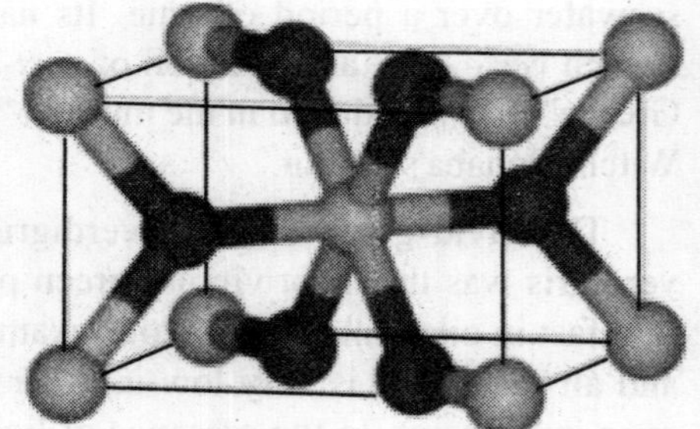

- rutile, a tetragonal mineral usually of prismatic habit, often twinned;
- anatase or octahedrite, a tetragonal mineral of dipyramidal habit;
- brookite, an orthorhombic mineral. Both anatase and brookite are relatively rare minerals;
- Titanium dioxide (B) or TiO_2(B), a monoclinic mineral.
- Titanium Oxide or TiO, as present in K and M spectral type stars.

Titanium dioxide occurrences in nature are never pure; it is found with contaminant metals such as iron. The oxides can be mined and serve as a source for commercial titanium. The metal can also be mined from other minerals such as ilmenite or leucoxene ores, or one of the purest forms, rutile beach sand.

Applications

As a pigment of high refringence Titanium dioxide is the most widely used white pigment because of its brightness and very high refractive index (n=2.4), in which it is surpassed only by a few other materials.

When deposited as a thin film, its refractive index and color make it an excellent reflective optical coating for dielectric mirrors and some gemstones, for example "mystic fire topaz". TiO_2 is also an effective opacifier in powder form, where it is employed as a pigment to provide whiteness and opacity to products such as paints, coatings, plastics, papers, inks, foods, and most toothpastes. Used as a white food coloring, it has E number E171. In cosmetic and skin care products, titanium dioxide is used both as a pigment and a thickener. It is also used as a tattoo pigment and styptic pencils.

This pigment is used extensively in plastics and other applications for its UV resistant properties where it acts as a UV reflector.

In ceramic glazes titanium dioxide acts as an opacifier and seeds crystal formation. In almost every sunscreen with a physical blocker, titanium dioxide is found both because of its refractive index and its resistance to discoloration under ultraviolet light. This advantage enhances its stability and ability to protect the skin from ultraviolet light.

As a photocatalyst

Titanium dioxide, particularly in the anatase form, is a photocatalyst under ultraviolet light. Recently it has been found that titanium dioxide, when spiked with nitrogen ions, is also a photocatalyst under visible light. The strong oxidative potential of the positive holes oxidizes water to create hydroxyl radicals. It can also oxidize oxygen or organic materials directly. Titanium dioxide is thus added to paints, cements, windows, tiles, or other products for sterilizing, deodorizing and anti-fouling properties and is also used as a hydrolysis catalyst. It is also used in the Graetzel cell, a type of chemical solar cell.

Titanium dioxide has potential for use in energy production: as a photocatalyst, it can carry out hydrolysis, i.e., break water into hydrogen and oxygen. Were the hydrogen collected, it could be used as a fuel. The efficiency of this process can be greatly improved by doping the oxide with carbon, as described in "Carbon-doped titanium dioxide is an effective photocatalyst".

As TiO_2 is exposed to UV light, it becomes increasingly hydrophilic; thus, it can be used for anti-fogging coatings or self-cleaning windows. TiO2 incorporated into outdoor building materials can substantially reduce concentrations of airborne pollutants such as volatile organic compounds and nitrogen oxides.

For Wastewater Remediation

TiO_2 offers great potential as an industrial technology for detoxification or remediation of wastewater due to several factors:

1. The process occurs under ambient conditions.
2. The formation of photocyclized intermediate products, unlike direct photolysis techniques, is avoided.
3. Oxidation of the substrates to CO_2 is complete.
4. The photocatalyst is inexpensive and has a high turnover.
5. TiO_2 can be supported on suitable reactor substrates.

Other Applications

It is also used in resistance-type lambda probes (a type of oxygen sensor).

Titanium yellow, also nickel antimony titanium yellow, nickel antimony titanium yellow rutile, CI Pigment Yellow 53, or C.I. 77788, is a yellow pigment with the chemical composition of $NiO.Sb_2O_5.20TiO_2$. Its CAS number is [8007-18-9]. It is a complex inorganic compound. Its melting point lies above 1000 °C, and it is extremely little soluble in water. While it contains antimony and nickel, their bioavailability is very low, so the pigment is relatively safe.

The pigment has crystal lattice of rutile, with 2-5% of titanium ions replaced with nickel(II) and 9-12% of them replaced with antimony(V). Titanium yellow is manufactured by reacting fine powders of metal oxides, hydroxides, or carbonates in solid state in temperatures between 1000-1200°C, either in batches or continuously in a pass-through furnace. Titanium yellow is used primarily as a pigment for plastics and ceramic glazes, and in art painting.

Ultramarine is a blue pigment consisting essentially of a double silicate of aluminium and sodium with some sulfides or sulfates, and occurring in nature as a proximate component of lapis lazuli. Also known in the past as *azzurrum ultramarine, azzurrum transmarinum, azzuro oltramarino, azur d'Acre, pierre d'azur, Lazurstein*. Current terminology for ultramarine include natural ultramarine (English), *outremer lapis* (French), *Ultramarin echt* (German), *oltremare genuino* (Italian), and *ultramarino verdadero* (Spanish). The pigment color code is P. Blue 29 77007. Ultramarine is the most complex of the mineral pigments, a complex sulfur-containing sodio-silicate ($Na_{8-10}Al_6Si_6O_{24}S_{2-4}$), essentially a mineralized limestone containing a blue cubic mineral called lazurite. Some chloride is often present in the crystal lattice as well. The blue color of the pigment is due to the S_3^- radical anion, which contains an unpaired electron.

Uses

The first noted use of lapis lazuli as a pigment can be seen in the 6th- and 7th-century AD cave paintings in Afghanistan temples, near the most famous source of the mineral. Lapis lazuli has also been identified in Chinese paintings from the 10th and 11th centuries, in Indian mural paintings from the 11th, 12th, and 17th centuries, and on Anglo-Saxon and Norman illuminated manuscripts from c.1100. Natural ultramarine is the most difficult pigment to grind by hand, and for all except the highest quality of mineral sheer grinding and washing produces only a pale grayish blue powder. At the beginning of the 13th century an improved method came into use, described by the 15th century artist Cennino Cennini. This process consisted of mixing the ground material with melted wax, resins, and oils, wrapping the resulting mass in a cloth, and then kneading it in a dilute lye solution. The blue particles collect at the bottom of the pot, while the impurities and colorless crystals remain in the mass. This process was performed at least three times, with each successive extraction generating a lower quality material. The final extraction, consisting largely of colorless material as well as a few blue particles, brings forth ultramarine ash which is prized as a glaze for its pale blue transparency.

The pigment was most extensively used during the 14th through 15th centuries, as its brilliance complemented the vermilion and gold of illuminated manuscripts and Italian panel paintings. From the beginning of the 16th century this pigment began to be imported into Europe from "over the sea," as azurrum ultramarinum. As lapis lazuli only yields from 2 to 3% of the pigment, it is not surprising to learn that the pigment used to be weighed up with gold. It was valued chiefly on account of its brilliancy of tone and its inertness in opposition to sunlight, oil, and slaked lime. It is, however, extremely susceptible to even minute and dilute mineral acids and acid vapors. Dilute HCl, HNO_3, and H_2SO_4 rapidly destroy the blue color, producing hydrogen sulfide (H_2S) in the process. Acetic acid attacks the pigment at a much slower rate than mineral acids. Because of this susceptibility, ultramarine was only used for frescoes when it was applied “secco”, in which the pigment was mixed with a binding medium and applied over dry plaster (such as Giotto di Bondone's frescos in the Cappella degli Scrovegni or Arena Chapel in Padua).

European artists used the pigment sparingly, reserving their highest quality blues for the robes of Mary and the Christ child. As a result of the high price, artists sometimes economized by using a cheaper blue, azurite, for under painting. Most likely imported to Europe through Venice, Italy, the pigment was seldom seen in German art or art from countries north of Italy. Due to a shortage of azurite in the late 16th and 17th century the demand for the already-expensive ultramarine increased dramatically. In 1814 Tassaert observed the spontaneous formation of a blue compound, very similar to ultramarine, if not identical with it, in a lime kiln at St. Gobain, which caused the Societé pour l'Encouragement d'Industrie to offer, in 1824, a prize for the artificial production of the precious colour. Processes were devised by Jean Baptiste Guimet (1826) and by Christian Gmelin (1828), then professor of chemistry in Tubingen; but while Guimet kept his process a secret Gmelin published his, and thus became the originator of the "artificial ultramarine" industry.

Chemistry and Manufacture

The raw materials used in the manufacture are: (1) iron-free kaolin, or some other kind of pure clay, which should contain its silica and alumina as nearly as possible in the proportion of SiO_2:Al_2O_2 demanded by the formula assigned to ideal kaolin (a deficit of silica, however, it appears can be made up for by addition of the calculated weight of finely divided silica); (2) anhydrous sulfate of soda; (3) anhydrous carbonate of soda; (4) sulfur (in the state of powder); and (5) powdered charcoal or relatively ash-free coal, or colophony in lumps. "Ultramarine poor in silica" is obtained by fusing a mixture of soft clay, sodium sulfate, charcoal, soda and sulfur. The product is at first white, but soon turns green ("green ultramarine") when it is mixed with sulfur and heated. The sulfur fires, and a fine blue pigment is obtained. "Ultramarine rich in silica" is generally obtained by heating a mixture of pure clay, very fine white sand, sulfur and charcoal in a muffle-furnace. A blue product is obtained at once, but a red tinge often results. The different ultramarines-green, blue, red and violet-are finely ground and washed with water.

Synthetic alternatives

Synthetic ultramarine is not as vivid a blue as natural ultramarine, since the particles in synthetic ultramarine are smaller and more uniform than natural ultramarine and therefore diffuse light more evenly. Synthetic ultramarine is also not as permanent as natural ultramarine.

Artificial, like natural, ultramarine has a magnificent blue colour, which is not affected by light nor by contact with oil or lime as used in painting. Hydrochloric acid immediately bleaches it with liberation of hydrogen sulfide. It is remarkable that even a small addition of zinc-white (oxide of zinc) to the reddish varieties especially causes a considerable diminution in the intensity of the colour, while dilution with artificial precipitated sulfate of lime ("annalin") or sulfate of baryta ("blanc fix") acts pretty much as one would expect. Ultramarine being very cheap, it is largely used for wall painting, the printing of paperhangings and calico, etc., and also as a corrective for the yellowish tinge often present in things meant to be white, such as linen, paper, etc. Bluing or "Laundry blue" is a solution of synthetic ultramarine that is used for this purpose when washing white clothes. Large quantities are used in the manufacture of paper, and especially for producing a kind of pale blue writing paper which is popular in Britain.

Ultramarine is based on the sodalite structure which is a three dimensional aluminosilicate cage containing three sulphur atoms bonded together to form an ion. These ions are charge balanced by cations of sodium in the natural material. The sodium ions can be ion exchanged with lithium and potassium as described above. The modification of the ions has a dramatic effect on the structure of the cages. Lithium being smaller than sodium causes the cage to contract whilst potassium being large causes the cage to expand. The modification of the cage structure and the interaction of the different cations with the central sulphur species modifies the colouration of the final pigment.

By treating blue ultramarine with silver nitrate solution, "silver-ultramarine" is obtained as a yellow powder. This compound gives a blue potassium- and lithium-ultramarine when treated with the corresponding chloride, and an ethyl-ultramarine when treated with ethyl iodide. Selenium- and tellurium-ultramarine, in which these elements replace the sulfur, have also been prepared.

Vermilion, also spelled vermillion, when found naturally-occurring, is an opaque reddish orange pigment, used since antiquity, originally derived from the powdered mineral cinnabar. Chemically the pigment is mercuric sulfide, HgS. Like all mercury compounds it is toxic. Vermilion is not on the colour wheel since the colour is mixed with a slight amount of gray.

Today vermilion is most commonly artificially produced by reacting mercury with molten sulfur. Most naturally produced vermilion comes from cinnabar mined in China, giving rise to its alternative name of *China red*.

As pure sources of cinnabar are rare, natural vermilion has always been extremely expensive. In painting, vermilion has largely been replaced by the pigment cadmium red, a pigment that is less reactive due to the replacement of mercury with cadmium, especially in certain applications such as watercolors. The last mainstream commercial source in watercolors was from the Belgian artist's materials company Blockx, although the pigment can still be obtained in oils, where it is considered more stable. Unlike mercuric sulfide, cadmium sulfide is available in a large range of warm hues, including hues obtained by the addition of selenium or zinc. The range is from lemon yellow to a dull deep red, sometimes referred to as "cadmium purple".

Vermilion is also the name of the typical colour of the natural ground pigment, which is a bright red tinged with orange. It is somewhat similar to the colour scarlet. As with cadmium sulfide, mercuric sulfide can be found in a range from a bright orange-toned red to a duller slightly bluish red. The differences in hue are due to the range in the size of the ground particles. The larger the average crystal is, the duller and less orange-toned it appears. It has been theorized that the more coarsely ground

“Chinese” form of vermilion is more permanent than the more orange "French" variety. It is also theorized that purification leads to increased stability, as with many other pigments.

History

There is evidence of the use of cinnabar pigment in India and China since prehistory; It was known to the Romans; Pliny the Elder records that it became so expensive that the price had to be fixed by the Roman government. The synthesis of vermilion from mercury and sulfur may have been invented by the Chinese; the earliest known description of the process dates from the 8th century.

The synthetically-produced pigment was used throughout Europe from the 12th century, mostly for illuminated manuscripts, although it remained prohibitively expensive until the 14th century when the technique for synthesizing vermilion was widely known in Europe. Synthetic vermilion was regarded early on as superior to the pigment derived from natural cinnabar. Cennino Cennini mentions that vermilion is

> “made by alchemy, prepared in a retort. I am leaving out the system for this, because it would be too tedious to set forth in my discussion all the methods and receipts. Because, if you want to take the trouble, you will find plenty of receipts for it, and especially by asking of the friars. But I advise you rather to get some of that which you find at the druggists' for your money, so as not to lose time in the many variations of procedure. And I will teach you how to buy it, and to recognize the good vermilion. Always buy vermilion unbroken, and not pounded or ground. The reason? Because it is generally adulterated, either with red lead or with pounded brick.”

Another red mercury pigment, mercuric iodide, briefly sold in the 19th century to artists as “Scarlet Lake” and “Iodide Scarlet”, was more vivid than either vermilion or cadmium red, but it is very light sensitive and few artists used it. One who did was J. M. W. Turner, an artist infamous for his use of even the most fugitive paints. His response to criticism from a paint dealer was to point out that he was not the one who produced the paints.

Vermilion was frequently adulterated due to its high price, usually with red lead, an inexpensive bright lead oxide pigment that was too reactive to be trustworthy enough for use in art.

"American Vermilion" is the name for a historical vermilion imitation. The Portuguese word for the colour red (vermelho) derives from this term.

Zinc oxide is a chemical compound with formula ZnO. It is nearly insoluble in water but soluble in acids or alkalis. It occurs as white hexagonal crystals or a white powder commonly known as **zinc white**. It remains white when exposed to hydrogen sulfide or ultraviolet light. Crystalline zinc oxide exhibits the piezoelectric effect and is thermochromic (it will change colour from white to yellow when heated, and back again when cooled down). Zinc oxide decomposes into zinc vapor and oxygen at around 1975°C. High-quality single-crystalline ZnO is almost transparent. Fumes of zinc oxide are generated when melting brass, because the melting point of brass is close to the boiling point of zinc. Exposure to zinc oxide in the air (also while welding) can result in a nervous malady called metal fume fever.

Zinc oxide occurs in nature as the mineral zincite.

Zincum Oxydatum is the latin name for Zinc Oxide commonly listed on homeopathic medicines.

Applications

Zinc oxide in a mixture with about 0.5% iron(III) oxide (Fe_2O_3) is called calamine and is used in calamine lotion. There are also two minerals, smithsonite and hemimorphite, which have been called calamine historically.Zinc peroxide, $ZnO_2 .½ H_2O$, is a white to yellow powder that is used in antiseptic ointments. Zinc white is used as a pigment in paints and is more opaque than lithopone, but less opaque than titanium dioxide. It is also used in coatings for paper. Chinese white is a special grade of zinc white used in artists' pigments. Because it reflects both UVA and UVB rays of ultraviolet light, zinc oxide can be used in ointments, creams, and lotions to protect against sunburn and other damage to the skin caused by ultraviolet light. Zinc oxide and stearic acid are important ingredients in the commercial manufacture of rubber goods. A proper mixture of these two compounds allows a quicker and more controllable rubber cure. Zinc oxide can also be used as a filler in some rubber mixtures. Zinc oxide is a semiconductor with a direct band gap of 3.37 eV (368 nm at room temperature, deep violet/borderline UV). A common application is in gas sensors. As of 2003, it has been utilized in recent research to build blue LEDs and transparent TFTs. *n*-type doped films are often used in thin film technology, where zinc oxide serves as a TCO (transparent conducting oxide). n-type doping is possible by introduction of aluminum, indium, or excess zinc. Oxygen vacancies generate states in the band gap and hence also cause an increase in conductivity. *p*-type doping is difficult and is currently an active area of research, with arsenic as the leading candidate dopant Thin-film solar cells, LCD and flat panel displays are typical applications of this material. Appropriately doped Zinc oxide may be transparent and conductive, and can therefore be used as a transparent electrode. Indium tin oxide (ITO) is another transparent conducting oxide often used in microelectronics. ZnO has also been considered for spintronics applications because of theoretical predictions of room temperature ferromagnetism. Unsubstantiated reports of ferromagnetism have been made, but presence of dilute magnetic semiconductors remains a large unanswered question in physics. ZnO layers are mainly deposited by sputter deposition and chemical vapor deposition (CVD). The latter method allows the growth of a rough layer, which can diffuse the incoming light by scattering, increasing the efficiency of solar cells. Recently, ZnO has been observed to act as a chemical reagent for Friedel Crafts reaction. Zinc oxide - recognized as a mild antimicrobial, wound healing and sunscreen agent. Primarily absorbs UVA light rather than scattering or reflecting, non-irritating, non-comedogenic, and micronized by forming many small micro particles for cosmetic use.

Zinc oxide has been studied as a treatment for cold sores and appears to shorten their duration.

When mixed with eugenol, zinc oxide eugenol forms which has restorative and prosthodontic applications in dentistry.

Pyroelectric coefficient

- Primary pyroelectric coefficient: -6.8 $\mu C/m^2 \cdot K$
- Secondary pyroelectric coefficient: -2.5 $\mu C/m^2 \cdot K$
- Total pyroelectric coefficient: -9.4 $\mu C/m^2 \cdot K$

Production Methods

Zinc oxide is produced by oxidation of metallic zinc vapor at elevated temperatures. There are two methods. Direct method and indirect method.

Indirect method

Metallic zinc is vaporized in suitable containers. Zinc vapor reacts with oxygen that is in the air to form zinc oxide.

Direct method

Zinc ores or roasted sulfide concentrates are mixed with coal. In a reduction furnace, ore is reduced to metallic zinc and vaporized zinc reacts with oxygen to form zinc oxide.

ORGANIC PIGMENTS

Pigment red 170 is an organic pigment extensively used in automotive coatings. It is produced industrially by converting p-aminobenzamide into the corresponding diazonium compound followed by diazotation with 3-hydroxy-2-naphththoic acid (2-ethoxy)anilide. In the solid state the hydrazo tautomer forms and several crystal structures exist.

In the initial α polymorph the molecules are arranged in a herringbone pattern with extensive hydrogen bonding. The φ polymorph is more dense and more stable and produced industrially by thermal treatment in water at 130°C under pressure. In this phase the molecules are planar and arranged in layers. Extensive hydrogen bonding exists within the layer but between layers the only interactions are Van der Waals forces. Dense crystal structures are preferred for pigments used in coatings because in the event of Photochemical decomposition the fragments are locked in place and are able to recombine. Research shows that by replacing the ethoxy group in this compound by a methoxy group the crystal structure is less stable and in the final application and the color fades more easily. By careful selection of substituents it is possible to optimize crystal structure and improve optical properties.

Indian Yellow, also called euxanthin or euxanthine, is a transparent yellow pigment used in oil painting. Chemically it is a magnesium euxanthate, a magnesium lake of euxanthic acid. It is a clear, deep and luminescent yellow pigment. Its color is deeper than gamboge but less pure than cadmium yellow.

Euxanthic acid itself is a glycoside, a conjugate of the aglycone euxanthone with a molecule of oxidized glucose derivative.

COOH O OH O HO OH OH

Indian yellow was used by artist painters in both oil paints and watercolors. Due to its luminescence, it faded in darkness and artificial light, but was vivid and bright in sunlight. It was likely first used by Dutch artists, and before the end of 18th century it was commonly used by artists across Europe. Its origin was unknown until an investigation in the year 1883.

Indian Yellow pigment was originally manufactured in rural India from the urine of cattle fed only on mango leaves and water. The urine was collected and dried, producing foul-smelling hard dirty yellow balls of the raw pigment. The process was declared inhumane and outlawed in 1908, as the cows were extremely undernourished, partly due to the fact that the leaves contain a mild toxin related to that found in poison ivy.The replacement, synthetic Indian Yellow Hue, is a mixture of Nickel Azo, Hansa Yellow and Quinacridone Burnt Orange. It is also known as **Azo Yellow** "Light" and "Deep", or **Nickel Azo Yellow**.

Lapis lazuli, also known as just **lapis**, is a stone with one of the longest traditions of being considered a gem, with a history stretching back to 7000 BC in Mehrgarh of Indian subcontinent, situated in modern day Balochistan, Pakistan. Deep blue in color and opaque, this gemstone was highly prized by the pharaohs of ancient Egypt, as can be seen by its prominent use in many of the treasures recovered from pharaonic tombs. It is still extremely popular today.Lapis is a rock and not a mineral because it is made up from various other minerals. To be a true mineral it would have one constituent only. The first part of the name is the Latin lapis, meaning stone. The second part, lazuli, is the genitive form of the medieval Latin *lazulum*, which came from Arabic *(al-)lazward*, which came from Persian *zhward*. This was originally a place-name, but soon came to mean blue because of its association with the stone. The English word azure, the Spanish and Portuguese *azul*, and the Italian *azzurro* are cognates. Taken as a whole, *lapis lazuli* means "stone of azure"

Sources

The finest lapis comes from the Badakshan area of Afghanistan. This source of lapis may be the oldest continually worked set of mines in the world, the same mines operating today having supplied the lapis of the pharaohs and ancient Sumerians. Using this ancient source, the Indus Valley Civilization's artists used to make beautiful carvings and traders used to trade them to distant places. More recently, during the 1980s conflict with the USSR, Afghanistan resistance fighters disassembled unexploded Soviet landmines and ordnance and used the scavenged explosive to help mine lapis to further fund their resistance efforts.

In addition to the Afghan deposits, lapis has been found in the Andes near Ovalle, Chile, where it is usually pale rather than deep blue. Other less important sources are the Lake Baikal region of Russia, Siberia, Angola, Burma, Pakistan, USA (California and Colorado), Canada and India.

Uses

Lapis takes an excellent polish and has been made into jewelry, carvings, boxes, mosaics, ornaments and vases. In architecture it has been used for cladding the walls and columns of palaces and churches.It was also ground and processed to make the pigment Ultramarine for tempera paint and, more rarely, oil paint. Its usage as a pigment in oil paint ended in the early 19th century as a chemically identical synthetic variety, often called French Ultramarine, became available.

DYE AND ITS CHROMATICITY

The Different Colours of White Light

Everyone is familiar with rainbows- see the top picture for a well known example! Sunlight is refracted by atmospheric water, producing bands of red, orange, yellow, green, blue, indigo and violet. These combined make up white light. If a light source is deficient in any colour band, the light appears to be coloured in the complementary colour. The table below shows wavelength, the corresponding colour, and its complementary colour.

Wavelength Range/nm	Colour	Complementary Colour
400-435	Violet	Green-yellow
435-480	Blue	Yellow
480-490	Blue-blue	Orange
490-500	Blue-green	Red
500-560	Green	Purple
560-580	Yellowy-green	Violet
580-595	Yellow	Blue
595-605	Orange	Green-blue
605-750	Red	Blue-green

For examples, see below:

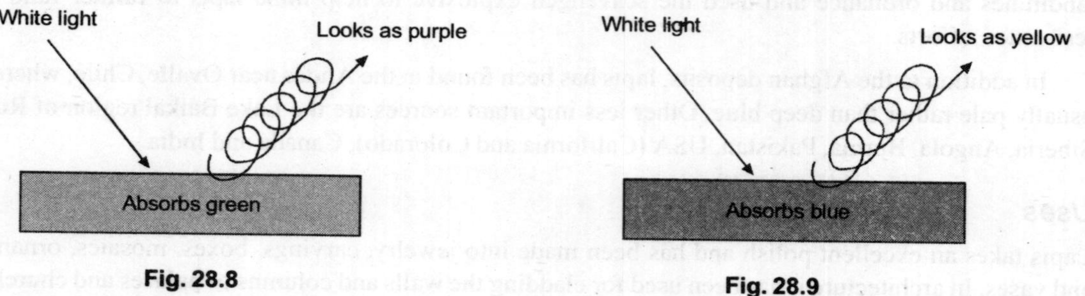

Fig. 28.8

Fig. 28.9

Fig. 28.8 and 28.9. The deflection of light and absorption of coloured body

Figure 28.8 shows the effect on white light reflected off a solid object.

Figure 28.9 shows the effect on white light transmitted through a solution, or other transparent article.

So what exactly causes an object to be coloured?

The Effect of Molecular Energy Levels

There are several ways in which molecules can appear to be coloured

- Simple molecular excitation, such as in a neon tube, may cause the appearance of colour. This is due to rotation and/or vibration of the molecules.
- Transition metal complexes are coloured due to the distortion of the metal's d-electron shell caused by ligands surrounding the metal ion.
- Electronic motion in conjugated organic systems, and charge transfer.
- Colour in crystalline solids arises from band theory- the blurring of many orbitals through-out the solid. Solids are only coloured if the gap between the Highest Occupied Molecular Orbital (HOMO, the Fermi level) and the Lowest Unoccupied Molecular Orbital (LUMO) is small enough.
- Colour due to refraction, scattering, dispersion and diffraction- these are all due to the geometrical and physical dimensions of a solid or a solution.

The first four mechanisms all rely on some form of energy transfer to move either molecules or electrons from their ground state into some excited state. However, only one of these effectively applies to dye molecules, since dye molecules are almost without exception organic conjugated systems. The overlapping p-orbitals effectively mean that no one electron absorbs more energy than another, since all p-electrons in the conjugated system are smeared above and below the molecule. Conjugated organic molecules absorb specific wavelengths of electro-magnetic radiation. If this absorption falls within the visible region, then the light reflected or transmitted is deficient in a particular colour, and the solid (or solution) appears coloured:

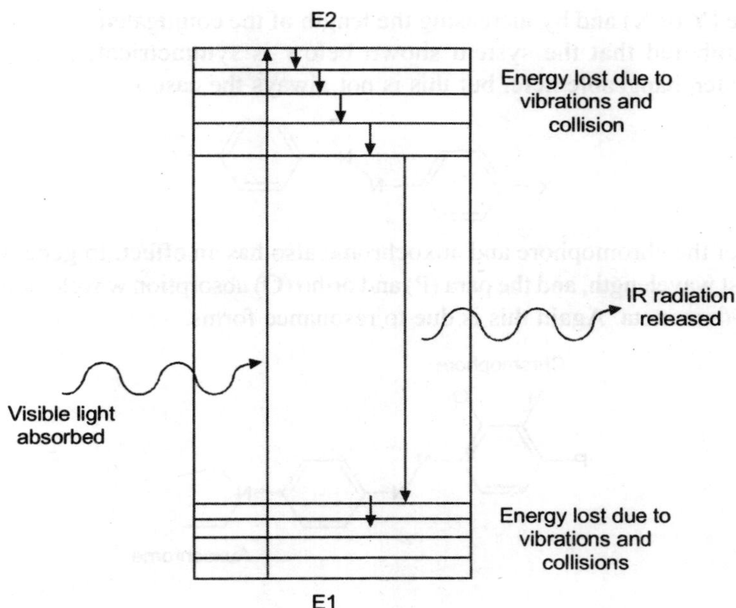

Fig. 28.10. The Energy state of the photons

The energy of the electronic transition can be calculated from

$$\Delta E = hn$$

where ΔE is the difference between the two electronic levels, h is Planck's constant and n is the frequency of the absorbed radiation.

What Affects the Colour Quality of the Dye?

Hue

The hue of a dye depends on the wavelength it absorbs. Since the wavelengths) the dye absorbs depends on its structure, we can see that any change which affects the p-system will affect the hue. A structural change which causes the absorption band to longer wavelengths (i.e. yellow → orange → red → violet → blue → green) is called a bathochromic shift. The reverse shift, towards shorter wavelengths is known as a hypsochromic effect.

Chromophores and Auxochromes

In 1876, Witt proposed that dye molecules contain two groups; the chromophore and the auxochrome. The chromophore is a group of atoms which control the colour of the dye. At that time, Witt suggested that the auxochrome was a salt-forming group, which helped to improve the colour of the dye. His theory was later modified when it was discovered that the chromophore is usually electron-withdrawing, and auxochromes are normally electron-donating. The two groups are connected by a conjugated system.So how can a dye molecule's hue be altered? A bathochromic shift may be caused by increasing the electron-withdrawing power of the chromophore (X or Y), increasing the electron-donating power of the auxochrome (Y or X) and by increasing the length of the conjugated system connecting the two. (It must be remembered that the system shown below is symmetrical, so the chromophore and auxochrome are interchangeable here, but this is not always the case.)

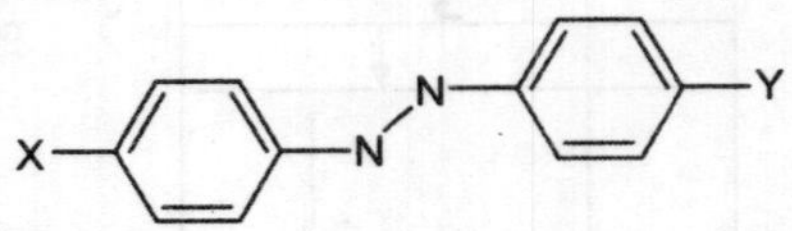

The position of the chromophore and auxochrome also has an effect. In general, the meta position (M) has the shortest wavelength, and the para (P) and ortho (O) absorption wavelengths are approximately equal, and longer than meta. Again this is due to resonance forms.

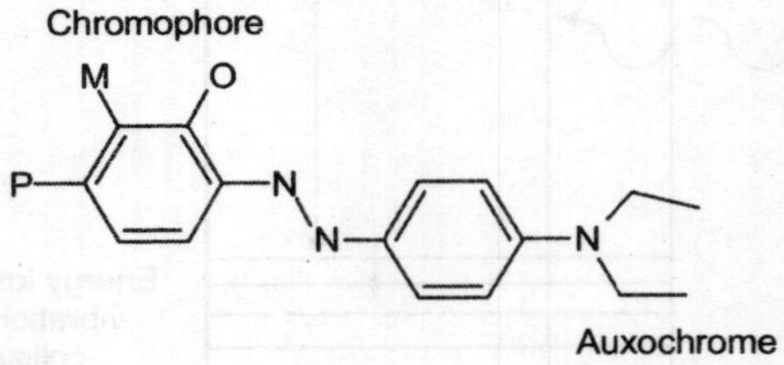

Chromophore

Auxochrome

Strength

The *molar extinction co-efficient* indicates the strength of a dye at low concentrations. It can be calculated using the Beer-Lambert Law

$$A = ecl$$

where A is the absorbance of the dye at a particular wavelength, e is the molar extinction co-efficient, c is the concentration of the dye and l is the path-length through the cell.

Brightness

The brightness is best described by the shape of the absorbance band; if the band is narrow and sharp, then the dye is bright, if broad, then the dye is dull. This is due to other wavelengths absorbed by the dye, other than the one which causes the hue. The superposition of these determines how broad an absorption band is. The components involved in histological staining are dyes and proteins. The fundamental process involved is the chemical bonding between the carboxyl groups of one and the amino groups of the other. The commonest bonds involved are ionic bonds, although there are exceptions especially in the case of nuclear staining of DNA. The use of colour to identify individual components of tissue sections is accomplished mainly with chemical dyes, although other means are occasionally used. Dyes, however, are the largest group that we can manipulate.

What Dyes Are?

In short, dyes are coloured, ionizing, aromatic organic compounds. It must be appreciated that they are individual chemicals, and like all chemicals, they are similar in their reactions to some other chemicals, and distinctly different from others. It may seem that this is a statement of the obvious, but we sometimes appear to view dyes as something other than ordinary chemicals, and I want to stress that the same rules that apply to sodium chloride, acetic acid, benzidine and a host of others also apply to dyes. This includes the possibility that they are toxic. They may be carcinogenic or mutagenic, or harmful to your health in some other way. Just because we call something by an appealing name and because it has an appealing colour, does not make it any the less harmful. Handle dyes with care! Put your own safety first!

What Colour is?

Dyes are aromatic organic compounds, and as such are based fundamentally on the structure of benzene. To us, benzene appears to be a colourless fluid. In fact it absorbs electromagnetic radiation just as dyes do, but it does so at about 200 nm so that we do not see it. The perception of colour is an ability of some animals, including humans, to detect some wavelengths of electromagnetic radiation (light) differently from other wavelengths. Normal daylight, or white light, is a mixture of all the wavelengths to which we can respond and some to which we cannot, in particular the infra-red and ultra-violet rays. We respond to wavelengths between about 400-700 nm. When an object absorbs some of the radiation from within that range we see the waves that are left over, and the object appears coloured. In reality this range we see makes up only a very small fraction of the electromagnetic spectrum.

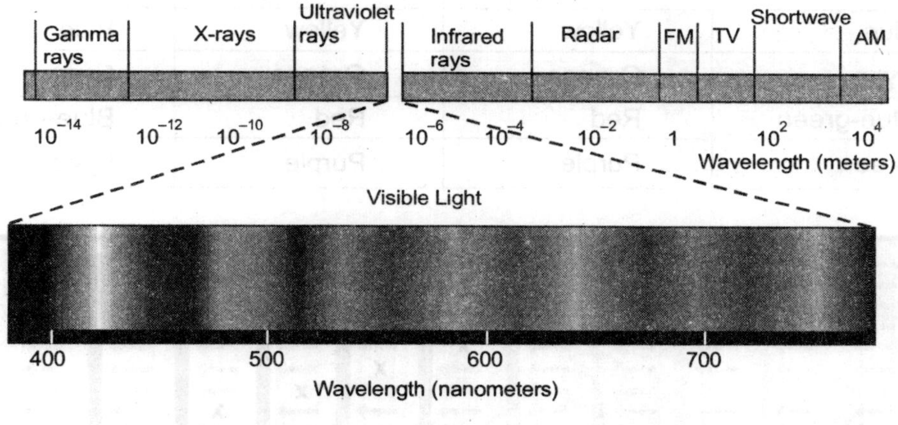

Spectral colours

Although the spectrum is continuous and therefore there are no clear boundaries between one color and the next, the following ranges may be used as an approximation.

violet	380-450 nm
blue	450-495 nm
green	495-570 nm
yellow	570-590 nm
orange	590-620 nm
red	620-750 nm

In scientific terms there is nothing special about the wavelengths in the visible range, other than being the major components of sunlight which are not removed by the earth's atmosphere. Their special importance is based exclusively on the ability of human retinas to respond to them, and to discriminate between them to a significant degree. These discriminations are what we call colour. Wavelengths just outside the visible range are considered colourless, even though there is no substantive difference between them and the limiting wavelengths inside the range. Some animals (bees, for example) can see these other wavelengths but, because humans do not, we consider them colourless. The point is that colour is a subjective phenomenon, and thinking of colour as something objective is misleading. For that reason, we should refer to the wavelengths involved rather than describe the human response to them. When some of the wavelengths found in white light are absorbed, then we see what is left over as coloured light. The colour that we see is referred to as the complementary colour of the colour that was removed. For instance, if the red rays are removed from white light, the colour we detect is blue-green. Blue-green is complementary to red, and red is complementary to blue-green

Complementary Colours

Removed	Observed	Removed	Observed
Violet	Yellow-green	Yellow-green	Violet
Blue	Yellow	Yellow	Blue
Cyan	Orange	Orange	Cyan
Blue-green	Red	Red	Blue-green
Green	Purple	Purple	Green

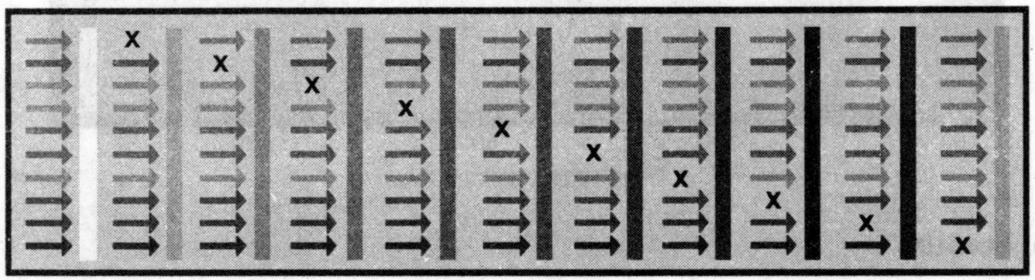

Absorption of light leaving the complementary colour

The perception of colour is merely a human evolutionary adaptation to the absence of some wavelengths in white light. Suppose however, that the same thing happens outside the range to which our eyes respond. Suppose a chemical removes radiation which has a wavelength about 200 nm, as benzene does. Is this coloured?

Well, obviously it is not. There has been no impact on the wavelengths of white light, so it is not coloured. However, what has happened to the ultra-violet radiation that was absorbed by the benzene is no different to what happens to the blue radiation that is removed when we look through a solution of acid fuchsin. The difference is solely whether we can detect it visually. Colour is not an objective phenomenon, it is the human detection and perception of electromagnetic radiation.

Why Dyes are Coloured?

Colour in dyes is invariably explained as a consequence of the presence of a ***Chromophore***. Since, by definition, dyes are aromatic compounds their structure includes aryl rings which have delocalised electron systems. These are responsible for the absorption of electromagnetic radiation of varying wavelengths, depending on the energy of the electron clouds. For this reason, chromophores do not make dyes coloured in the sense that they confer on them the ability to absorb radiation. Rather, chromophores function by altering the energy in the delocalised electron cloud of the dye, and this alteration results in the compound absorbing radiation from within the visible range instead of outside it. Our eyes detect that absorption, and respond to the lack of a complete range of wavelengths by seeing colour.

Chromophores are atomic configurations which can alter the energy in delocalised systems. They are composed of atoms joined in a sequence composed of alternating single and double bonds. Double bonds in organic compounds can be of two types. If the atoms with double bonds are not adjacent, they are termed ***isolated*** double bonds, and exist independently of other double bonds in the same molecule. If adjacent atoms have double bonds they are termed conjugated double bonds and the bonds interact with each other. Chromophore configurations often exist as multiple units, having ***conjugated*** double bonds, and are more effective when they do so. This is due to the interaction between the double bonds, which causes partial delocalisation of the electrons involved in the bonds. In this case, although specific atoms are involved in the bonds, the electrons are distributed over a larger area than the specific atoms and also involve adjacent atoms that have double bonds.

The point of this is that conjugated systems have partially delocalised electrons, and the energy in these delocalised electrons can impact on the energy of the delocalised electrons of the parent aromatic compound by extending the number of electrons involved in the system and the energy needed to keep the whole system in place.

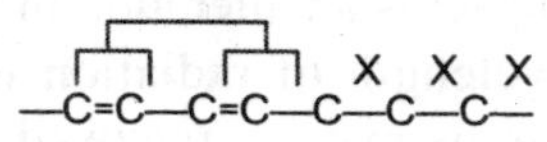

X X X
—C=C—C=C—C—C—C—

Conjugated Double Bonds (Red)

Another common chromophore is the nitro group. This chromophore is a nitrogen with two oxygen atoms attached. One oxygen is shown attached with a single bond, the other with a double bond. In fact, like the carbon atoms in benzene, these two oxygen atoms are attached to the nitrogen with bonds of equal strength. The extra electrons are delocalised between the three atoms.

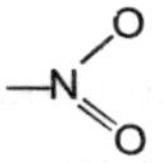

O
—N
O

Nitro group

The quinoid ring is found in many dyes. It is a ring structure with two points at which chromophores can attach. It should be thought of as a closed system of conjugated double bonds. The attachment of configurations which add delocalised electrons to the system at one point, and the attachment of configurations which extend the atoms involved in the delocalisation at the other, causes very dramatic shifts in the wavelengths which these compounds absorb. For this reason, quinoid ring configurations are considered to be extremely powerful chromophores, producing very intensely coloured compounds.

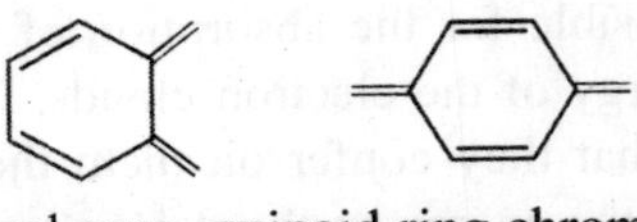

Ortho- and para- quinoid ring chromophores

To sum this up, chromophores are atomic configurations that contain delocalised electrons. Usually they are represented as nitrogen, carbon, oxygen and sulphur that have alternate single and double bonds. By incorporating the delocalised electrons in these configurations into the delocalised electrons in the aryl rings of aromatic compounds, the energy contained in the electron cloud can be modified. If the energy incorporated into the electron cloud is changed, then the wavelength of the radiation it absorbs will also change. If this change in the wavelength to be absorbed is sufficient to cause any absorption at all within the visible range, then the compound will be coloured.

Below are the usual chromophores seen in histological dyes:

– C = C – – C = N – – C = O –

– N = N – – NO_2 Quinoid rings

Other Effects

There is more than one effect when chemical groups are attached to aryl rings. Any delocalised electrons and their energy can simply be added to that already present, thus increasing it. Also, the delocalised electrons may be shared by more atoms than those in the original structure, by adding in to the delocalised system the atoms in the chromophore and any modifiers that may be present. Whether these effects occur jointly or separately, the final impact is an alteration in the overall energy of the electron cloud with a subsequent effect on the wavelength of radiation to which the whole molecular system reacts.Another possibility is that electrons may be removed from the electron cloud, and this may result in loss of colour. Removal of electrons may cause the remaining electrons to revert to local orbits. A good example would be Schiff's reagent. When sulphurous acid reacts with pararosanilin, a sulphonic group attaches to the central carbon atom of the compound. This disrupts the conjugated double bond system of the quinoid ring, causing the electrons to become localised and the ring to cease being a chromophore. Consequently, the dye becomes colourless.

Auxochrome

Below are the usual auxochromes found in histological dyes:-

$-NH_3$ –COOH $-HSO_3$ –OH

Auxochromes are groups which attach to non ionising compounds yet retain their ability to ionise. While this definition is largely correct, it is also inadequate. This is because it restricts the definition of the auxochrome to ionisation, and does not comment on the effect of auxochromes on the absorbance of the resulting compound.

The word ***auxochrome*** is derived from two roots. The prefix ***auxo*** is from ***auxein***, and means ***increased***. The second part, ***chrome*** means ***colour***, so the basic meaning of the word auxochrome is ***colour increaser***. This word was coined because it was noted originally that the addition of ionising groups resulted in a deepening and intensifying of the colour of compounds.

	Colour enhancing by an auxochrome
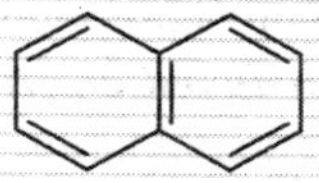	To the left is *naphthalene*, a colourless compound.
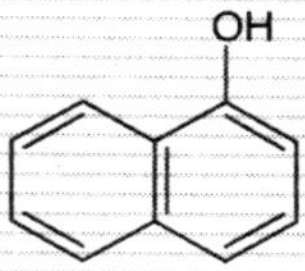 OH	The addition of a single hydroxyl group to naphthalene produces *1-naphthol* which is also a colourless compound, but one which can ionise.
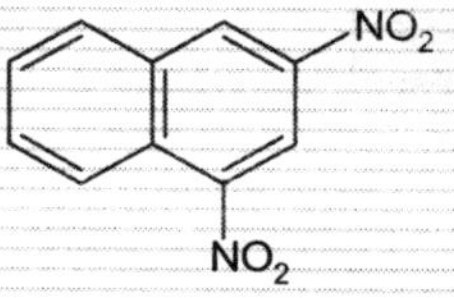 NO_2, NO_2	If instead of a hydroxyl group we add the nitro group, which is a chromophore, we get the compound *2,4-dinitronaphthalene*. The addition of this chromophore has caused it to become pale yellow.
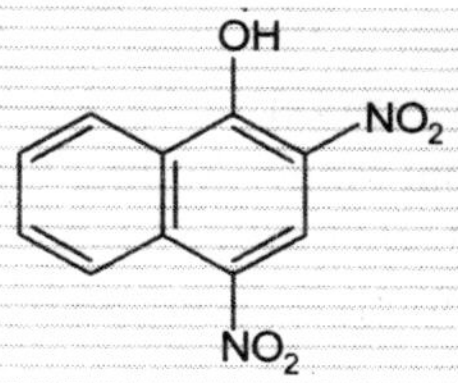 OH, NO_2, NO_2	If instead of a hydroxyl *or* nitro groups, *both* a hydroxyl and nitro groups are added, we get the deep yellow dye, *martius yellow*.
The addition of *both* an auxochrome *and* a chromophore results in a much stronger alteration of the absorption maximum of the compound. The hydroxyl group must have deepened the colour, showing that auxochromes are also chromophores.	

Sometimes the term ***auxochromophoric*** is used to denote the action of an auxochrome that modifies the colour as well as permitting ionisation. This term infers that the colour modifying effects of auxochromes are rare, but this is not the case. The effect on absorption should not be considered an incidental aspect of the auxochrome's action, but an integral and fundamental part of it. The word auxochromophoric is redundant.

Edward Gurr proposed the terms ***colligator*** and ***non-colligator*** to distinguish between ionising auxochromes and colour modifying effects.

Auxochromes are of two types. They may have a positive charge as the amino group and its substituted variants. Or they may be negatively charged as the carboxyl and hydroxyl groups, and the sulphonic group. This last is commonly used to convert basic dyes to acid dyes. Both negatively charged and positively charged auxochromes may be present on a single molecule.

Resonance

The process by which electrons are stimulated by radiation is ***resonance***. It should be made clear at the outset that resonance is not the same as vibration.

Resonance is the induction of a response in one energy system from another energy system in close proximity, which is operating at the same energy level (frequency). In aromatic organic compounds, including dyes, the two energy systems are ***electromagnetic radiation***, and the ***delocalised electron cloud***. Do not confuse resonance with the imaginary resonance hybrids used in an older explanation for the structure of benzene.

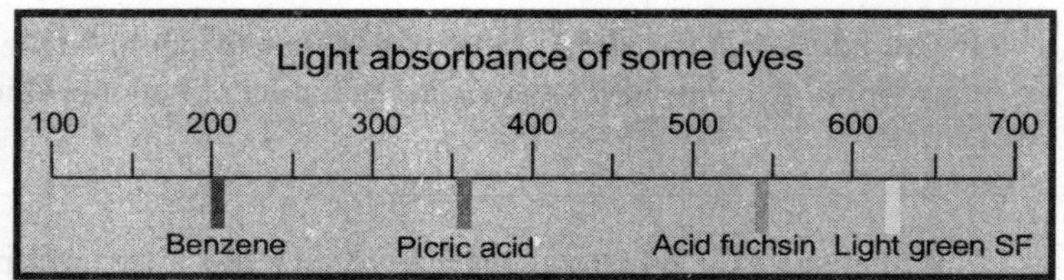

Absorption of benzene compared to some dyes

Modifiers

Colour modifiers such as methyl or ethyl groups alter the colour of dyes by altering the energy in the delocalised electrons. By themselves they cannot do this enough to cause absorption in the visible range, but they can affect the shade significantly when absorption is already in that range. Adding more of a particular modifier results in a progressive alteration of colour. Compounds that differ from each other in this kind of regular fashion are called homologues. A very good example is seen with the Methyl violet series.

	Alteration of colour by modifiers
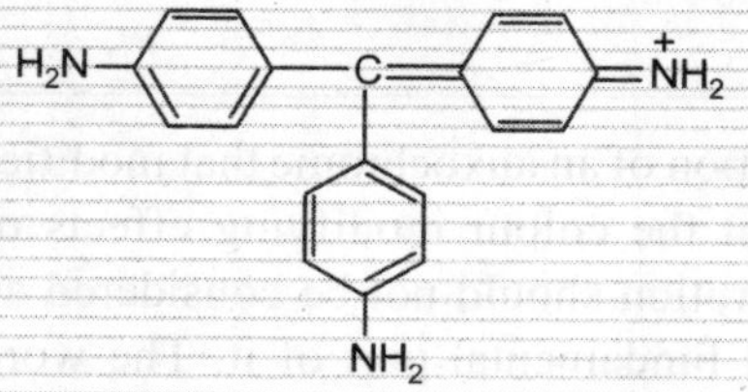 H_2N, C, $\overset{+}{N}H_2$, NH_2	Without any methyl groups the parent dye is called ***pararosanilin*** and is red.

$(CH_3)_2N$ — C = $\overset{+}{N}H_2$; $N(CH_3)_2$	When four methyl groups are added we get the reddish purple dye ***methyl violet***.
$(CH_3)_2N$ — C = $\overset{+}{N}(CH_3)_2$; $N(CH_3)_2$	As more methyl groups are added we get the purple blue dye ***crystal violet*** which has six such groups.
$(CH_3)_2N$ — C = $\overset{+}{N}(CH_3)_2$; $N^+(CH_3)_2$	If a seventh methyl group is added, the resulting dye is ***methyl green***.

The Colour

The colour of the dye is caused by the absorbance of electromagnetic radiation. We have constantly referred to the ***wavelength that is absorbed*** in the singular, but a simple scan of a dye solution with a spectrophotometer shows that dyes do not remove a single wavelength. Rather they absorb radiation on either side of the wavelength most completely removed (the absorption maximum). Plotting the wavelength absorbed against the degree of absorbance usually results in a display resembling a bell curve. If any part of this curve is in the visible range, the dye will appear coloured. White light is a mixture of wavelengths. Some of these have a relationship to the energy in the delocalised electron cloud of the dye molecule. By the process of resonance, previously described, the electron cloud will respond to the energy contained in that radiation by absorbing it, and removing it from the spectrum. As a consequence the white light will cease to be white and will display the colours of the wavelengths left over. The transmitted light will have the complementary colour to the wavelengths removed. The Effect. When light illuminates a dye, some of it is absorbed as energy. Since energy is not destroyed, something must then happen. We could use an analogy of heating water - the water's temperature rises, molecular vibration increases. However, we do not see anything else, as the water just sits there being water.

The same can happen with dyes, we may not observe anything particular as the effect may be at the atomic level. There are several possibilities, however.

A. The energy level in the electrons in an unaffected dye is called the ground state. When electromagnetic radiation, as light energy, is absorbed the electrons become more energised.

B. With most dyes, there is then a gradual decay and the electrons return to the ground state. We do not see anything. Nevertheless, something may happen that we do not see. Possibly there is an increase in temperature, or some chemical changes occur which disrupt the dye's structure and cause it to lose colour - fading.

C. Another possibility is that the return to the ground state is not gradual, but sudden. If this is accompanied by emission of any residual energy in the form of light, we observe the dye glowing - fluorescence. Since the emitted light must always contain less energy than the absorbed light, as some was used to energise the electrons, the emitted radiation is always at longer wavelengths than the absorbed radiation. By manipulating the light available, we can cause ultra-violet light to be absorbed and visible light to be emitted.

D. A third possibility is that the electrons stabilise in their newly energised state. After a passage of time they then return to the ground state. If this happens gradually, we may observe nothing, with the same possibilities regarding fading and temperature increase as before.

E. If return to the ground state happens suddenly, and the residual energy is emitted as light, we once again see the dye glowing - phosphorescence. As with fluorescence, the light emitted is always a longer wavelength than the light absorbed, but the disparity is greater with phosphorescence due to the greater energy consumed in keeping the electrons in the excited state.

Note that the difference between fluorescence and phosphorescence is in whether the electrons stabilise in the excited state before returning to the ground state. With any stability, no matter how long (or short), it is considered phosphorescence.

Conclusion

The explanation of the relationship between structure and colour depends on the basic atomic structure of the aryl ring, and the shared or delocalised electrons that this atomic arrangement has. The ability to absorb radiation is inherent in this structure. The effect of other atomic configurations is to modify the energy contained in the delocalised electron cloud so that the compound absorbs electromagnetic radiation at a wavelength in the visible range. Some also ionise, enabling the compound to chemically react with ionising tissue groups. Colour, fading, fluorescence and phosphorescence are all seen to be different effects of the same fundamental process.

THEORY OF CHROMATICITY

Armstrong's Theory

It is sometimes called 'quinonoid theory'. This theory states that all the coloured compounds must have a quinonoid structure in its chemical constitution. This postulation is true because, when we look at the structure of benzene and benzoquinone, benzene is colourless and benzoquinone is coloured.

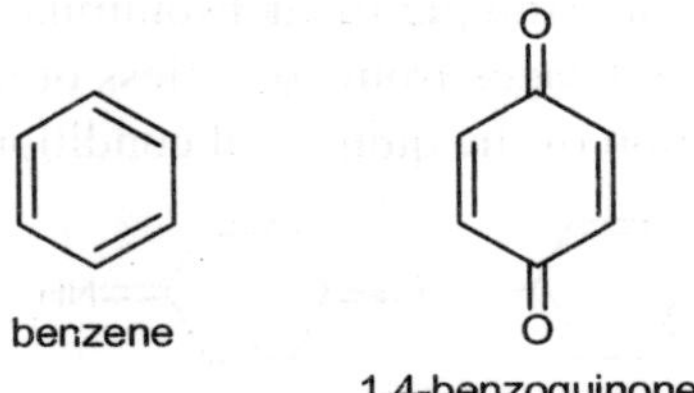

Secondly, the addition of phenolphthalein to the bezinoid structure makes it colourless, but added to the benzoquinonoid structure makes it coloured.

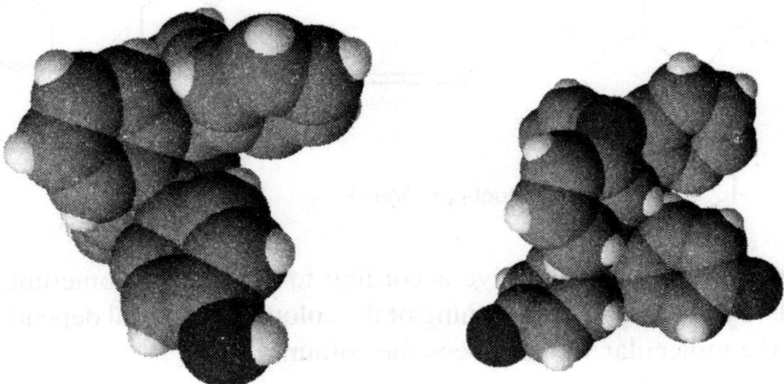

Phenolphthalein (colourless) Phenolphthalein, (coloured) quinonoid

This theory does not have any ratiocination for all the compounds also imminoquinones have quinonoid structure, still they are not coloured. On the other hand azobenzene, fluvene etc are all having coloured dyes in their family.

Although this theory has many limitations, but it has opened a scope for research for the colour and dyes.

Hantzsch Modification of Armstrong Theory

Hantzsch postulated about the Armstrong's theory, he stated that colour is a additive property, it means that the whole molecule is responsible for the colour of the compound, for example when -NO2 and -ONa gropu both are present in a molecule then only produces colour otherwise not, thus Hantzsch stated that two groups must be interacting to derive a new structure of the coloured compound. He called it chromoisomerism.

Baeyer's Theory

In 1907, Baeyer, said that structureal oscillation of the quinonoid condition between the benzene rings is responsible for the colour. The change from one form to another involves the movment of electrons and not the whole atom.

This theory is best explained by an example of fuchsonimine, which is colourless, but when an amino group (from aniline) is added, it chages from colourless ot deep violet (Doebner's violet). This colour is due to the structural oscillation of the quinonoid condition between the two benzene rings.

Fuchhsonimine

Doebner's Violet

Neitzki's Theory: The colour of the dye according to this theory sometimes depends upon the addition of substituents. Further also the deepening of the colouring material depends upon the molecular weight, increasing in the molecular weight deeps the colour.

For example when benzene is replaced by naphthalein in azo dyes colour shifts from yellow to red.

This rule is ture in genral but in the case of deepening of colour the nature of the groups added is very important and not the weight of it. This is well understood by theexample of the following compound.

$$CH_3 - (CH_2)_a - X - (CH_2)_b - CH_3$$

Where X is chromophore, here if X remains same and that of a and b have different values, then it wuld is interesting to stay that change of colour does depent upon the value of a and b.

Modern Theories of Chromaticity

There are some modern thinking about the theory of the chromaticity and so thay have been cited briefly.

Quantization of Light Energy: Light is the composition of energy, or quantized energy; these are called light quanta or photons. The coherence or con-coherence of these photons produces the monochromatic or muti-chromatic light after getting reflected or absorbed from the material.

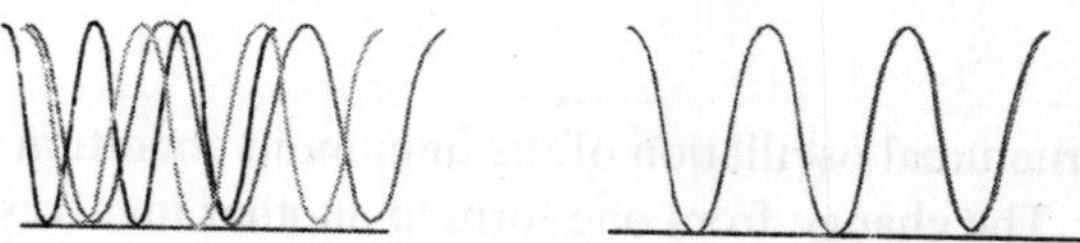

Fig. 28.11. The coherency and non-coherency of the photon

The Absorption of Radiation by Molecules: As discussed above, (see page 1319 - 321)a compound then only appears coloured when it systamatically and that also slectively absorbs light, in the visible region and reflects the rest of the light in the visible region. This should be carefully induced in mind that the chromaticity of any substance depends only on the amount of light energy absorbed in the visible spectrum.

The main function of the absorbed energy level is to raise the molecule from the ground state (E_0) or the zero level of the energy to the higher energy level E_1. The difference of the energy level is denoted by ΔE, this is the quantum of the energy. This can be mathematically expressed by the folloing equation.

$$\Delta E = E_1 - E_0 = hv = h\frac{c}{\lambda}$$

The amount of energy absorbed ΔE depends upon the structural configuration of that dye.

Where h = Plank's Constant

c = Velocity of Light

λ = Wavelenth of the absorbed radiation

On the other way ΔE depends upon how tightly the electrons are placed in the bands and according to that absorption will occur in the Ultra violet or the visible region, for example:

If the electrons of a molecule are tightly bound as in saturated compounds no light of visible region will be absorbed but only light of ultraviolet region will be absorbed and hence the compound will appear colourless. On the other hand, electrons of a molecule are if loosely bound as in saturated compounds, the absorption may occur in the visible region and the substance will then appear coloured. Fore a substance the values of E lies between 71 and 35 Kcal/mole as wee pass from 4000-8000 Å. It means that if an organic substance is to be coloured, it must have free electrons which can be raised to exited state from ground state.

It is known that elctrons occupy definite orbitals, so the E as the frequency of ligt absorbed will certainly have a definite value, but each values of the frequency of light is associated with the particular line in the spectrum (band of colour). Therefore, the spectrum of a compound will consist a large number of excited states of the large number of molecules consisting the compound.

Dipole Moment: The dipole moment playes a very important role in the transition of the molecule, this is because the colour will only change if the dipole moment changes for example a symmetrical molecule gives a lesser scope for the trransition of the dipole, the greater the absorption dipole. Thus, if the group, which is going to be introduced changes the symmetry by decreasing it, therby decreasing the transition dipole, and herby increasing the absorption of light.

We will now use the above facts in the theories of chromaticity in the following subheads.

The Resonance Theory

This is also known as valence band theory. The following are the different aspects of the theory.

(*i*) In chromaticity the groups of atoms are chromophores, among these π - electrons of the atomatic ring. This increase in resonance increases the intensity of absorption of light and also shifts the absorption band to longer wavelength, responsible for deepening of the colour. From this it is converged that the increase in resonance must deeepn the colour and actuaklly it is the fact.

(*ii*) The oscillation of electron pair changes the dipole moment, the under mentioned order is obserbed for the case of exciattion of different gropus.

increase of dipole moment →

N=O > $C\equiv S^+$ > N≡N > $C\equiv O^+$ > C≡N > C≡C

nitroso methylidynesulfonium azo methylidyneoxonium cyano carbon

← deepening of colour

(*iii*) Resonance theory explains also the relation between the colour and symmetry of themolecule or the transition dipole of the molecule, because as the number of charged canonical structure increase the colour of the compound deepens. Also lesser the possibility of oscillation of the charge and also the path of oscillation is shorter lesser the wavelength of light will be absorbed therefore the lighter will be the colour of the compound.

A very simple example is benzene, nitrobenzene and *p*-nitroaniline.

Benzene is colourless seen from its two Kekule structure which are the two possible resonance hybrids, I and II a third canonical structure can be drawn, this structures contribute less or no to the ground or excited state, or we can say the E is much higher.Secondaly the benzene molecule is symmetrical in shape and thus making the absorption weak.

Nitrobenzene is light yellow. In nitrobenzene, charged structure contribute much more therefore resulting the absorption is sifted to much higher wavelength the blue region, thus visible in the yellow the somplementary region. This is also caused due the dissymmetry of the nitrobenzene molecule due to the introduction of the -NO_2 group.

Nitroaniline is deep yellow

The comparison of the dipole moment for the nitrobenzene (4.21D) the aniline (1.48D) and nitroaniline (6.1D) also unconclaves the fact that a reality high contribution of the charged structure to the resonance hybrid of *p*-nitroaniline. The larger value (6.1D) for *p*-nitroaniline over the summationof the values of aniline and nitro benzene (4.21 + 1.48 = 5.69) D is due to the fact that each of the amino group and also the nitro groups comparatively helps the other in sifting the actual state of the molecule in the direction of charged structure.

Relation between Conjugation and Resonance. The longer the conjugation in a molecule, deeper will be the colour because the conjugated system of double bonds (or of chhromophores) provides a long path resonance. Further, the absorption shifts progressively to longer and longer wavelength.

The increase in the number of C = C bonds in conjugated system (see Fig 28.11). This has been obserbed in the case of a series of diphenylpolenes in benzene solution. In this case, N can have different values from 1, 2, 3, 4, 5, upto 15. When n = 1 or 2 the compound is colourless and it is pure yellow when n = 3 and becomes orange when n = 5 and deepens as the value of n increases. The colour sequence is shwn in Fig. 28.12.

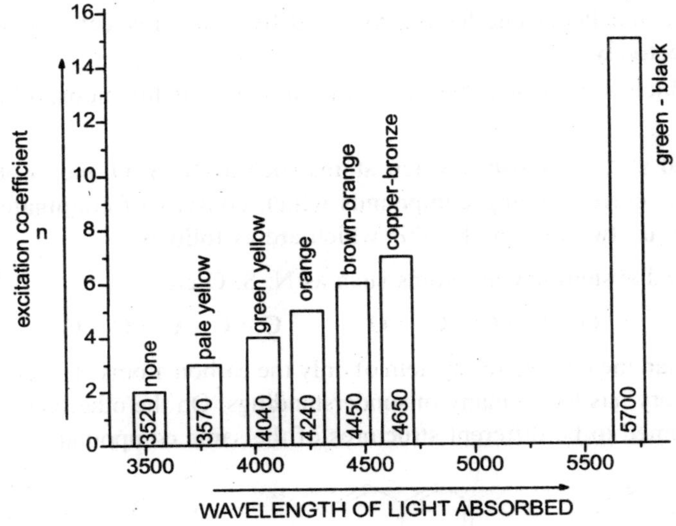

Fig. 28.12. Comparison of wavelength and exitation co-efficient.

According to Lewis and Calvin, the deepening of the colour with the increasing in the wavelength of the conjugation has been attributed to the increae in the number of electrons which are parcipating in conjugation. The increase in conjugated system is also able to increase the deep colour of some compounds which do not possess aromatic nucleus. For example - carotene, which contains a caonjugated system of ii elthylenic bonds, absorbs at 4510Å and is orange red in colour.

β - Carotene

As the number of fused rings increase, the colour deepens more and more. This can be proved by the following statements:

(*i*) Benzene having one ring is colourless,

(*ii*) Naphthalene and Anthracene both having two rings respectively are also colourless.

(*iii*) Naphthacene and Pentacene having four and five rings respectively are yellow and blue in colour respectively.

(*iv*) Graphite, which is a mound of benzene ring, fused in all directions, is black absorbs almost all the colours.

If the conjugated system is also haveign atoms such as N, S, O, etc, it absorbs light of longer wavelengths than the corresponding compounds which consists of conjugated system of only the carbon atoms. There are two reasons for this which are as follows.

(*i*) The conjugated system having atoms such as N, S, O etc.

$$O = C - C = C^{+} - O - \qquad C = C - C = C - C$$

is less symmetrical than the conjugated system of only the carbon atoms. The qualitative application of the valance bond theory has led to many misunderstandings. On the otherhand, mesomeric structures were repeatedly assumed to be different structures of the same compound.

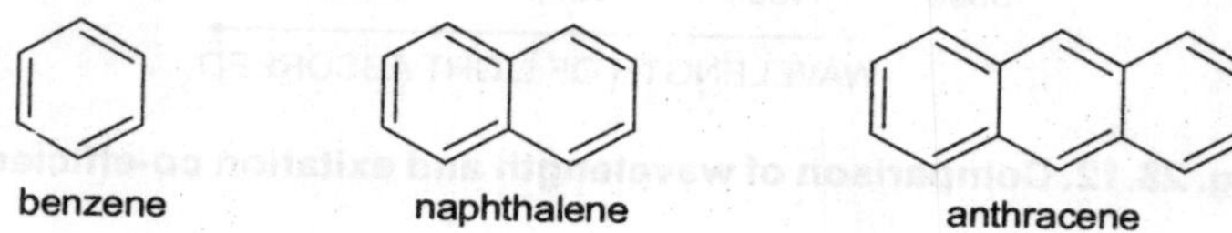

naphthacene pentacene Graphite

And on the otherhand the mesomeric structures were connected with excited states. However, the true low energy state lies between these two extremes.

Application of Molecular Orbital Theory on Chromaticity

As discussed in above of Architecture of chemistry an inorganic approach about this theory, it means the transference of one electron from an orbital of lower energy to thatof higher energy. These electrons may be , or n (non-bonding) electrons. The higher energy atates are commonly called antibonding orbitals.(see chapter 5 chemical bonding). The antibonding orbitals associated within (non-bonding) electrons because they do not form bonds. Fig 28.13 shows, the most essential types of energy levels.

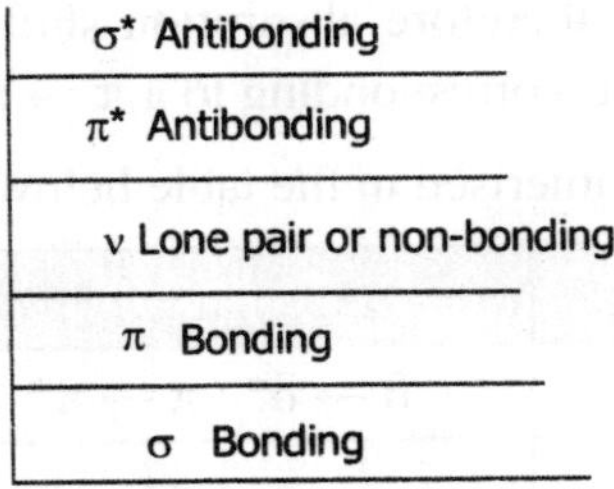

Fig. 28.13. The Molecular orbitals according to the energy levels.

Electronic Transitions

The electronic transitions can occur by the absorption of ultra violet and visible radiation. Although several transitions are possible but only the following transitions are allowed.

$\delta \rightarrow \delta^*$

This transition occurs when a bonding δ-electron is excited to an antibonding δ-orbital thst is δ*. This type of transition requires a very large amount of energy as δ-electrons are very tightly bound. Hence the compounds like saturated hydrocarbons which do not have any π or δ electrons may undergo only $\delta \rightarrow \delta^*$ transitions. However, these transitions do not take place by absorbing in the ordinary ultra violet region. e.g. ethane absorbs at 135 μ

$n \rightarrow \delta^*$

This transition occurs when one electron of a lone pair (i.e) a non-bonding pair of electron ion is excited to an antibonding δ-orbital i.e δ^*. Compounds having non-bonding electrons on hetero atoms like N, S, O etc can undergo $n \rightarrow \delta^*$ transitions.

But the non-bonding, electrons are much more loosely held then the δ-electrons. Therefore, the energy required for δ-δ* transitions. Hence the compounds having non-bonding electrons usually absorb in the ordinary U-V region. For example, methyl iodide shows λ_{max} at 258 mμ while trimethylamine at 227mμ.

Both transitions $n \rightarrow \pi^*$ and $\pi \rightarrow \pi^*$ transitions are of lower energy than the $n \rightarrow \delta^*$ transitions. Acetaldehyde shows two absorptions, one at 180 mμ corresponding to a $\pi \rightarrow \pi^*$ transition while the other at 290m corresponding to a $n \rightarrow \pi^*$ transition.

The various transitions may be arranged in their descending order of energy.

$$\delta \rightarrow \delta^* > n \rightarrow \delta^* > n \rightarrow n^* > n \rightarrow \pi^*$$

Let us now consider some possible examples:

(*a*) The $\pi \rightarrow \pi^*$ transitions for example for simple alkenes takes place in the vacuum untraviolet region. For example ethylene absorbs at 175 mμ. Conjugation of double bonds decreases the genergy required for $\pi \rightarrow \pi^*$ transition and therefore absorption shifts to longer wavelength. For example butadiene shows absorption at 217mμ corresponding to a $\pi \rightarrow \pi^*$ transition and is yellow in colour.

The above results have been summerised in the table below: -

Compound	Transition	Absorption band
$H_2C = CH_2$	$\delta \rightarrow \delta^* / \pi \rightarrow \pi^*$	175 mμ
$H_2C = CH_2 - CH - CH - CH_2$	$\pi \rightarrow \pi^*$	217 mμ
β - carotene	$\pi \rightarrow \pi^*$	451 mμ

Ultra-violet spectrum of a compound is mainly used for detecting the presence of conjugation and also for determining the nature of the conjugating system.

In order to explain the results of conjugating system of only carbon atoms; let us consider the compound butadiene $H_2C = CH - CH = CH_2$. This study will help us to know the relation between colour and orbital theory.

Let us have a pip in the Fig 28.14.

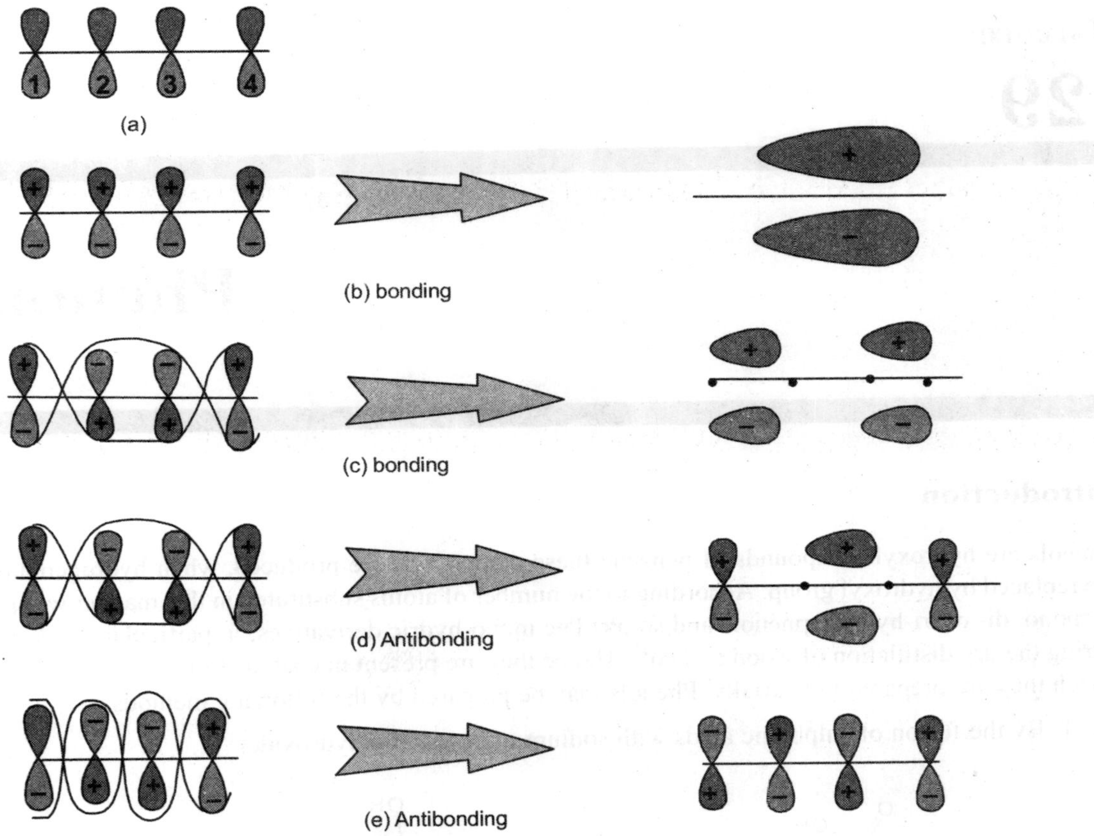

Fig. 28.14. The Bonding system in butadiene

Figure 28.14 shows the four P_3 prbitals associated with four sp^2 hybridised carbon atoms of butadiene. All the four P_3 otrbitals are forming molecular orbitals which cover all the four P_3 orbital atoms. Two of four P_3 electrons will occur one molecular orbital (see fig 28.14 b) and remaining two the second molecular orbital (28.14 c) Fig 28.14 d and e represent the excited state of molecule.

When the butadiene is excited four transitions from lower energy level to that of higher energy level are possible i.e (b) to (d), (b) to (e) (c) to (d) or (c) to (e). If all these four transition takes place four absorption bands will be produced. So if ΔE for (*c*) $|\rightarrow$ (*d*) transition is lowesr than all the three other transitions and thus, the absorption band of longest wavelength will corresponding to such a transition.

From the above discussions it is evident that the Molecular orbital theory is the best supporting theory behind colours.

CHAPTER

29

Phenols

Introduction

Phenols are hydroxyl compounds of benzene these compounds are produced, when hydrogen atoms are replaced by hydroxyl group. According to the number of atoms substituted in this manner we speak of mono, di- or tri-hydride phenols and so on. The mono-hydric derivatives, in particular, are formed during the dry distillation of wood and cola. Hence they are present in coal tar from the carbolic oil of which they are prepared industrially. Phenols may be prepared by the following methods:

1. By the fusion of sulphonic acids with sodium or potassium hydroxide.

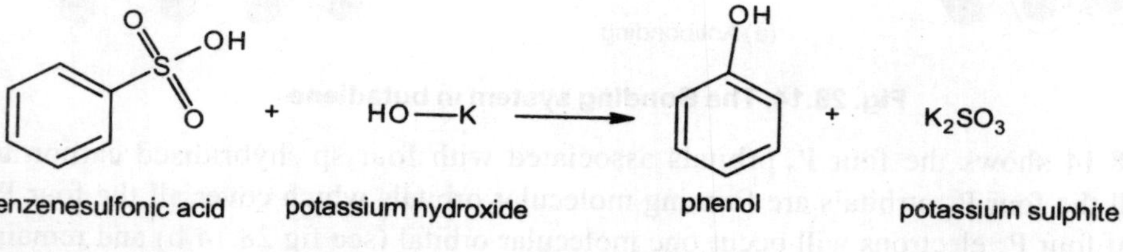

benzenesulfonic acid + HO—K (potassium hydroxide) → phenol + K_2SO_3 (potassium sulphite)

2. By heating diazomium salts with water, diazomium chloride salts are used because nitrates, sets free the nitric acid can produce nitro phenols instead.

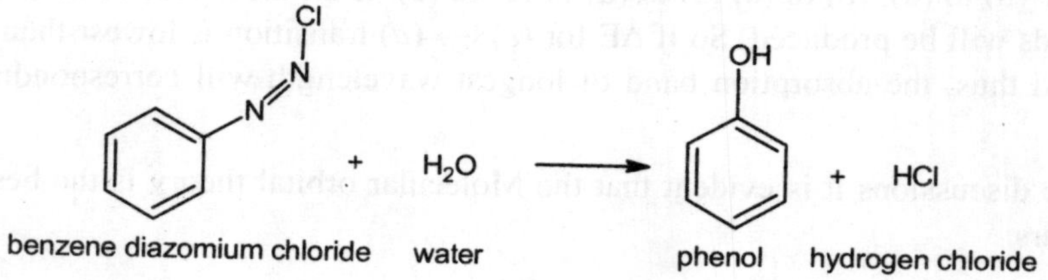

benzene diazomium chloride + H_2O (water) → phenol + HCl (hydrogen chloride)

3. By the action of oxygen on aromatic organo-magnesium compounds and decomposition of the resulting product with dilute hydrochloric.

Types of Phenols

The Monohydric Phenols and their derivatives

Phenol, Carbolic acid, Acidum Carbolicum: This is the chief constituent if that fraction of coal tar boiling at 170°C to 230°C and generally known as middle oil or carbol oil. It is prepared from this source after removal of naphthalene, by shaking out with dilute caustic soda. The aqueous layer is run off and phenol precipitated with sulphuric acid or carbon dioxide. Finally it is purified by distillation.

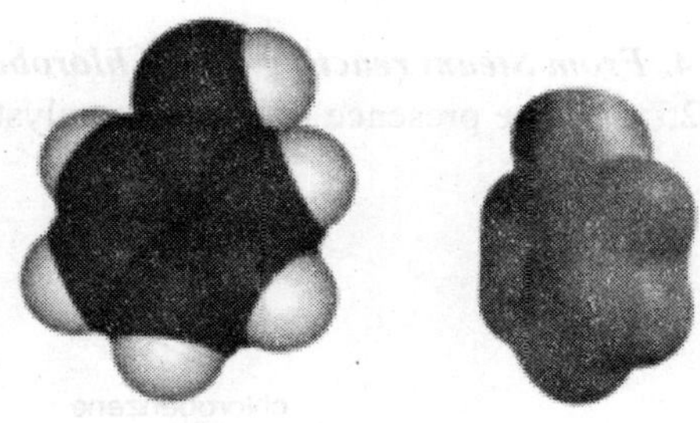

Preparation

This is the industrial process for the phenol preparation; it is recovered from the middle oil or carbol oil (so the name carbolic has been designated).

1. From Benzene: A certain amount of phenol is also prepared from benzene by converting it into alkali benzene sulphonated and fusing this with solid caustic alkali.

Benzene —(sulphonation, H_2SO_4)→ Benzenesulfonic acid —(NaOH, fused)→ Phenol

2. Oxidation of Benzene: Phenol can also be obtained from benzene by its oxidation in air at high temperature.

Benzene —(oxidation, 300 °C)→ Phenol

3. From Chlorobenzene: Chlorobenzene when heated with aqueous sodium hydroxide at high temperature (350°C) an under a pressure of 200-300 atmos in the pressure in the presence of copper salts forms phenols.

chlorobenzene + sodium hydroxide $\xrightarrow[\text{200 - 300 atmos}]{\text{350 °C}}$ phenol + sodium chloride

4. *From Steam reacting with Chlorobenzene*: Chlorobenzene can also be hydrolyzed with steam at 425°C in the presence of copper catalyst.

chlorobenzene + H_2O (steam) $\xrightarrow{CuCl_2}$ phenol + HCl

5. *From Benzene diazomium Chloride*: By heating it with water.

benzene diazomium chloride + H_2O (steam) $\xrightarrow{\text{Heat}}$ phenol + HCl + N_2

6. *From Grignard Reagent*: By the action of oxygen and subsequent hydrolysis.

Grignard's Reagent + O ⟶ ⟶ phenol ⟶ Mg(OH)Br

7. *From Salicylic Acid*: By heating it with soda lime

salicylic acid + CaO $\xrightarrow{\text{heat}}$ phenol + $CaCO_3$

Industrial Process of Phenol Preparation from Tar

A fraction boiling between 150-230°C is separated by either continuous or batch distillation of tar. Depending upon the origin of the tar, this fraction contains from 10 to 30% tar acids, and less that 1% tar bases. An installation for the continuous processing of about 7000 lb/hr is shown in Fig. 29.1. Column A is a steel extraction column, about 4 ft 6 inch in diameter, fitted with about 25 ft of 1 inch steel Rasching rings. About 7,000 lb/hr of the carbolic acid fraction containing about 15% tar acids are heated to about 100°C, enters the bottom of the column A at 1. About 4,800 lb/hr of about 10% sodium hydroxide solution, also near 100°C, enters the top of the column at 2. The caustic extracts the acid in continuous counter current flow. The oil leaves the column at 3 and goes top the acid extraction for recovery of tar bases. The alkaline extract, cresols and xylenols, over flow at 4 into column B, a steel column about 3 feet in diameter, containing about 25 ft of 1 inch steel Rasching rings. About 5000 lb/hr of steam is blown into the bottom of the column at 5. The steam drives the hydrocarbon and residual bases out. The vapours are condensed in 6 and go to an acid extraction. The alkaline extract is pumped to then top of the column C, where the acid are liberated, or "sprung" from their sodium salts by the action of about 10,000 ft^3/hr of lime kiln gas containing about 30% CO_2 entering at 7, coming from the bicarbonate column D. The bottoms from C separate into two phase, an aqueous layer of sodium carbonate solution, containing about 12% Na_2CO_3 saturated with coal tar acid and organic layer, containing the free tar acid water, and some sodium phenate. The phases are separated at 8. The organic layer

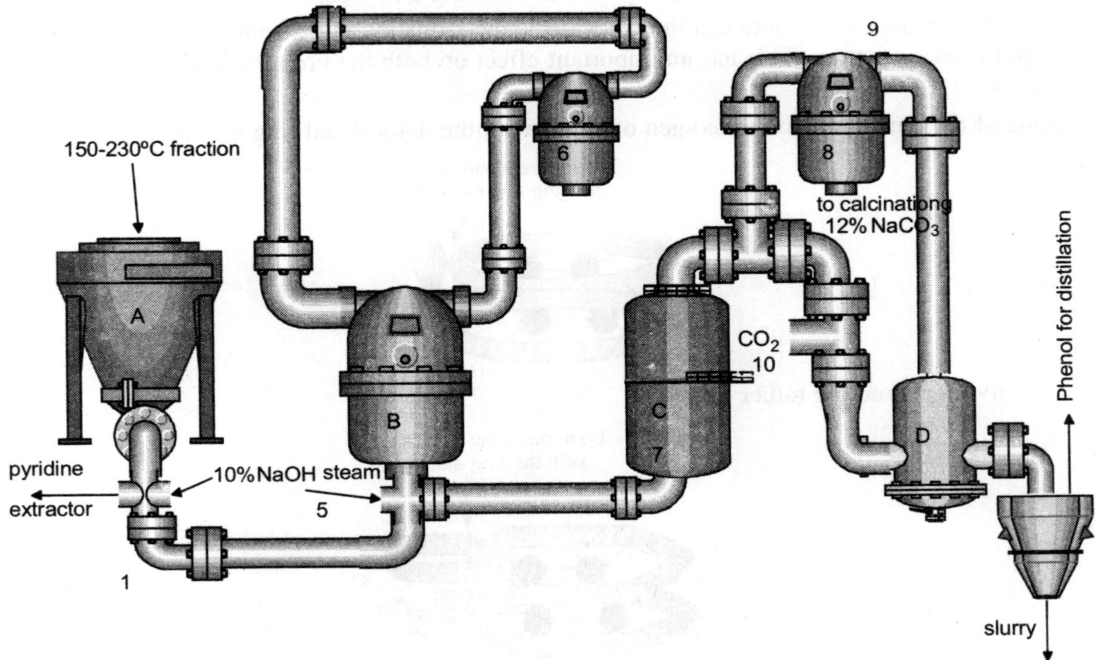

Fig. 29.1. The Preparation of Phenol from Tar

flows at 9 into the top of the bicarbonate column D, where it meets the lime kiln gas entering at 10. The large excess of carbon dioxide neutralizes all the unconverted salt dissolved in the acids, converts it into bicarbonate, and precipitates a portion of it as a solid. The slurry leaving the bottom D is pumped into the separator at 11 from which the free acids pass to a continuous distillation system, which separates them into their constituents, while the aqueous slurry returns to causticizing plant. The quantity of lime kiln gas entering at 10 is controlled so that

A few percent of the sodium phenate remains unconverted in the liquid leaving column T, because otherwise the precipitation of bicarbonate crystals in C would lead to difficulties. The carbonate solution from 8 is causticised in the normal way by means of lime. The presence of tar acids in the carbonate solution does not interfere with the causticizing sludge is washed properly there is no important loss of tar acids.

Structure

The simplest way to draw the structure of phenol is:

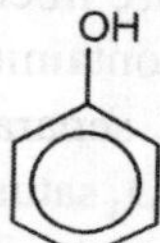

. . . but to understand phenol properly, you need to dig a bit deeper than this.*

There is an interaction between the delocalised electrons in the benzene ring and one of the lone pairs on the oxygen atom. This has an important effect on both the properties of the ring and of the -OH group.

One of the lone pairs on the oxygen overlaps with the delocalised ring electron system . . .

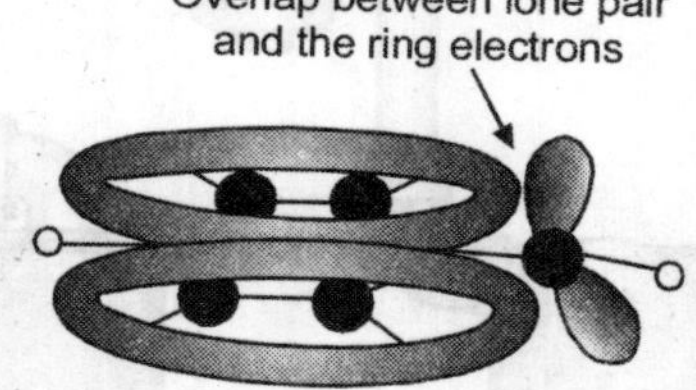

. . . giving a structure rather like this:

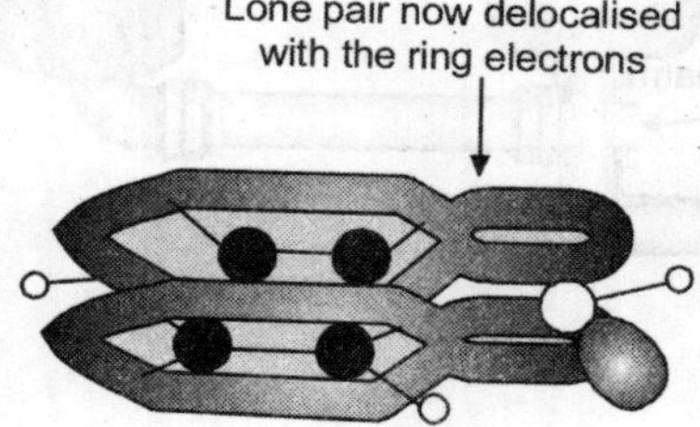

*Warning! You need to understand about the bonding in benzene in order to make sense of this next bit.

The donation of the oxygen's lone pair into the ring system increases the electron density around the ring. That makes the ring much more reactive than it is in benzene itself. That is explored in another page in this phenol section.

- The alcohol functional group consists of an O atom bonded to an sp^2-hybridized aromatic C atom and a H atom via s bonds.
- Both the C-O and the O-H bonds are polar due to the high electronegativity of the O atom.
- Conjugation exists between an unshared electron pair on the oxygen and the aromatic ring.
- This results in, compared to simple alcohols:
  - a shorter carbon-oxygen bond distance
  - a more basic hydroxyl oxygen
  - a more acidic hydroxyl proton (-OH)

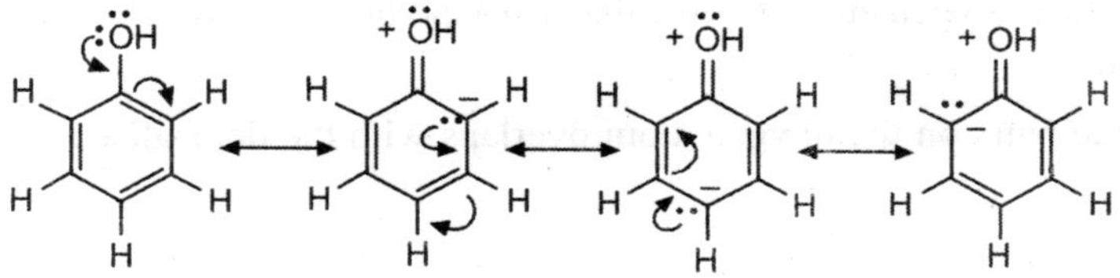

Physical Properties

Melting point	= 42°C
Boiling Point	= 183°C
Formula Weight	= 94.11124
Composition	= C(76.57%) H(6.43%) O(17.00%)
Molar Refractivity	= 28.13 ± 0.3 cm^3
Molar Volume	= 87.8 ± 3.0 cm^3
Parachor	= 222.2 ± 4.0 cm^3
Index of Refraction	= 1.553 ± 0.02
Surface Tension	= 40.9 ± 3.0 dyne/cm
Density	= 1.071 ± 0.06 g/cm^3
Polarizability	= 11.15 ± 0.5 $10^{-24}cm^3$
Monoisotopic Mass	= 94.041865 Da
Nominal Mass	= 94 Da
Average Mass	= 94.1112 Da
Solubility	= 8gm/100 ml of water

Chemical Properties

1. Acidity of Phenols: Unlike alcohols (which also contain an -OH group) phenol is a weak acid. A hydrogen ion can break away from the -OH group and transfer to a base.

For example, in solution in water:

OH + H_2O $\rightleftharpoons$ O^- + H_3O^+

A phenoxide ion

Phenol is a very weak acid and the position of equilibrium lies well to the left.

Phenol can lose a hydrogen ion because the phenoxide ion formed is stabilised to some extent. The negative charge on the oxygen atom is delocalised around the ring. The more stable the ion is, the more likely it is to form.

One of the lone pairs on the oxygen atom overlaps with the delocalised electrons on the benzene ring.

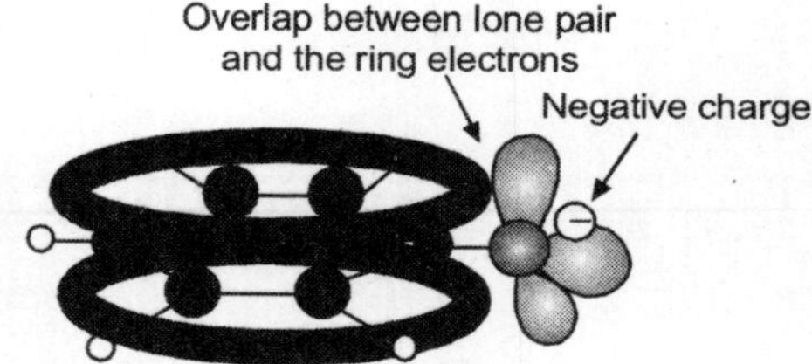

This overlap leads to a delocalisation which extends from the ring out over the oxygen atom. As a result, the negative charge is no longer entirely localised on the oxygen, but is spread out around the whole ion.

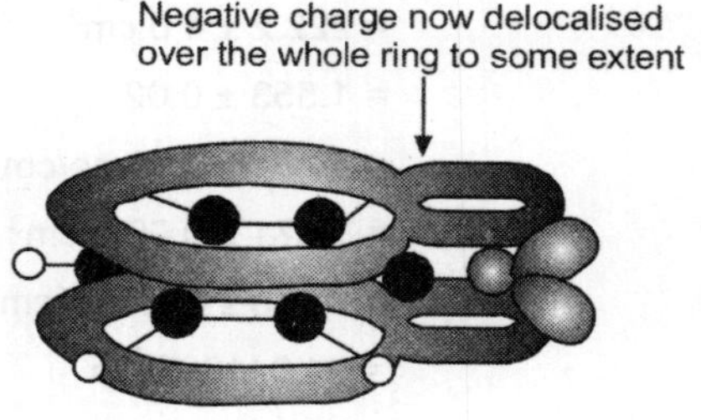

Spreading the charge around makes the ion more stable than it would be if all the charge remained on the oxygen.

However . . . oxygen is the most electronegative element in the ion and the delocalised electrons will be drawn towards it. That means that there will still be a lot of charge around the oxygen which will tend to attract the hydrogen ion back again.

That's why phenol is only a very weak acid.

Properties of Phenol as an Acid with Indicators

The pH of a typical dilute solution of phenol in water is likely to be around 5-6 (depending on its concentration). That means that a very dilute solution isn't really acidic enough to turn litmus paper fully red. Litmus paper is blue at pH 8 and red at pH 5. Anything in between is going to show as some shade of "neutral".

With Sodium Hydroxide Solution

Phenol reacts with sodium hydroxide solution to give a colourless solution containing sodium phenoxide.

$$C_6H_5OH + NaOH \longrightarrow C_6H_5O^-Na^+ + H_2O$$

Sodium phenoxide

In this reaction, the hydrogen ion has been removed by the strongly basic hydroxide ion in the sodium hydroxide solution.

With Sodium Carbonate or Sodium Hydrogen Ccarbonate

Phenol isn't acidic enough to react with either of these. Or, looked at another way, the carbonate and hydrogen carbonate ions aren't strong enough bases to take a hydrogen ion from the phenol.

Unlike the majority of acids, phenol *doesn't* give carbon dioxide when you mix it with one of these.

This lack of reaction is actually useful. You can recognise phenol because:

- It is fairly insoluble in water.
- It reacts with sodium hydroxide solution to give a colourless solution (and therefore must be acidic).
- It doesn't react with sodium carbonate or hydrogencarbonate solutions (and so must be only very weakly acidic).

With Metallic Sodium

Acids react with the more reactive metals to give hydrogen gas. Phenol is no exception - the only difference is the slow reaction because phenol is such a weak acid.

Phenol is warmed in a dry tube until it is molten, and a small piece of sodium added. There is some fizzing as hydrogen gas is given off. The mixture left in the tube will contain sodium phenoxide.

Warning! Under no circumstances should you try this without professional supervision, and good acess to medical help. The risks involved in careless handling of the hot phenol and sodium are too great.

OH

2 + 2Na ⟶ 2 + H_2

O^-Na^+

Sodium phenoxide

Resonance Effect on the Acidity of Phenols

What is the source of this enhanced stability? First, the phenolate anion is stabilized by resonance. Second, the polar effect of the benzene ring stabilizes the, negative charge. Both resonance stabilization and polar effects are the same effects that stabilize benzylic carbon ions Of course, a phenoxide anion is a benzylic anion in which the benzylic group is an oxygen instead of a carbon!

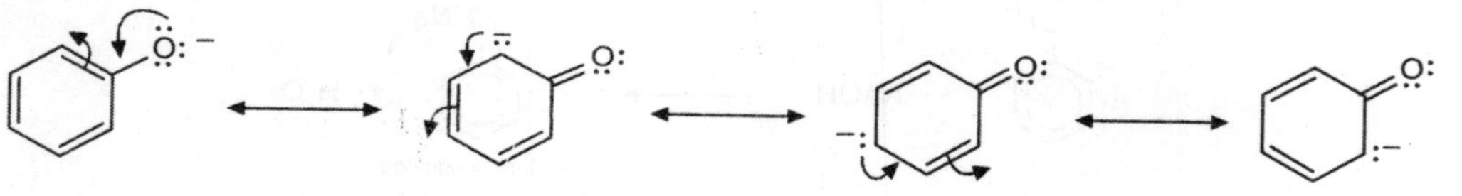

Alkoxides are stabilized neither by resonance nor by the polar effect of benzene rings or double bonds.

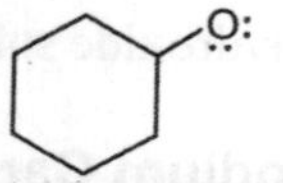

Because phenoxide ions are stabilized by both resonance and polar effects, less energy is required to form phenoxides from phenols than is required to form alkoxides from alco-hols. Because pKa is directly proportional to the standard free energy of ionization, phenols have lower pKa values, and are thus more acidic, than alcohols.

Cyclohexanoate anion
no resonance structures
no polar effect of double bond

Substituent groups can also affect phenol acidity by both polar and resonance effects. For example, the relative acidities of phenol, m-nitrophenol and p-nitrophenol reflect the operation of both effects.

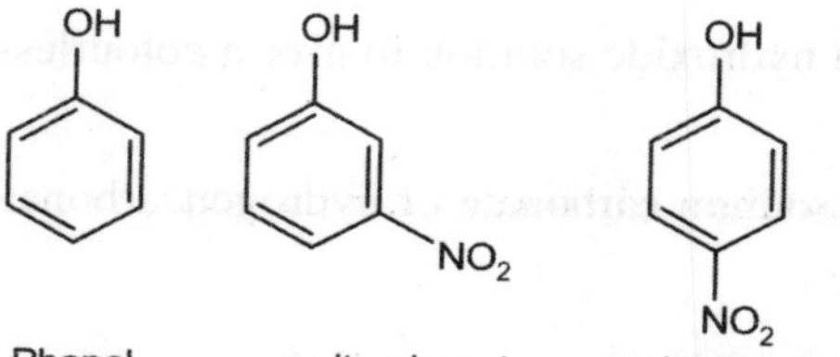

OH OH OH

NO_2 NO_2

Phenol
9.95

m-nitrophenol
8.32

p-nitrophenol
7.21

m-Nitrophenol is more acidic than phenol because the nitro group is very electronegative. The polar effect of the nitro substituent stabilizes the conjugate-base anion for the same rea-son that it would stabilize the conjugate base of an alcohol. Yet p-nitrophenol is more acidic than m-nitrophenol by more than one pKa unit, even though the p-nitro group is farther from the phenol oxygen. This cannot be the result of a polar effect, for polar effects on acidity decrease as the distance between the substituent and the acidic group in-creases. The reason for the increased acidity of p-nitrophenol is that the p-nitro group sta-bilizes the conjugate-base anion by resonance.

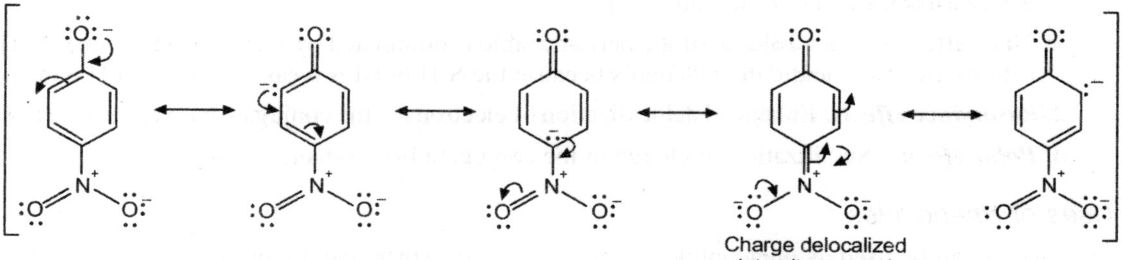

Charge delocalized into nitro ground

The colored structure is especially important because it places charge on the electronega-tive oxygen atom. In m-nitrophenol, however, it is not possible to draw a resonance struc-ture that delocalizes the negative charge into the nitro group.

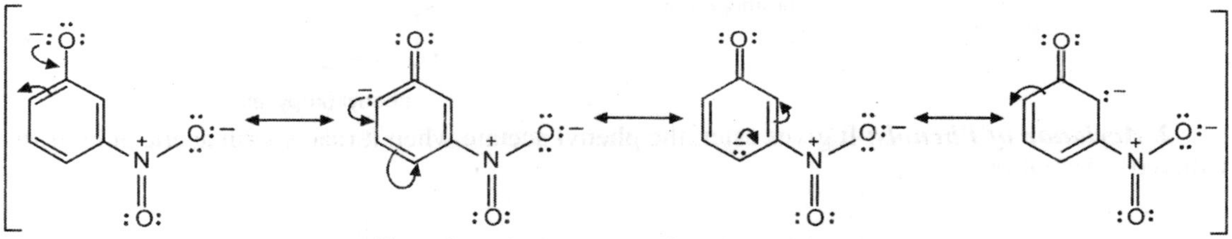

Fewer important structures than the *para*-isomer

Because p-nitrophenoxide has more resonance structures, it is more stable relative to its corresponding phenol than is m-nitrophenoxide. Hence p-nitrophenol is the more acidic of the two nitrophenols. The acid-strengthening resonance effect of *ortho-* and *para-nitro* groups is so large that 2,4,6-trinitrophenol (picric acid) is actually a strong acid.)

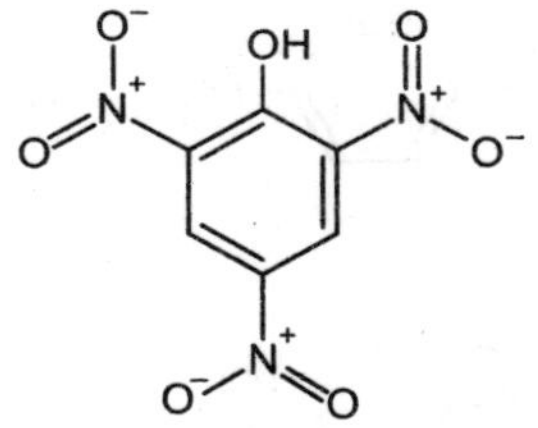

picric acid [2,4,6 trinitrophenol]

Let's summarize the factors that govern acidity as we've seen them in operation so far:

1. *Element effects:* Other things being equal, compounds are more acidic when the ele-ment to which the acidic hydrogen is bound has a higher atomic number within either a row or group of the periodic table.

a. The effect within a row (period) of the periodic table is dominated by relative elec- tronegativities (or electron affinities). Thus, water is much more acidic than methane, and phenol is much more acidic than toluene, because oxygen (higher atomic number) is more electronegative than carbon (lower atomic number).

b. The effect within a column of the periodic table is dominated by relative bond energies. Thus, thiols are more acidic than alcohols because the S-H bond is weaker than the 0-H bond.

2. *Resonance effects:* Enhanced delocalization of electrons in the conjugate base en-hances acidity.

3. *Polar effects:* Stabilization of charge in the conjugate base enhances acidity.)

Uses of Phenoxides

Phenoxides can be used as nucleophiles, for example, aryl ethers can be prepared by the reaction of phenoxides anion and an alkyl halide

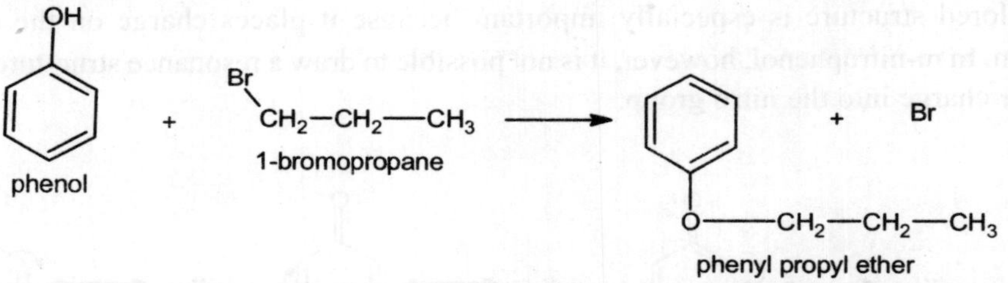

OH

phenol + Br–CH_2–CH_2–CH_3 (1-bromopropane) ⟶ O–CH_2–CH_2–CH_3 (phenyl propyl ether) + Br

2. *Acylation of Phenols*: It gives ester, the phenyl acetate when it reacts with acetic anhydride it is called o – Acylation.

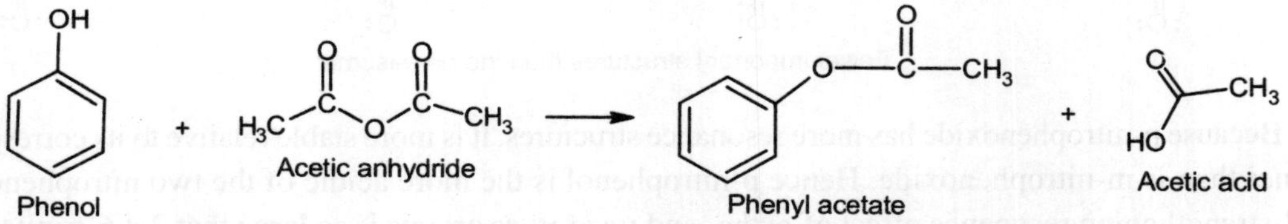

OH

Phenol + H_3C–C(=O)–O–C(=O)–CH_3 (Acetic anhydride) ⟶ O–C(=O)–CH_3 (Phenyl acetate) + HO–C(=O)–CH_3 (Acetic acid)

Type of Acylation in Phenols

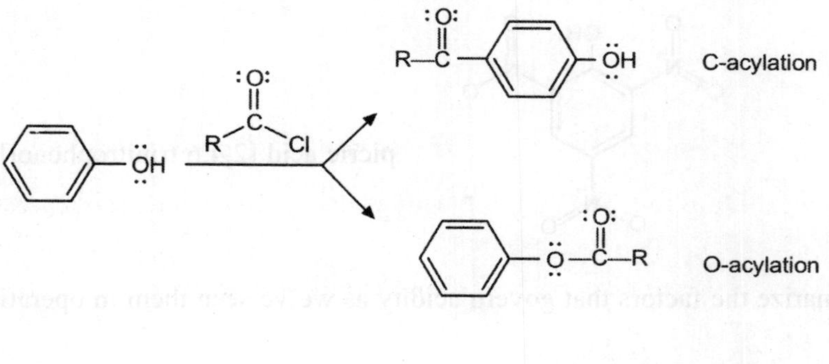

Ph–$\ddot{O}$H + R–C(=:O:)–Cl ⟶ R–C(=:O:)–C_6H_4–$\ddot{O}$H C-acylation

⟶ Ph–$\ddot{O}$–C(=:O:)–R O-acylation

Reaction type:

C-acylation = electrophilic aromatic substitution

O-acylation = nucleophilic acyl substitution

Summary

- Phenols are examples of bidentate nucleophiles, meaning that they can react at *two* positions:
  - on the aromatic ring giving an aryl ketone via **C-acylation**, a Friedel-Crafts reaction or,
  - on the phenolic oxygen giving an ester via **O-acylation,** an esterification
- Reagents :

  C-acylation : acylating agent (acyl chloride or anhydride) and $AlCl_3$

  O-acylation : acylating agent (acyl chloride or anhydride)
- The product of **C-acylation** is more stable and predominates under conditions of **thermodynamic control** (i.e. when $AlCl_3$ is present).
- The product of **O-acylation** forms faster and predominates under conditions of **kinetic control**
- **O-acylation** can be promoted by either:
  - acid catalysis via protonation of the acylating agent, increasing its' electrophilicity or
  - base catalysis via deprotonation of the phenol, increasing its' nucleophilicity.
- It is also known that aryl esters readily rearrange to aryl ketones in the presence of $AlCl_3$, a reaction known as the **Fries rearrangement:**

aryl ester —($AlCl_3$, heat)→ *o*-and *p*-aryl ketones

3. Carboxylation of Phenols (Kolbe-Schmitt reaction)

ONa (sodium phenoxide) —(1. CO_2/125°C, 100 atm; 2. H^+)→ OH / CO_2H (salicylic acid)

Reaction type: Electrophilic Aromatic Substitution

Summary

- Heating the nucleophilic phenolate salt with carbon dioxide under high pressure / temperature results in regioselective *ortho*-substitution.
- This process is also known as the **Kolbe-Schmitt synthesis**.

- o-hydroxybenzoic acid is more commonly known as *salicyclic acid.*

The geometry allows for an intramolecular hydrogen bond

Consider the phenolate to be an enolate, hence reactions at the a-C are typically favored.

In some of these reactions, the – OH group has special effects that are not common to other substituent groups.

Because the – OH group is a strongly activating substituent, phenol can be halogenated once under conditions that are titally ineffective for benzene itself.

phenol + Br_2 $\xrightarrow[\text{or } CS_2]{CCl_4}$ 4-bromophenol + HBr

It is to be noticed the mild conditions of this reaction.

A Lewis acid such as $FeBr_2$ is not required*. But when phenol reacts with Br_2 in H_2O (bromine water) more extensive bromination occurs and 2, 4, 6 tribromophenol

phenol + $3Br_2$ $\xrightarrow{H_2O}$ 2,4,6-tribromophenol + 3HBr

This more extensive bromination occurs for two reasons. First, bromine reacts with water to give protonated hypobromous acid. A more potent electrophile than bromine.

Br_2 + H_2O ⇌ Protonated Hypobromous acid ⇌ HBr + Br—OH (hypobromous acid)

Second in aqueous solutions near neutrality, phenol partially ionizes to its conjugate – base phenoxides anion. Although only a small amount of this anion is present, it very reactive and brominates instantly thereby pulling the phenol – phenolate equilibrium to the right

*A solution of Br_2 in CCl_4 is generally used to add bromine to alkenes).

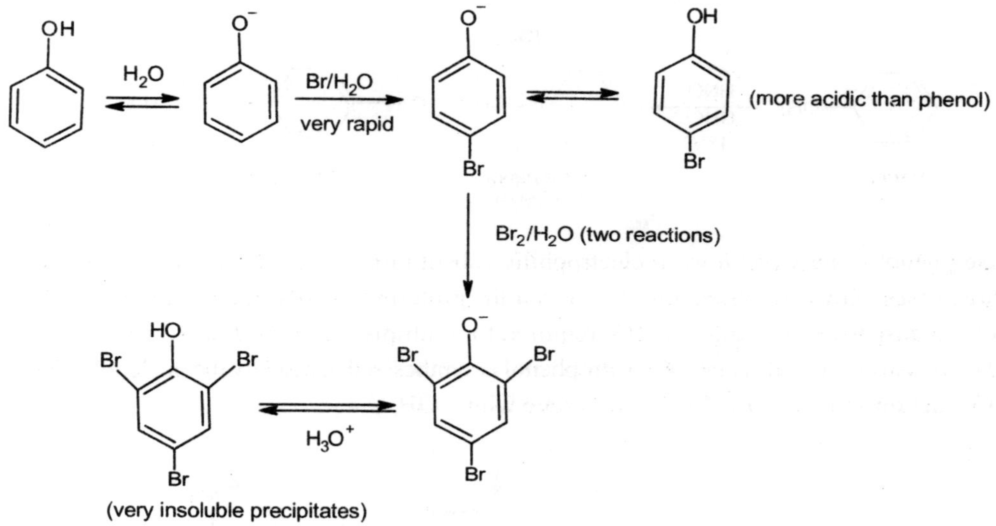

OH, H_2O, O⁻, Br/H_2O very rapid, Br, OH, Br (more acidic than phenol)

Br_2/H_2O (two reactions)

HO, Br, Br, Br, H_3O^+, O⁻, Br, Br, Br

(very insoluble precipitates)

Phenoxides ion is much more reactive than phenol because the reactive intermediate uis not a carbocation, but is instead a more stable neutral molecule.

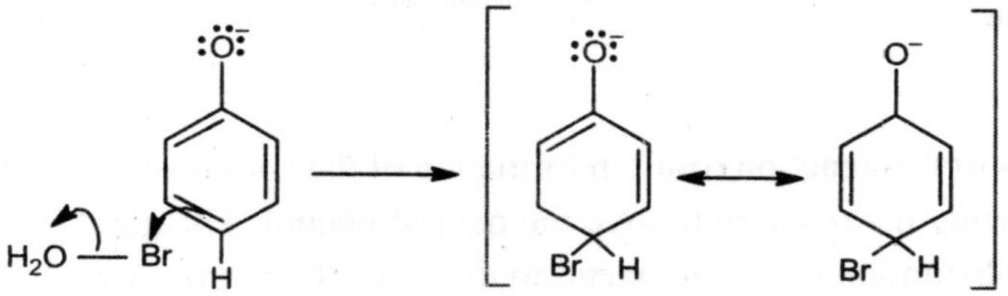

:Ö:⁻, H_2O — Br, H, :Ö:⁻, Br H, O⁻, Br H

p-Bromophenol is also in equilibrium with its conjugate base p-bromophenoxide anion, which brominates again until all ortho and para positions have been substituted. Notice in Eq. above that in the second and third substitutions the powerful *ortho, para-directing* and activating effects of the -2:- group override the weaker deactivating and directing effects of the bromine substituents. In strongly acidic solution, in which formation of the phenolate anion is suppressed, bromination can be stopped at the 2,4-dibromophenol stage.

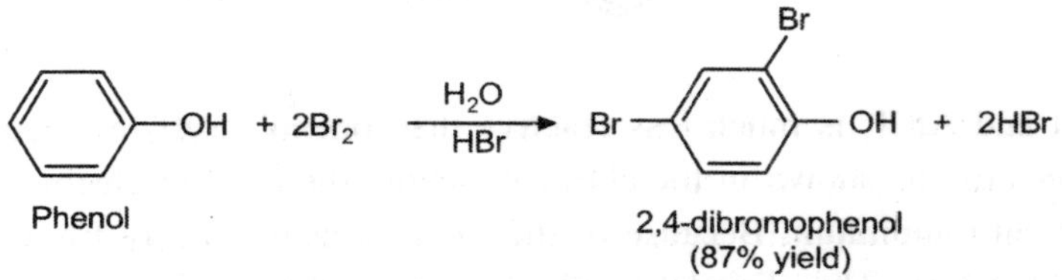

Phenol —OH + $2Br_2$ $\xrightarrow[HBr]{H_2O}$ Br—(Br)—OH + 2HBr

Phenol

2,4-dibromophenol (87% yield)

Phenol is also very reactive in other electrophilic substitution reactions, such as nitra-tion. Phenol can be nitrated once under mild conditions.

Phenol $\xrightarrow[\text{CHCl}_3,\ 15°C]{HNO_3}$ o-nitrophenol (26% yield) + p-nitrophenol (61% yield) + H_2O

Because phenol is activated toward electrophilic substitution, it is also possible to nitrate phenol two and three times. However, direct nitration is *not* the preferred method for syn-thesis of di- and tri-nitrophenol, because the concentrated HNO_3 required for multiple nitrations is also an oxidizing agent, and phenols are easily oxidized. Thus 2,4-dinitrophenol is synthesized instead by the nucleophilic aromatic substitution reaction of l-chloro-2,4-dinitrobenzene with – OH.

chlorobenzene $\xrightarrow[H_2SO_4]{HNO_3}$ 1-chloro-2,4-dinitrobenzene $\xrightarrow[2)\ H_3O^+]{1)\ ^-OH}$ 2,4-dinitrophenol

The basic conditions of this reaction result in formation of the conjugate-base anion of the product; the H_3O^+ is added following the reaction to give the neutral phenol. The great reactivity of phenol in electrophilic aromatic substitution does not extend to the Friedel – Crafts Acylation reaction, because phenol reacts rapidly with the $AlCl_3$ catalyst

+ $AlCl_3$ ⟶ + HCl

The adduct phenol and $AlCl_3$ is much less reactive than phenol itself in electrophilic aromatic substitution reaction because, as shown in the above equation, the oxygen electrons are delocalized onto the electron-deficient aluminium. Because of their delocalization away from the benzene ring during Friedel-Crafts Acylation. Thus Friedel-Crafts Acylation pf phenol occurs slowly, but can be carried out successfully at elevated temperatures. Because it is not activated, the ring is acylated only once.

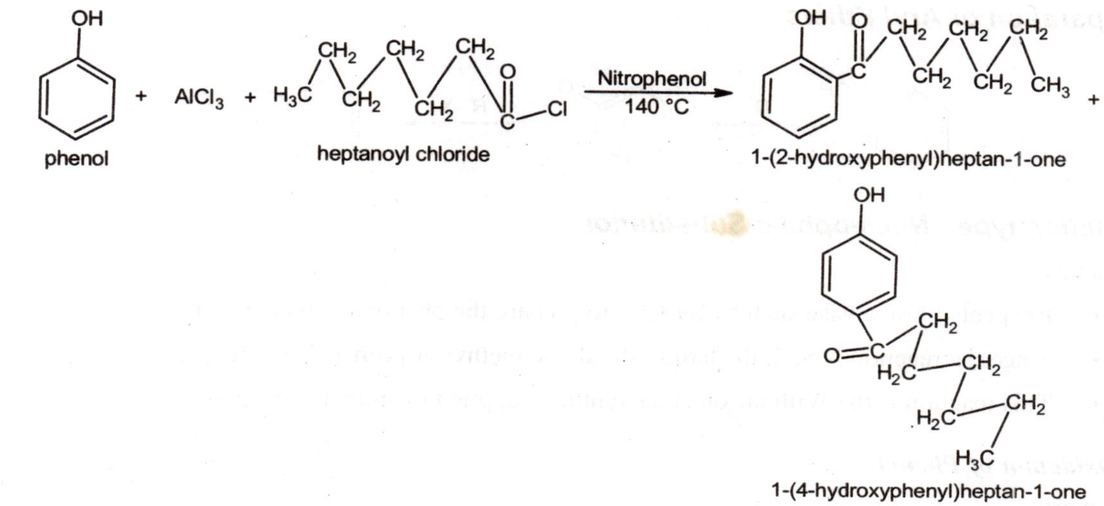

phenol + $AlCl_3$ + heptanoyl chloride $\xrightarrow[140\ °C]{\text{Nitrophenol}}$ 1-(2-hydroxyphenyl)heptan-1-one + 1-(4-hydroxyphenyl)heptan-1-one

Friedel-Crafts alkylation of phenol is also possible.

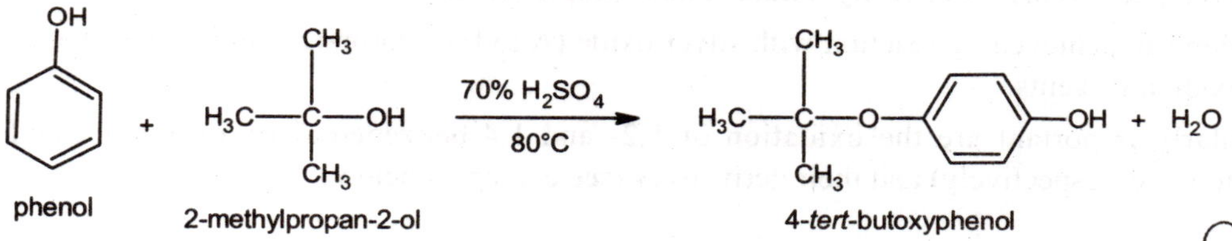

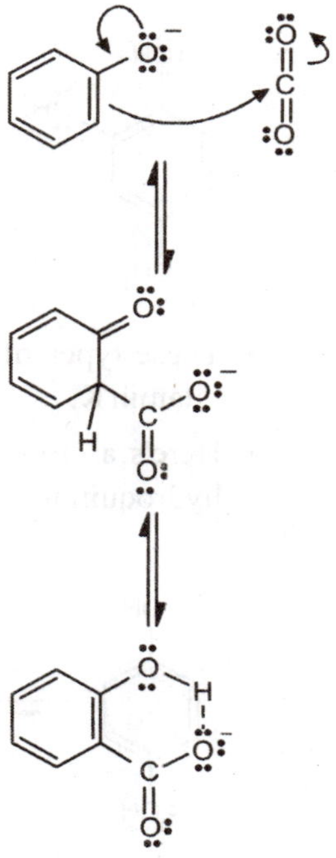

phenol + 2-methylpropan-2-ol $\xrightarrow[80°C]{70\%\ H_2SO_4}$ 4-*tert*-butoxyphenol + H_2O

Mechanism for Carboxylation of Phenols

Step 1: The nucleophilic phenolate (reacting like an enolate) reacts with the electrophilic carbon of carbon dioxide in the ortho position (compare this with an Aldol reaction).

Step 2: The non-aromatic cyclohexadienonecarboxylate intermediate tautomerises to the more stable aromatic enol which is further stabilized by an intramolecular hydrogen bond. An acidic work-up will generate the carboxylic acid.

Preparation of Aryl Ethers

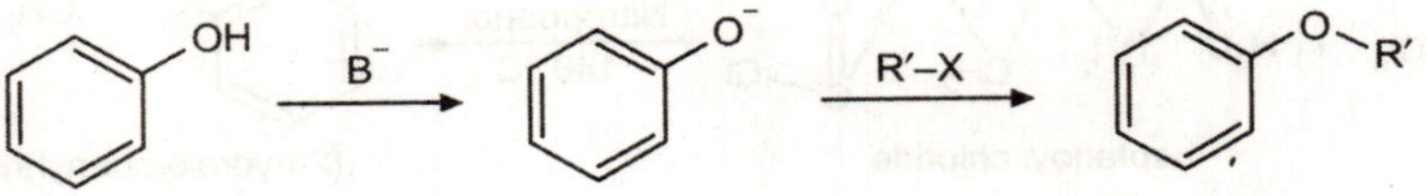

OH B⁻ → O⁻ R'–X → O–R'

Reaction type : Nucleophilic Substitution

Summary

- Reagents : Use a base such as Na_2CO_3 to prepare the phenoxide, then add the alkyl halide.
- Since the reaction is S_N2, the halide should be methyl or primary alkyl halides (or tosylates).
- This reaction is the Williamson ether synthesis applied to aromatic alcohols.

4. Oxidation of Phenols

Summary

- In general, phenols are more easily oxidized than simple alcohols.
- Oxidation can achieved by reaction with silver oxide (Ag_2O) or chromic acid ($Na_2Cr_2O_7$), or other oxidizing agents.
- Particularly important are the oxidation of 1,2- and 1,4-benzenediol (pyrocatechol and hydroquinone, respectively) and their derivatives (see examples below):

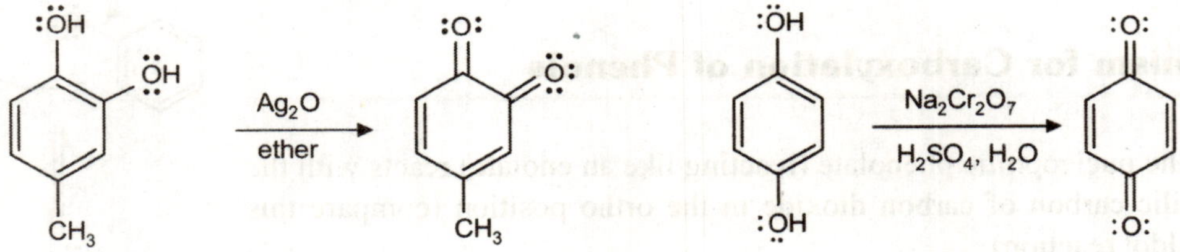

:ÖH ÖH CH3 — Ag_2O, ether → :O: :O: CH3 ; :ÖH :ÖH — $Na_2Cr_2O_7$, H_2SO_4, H_2O → :O: :O:

- These types of systems are important in biological redox-systems such as coenzyme Q and vitamin K.
- Here's a closer look at the two one electron transfers that are believed to take place when hydroquinone is oxidized to benzoquinone

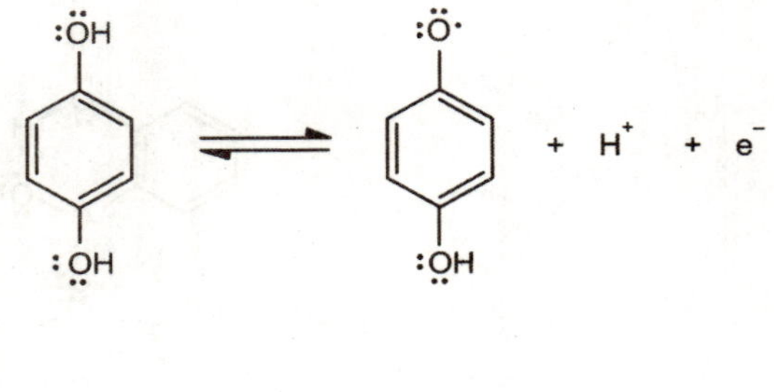

:ÖH :OH ⇌ :Ö· :OH + H^+ + e^-

Loss of a proton and an electron generates a phenoxy radical

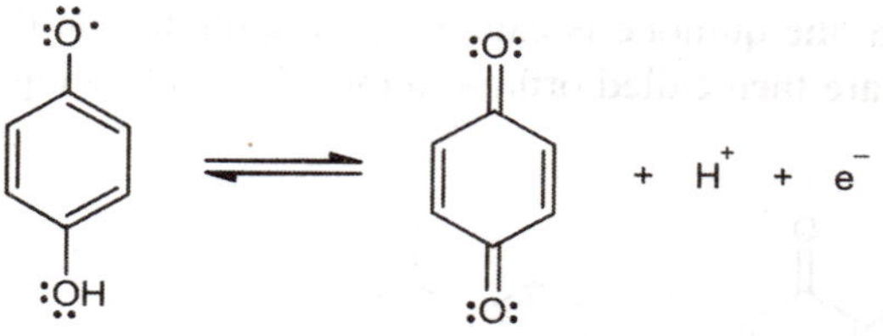

:Ö· ⇌ :O: + H^+ + e^-

:ÖH :O:

Loss of a second proton and a second electron completes the oxidation.

Can you draw a curly arrow scheme that shows how the quinone may be formed from a diradical?

Oxidation of Phenols and Transformation to Quinones

The most common quinones are explained below:

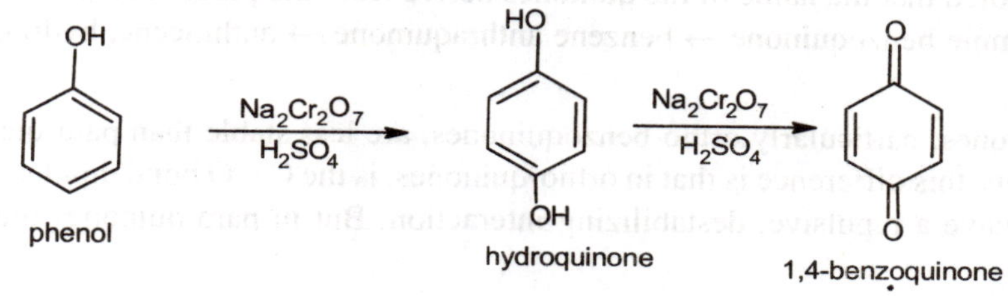

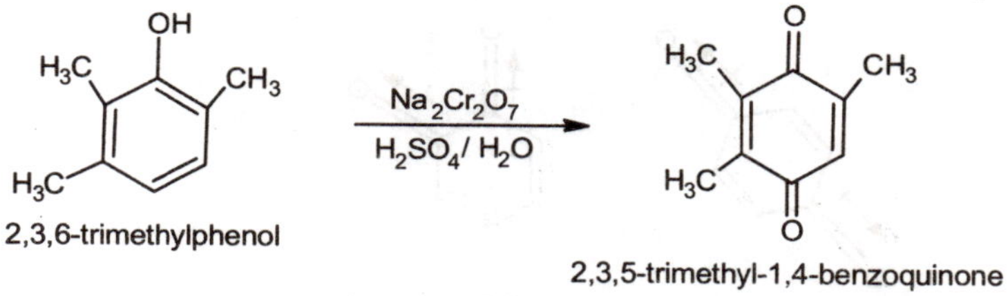

phenol $\xrightarrow[H_2SO_4]{Na_2Cr_2O_7}$ hydroquinone $\xrightarrow[H_2SO_4]{Na_2Cr_2O_7}$ 1,4-benzoquinone

2,3,6-trimethylphenol $\xrightarrow[H_2SO_4/\ H_2O]{Na_2Cr_2O_7}$ 2,3,5-trimethyl-1,4-benzoquinone

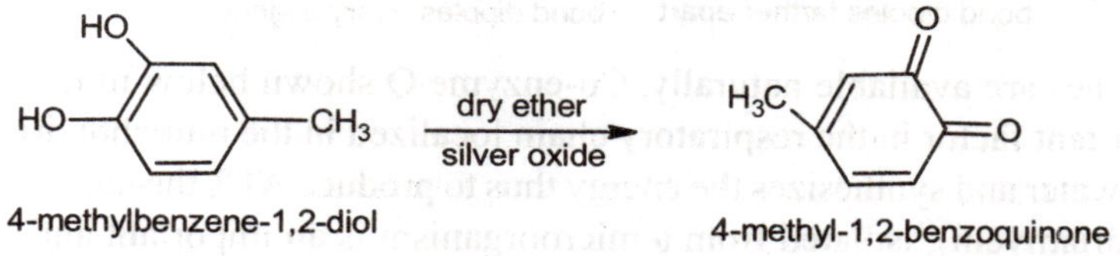

4-methylbenzene-1,2-diol $\xrightarrow[\text{silver oxide}]{\text{dry ether}}$ 4-methyl-1,2-benzoquinone

It is to be noted that hydroquinones, the 4 - methyl benzene 1, 2, diol and phenols with an undistributed position para to the hydroxyl group are oxidized to quinones.

As lime lighted from the above structure quinones must have the following structures

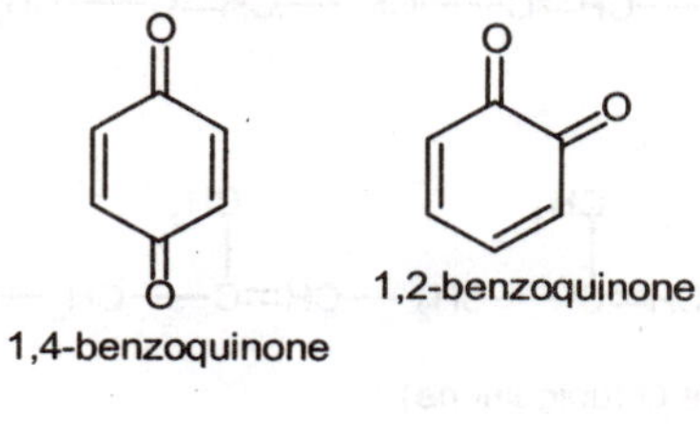

1,4-benzoquinone

1,2-benzoquinone

If the quinone oxygen have a 1, 4 i.e. para) relationship, the quinone is called a para-quinone the if oxygens are in a 1,2(ortho) arrangements, the quinones are then called ortho-quinone. The following compounds are typical quinones.

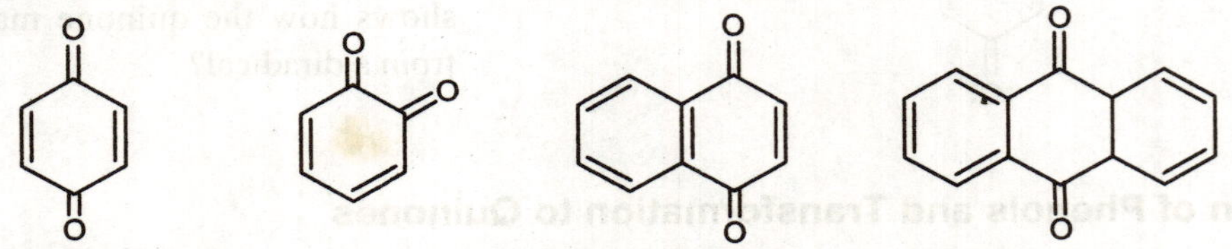

1,4-benzoquinone 1,2-benzoquinone 1,4-naphthoquinone anthraquinone

It is to be noted that the name of the quinones derive from the parent compounds from which they derive, for example benzoquinone → benzene anthraquinone → anthracene, hydroquinone → phenol and so on.

Ortho-quinones, particularly ortho-benzoquinones, are less stable than para-quinone isomers. On reason may be for this difference is that in ortho-quinones, is the C = O bond dipoles are nearly aligned, and therefore, have a repulsive, destabilizing interaction. But in para-quinones these dipoles are far apart.

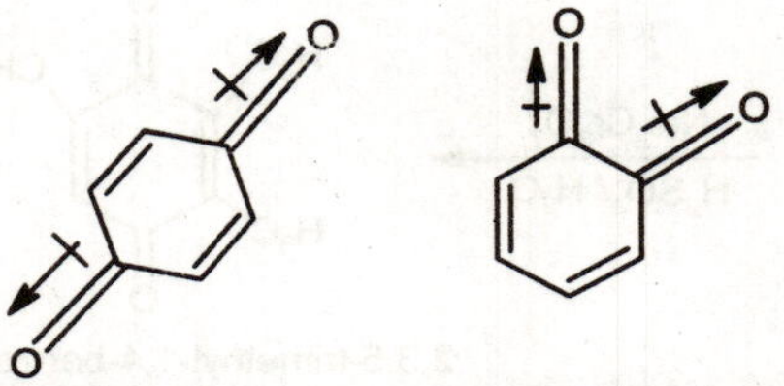

1,4-benzoquinone
bond dipoles farther apart

1,2-benzoquinone
bond dipoles neary aligned

A number of quinones are available naturally, Co-enzyme Q shown below in its oxidized state of ubiquinone, in an important factor in the respiratory chain localized in the mitochondrion that converts oxygen ultimately into water and synthesizes the energy thus to produce ATP, the universal "biochemical fuel" Dooxorubicin (adriamycin), isolated from a microorganism is an important anti-tumour drug.

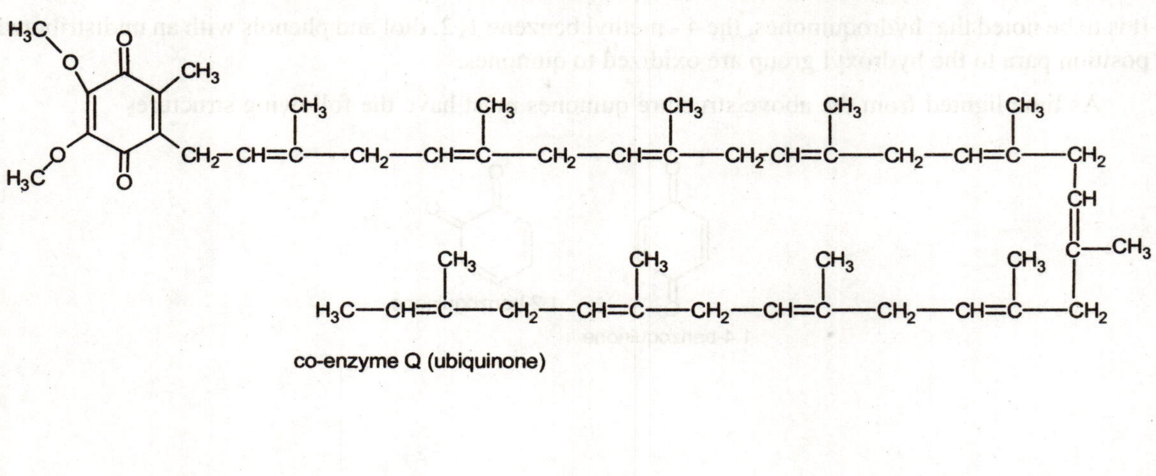

co-enzyme Q (ubiquinone)

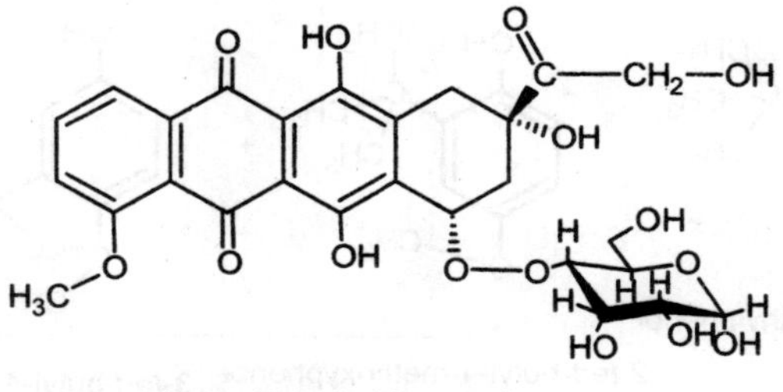

doxorubicin (adriamycin)

The oxidation of phenols by air (O_2) to coloured, quinone - containing products is the reaction responsible fort he darkening that is observed when some phenols are stored from a long time.

Practical Applications of Phenol Oxidation : The oxidation of phenols has several important practical applications. For example, phenols are sometimes used to inhibit free radical reactions that result in the oxidation of other compounds. The basis of this effect is that many free radicals. Abstract a hydrogen from hydroquinone to form a very stable radical called a semiquinone.

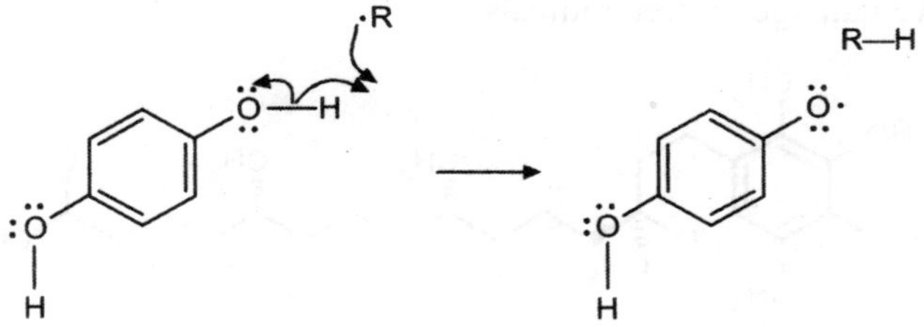

Semiquinone

(The semiquinone radical, like the benzyl radical, is resonance-stabilized, as shown in the following equation.) A second free radical can react with the semiquinone to complete its oxidation to quinone.

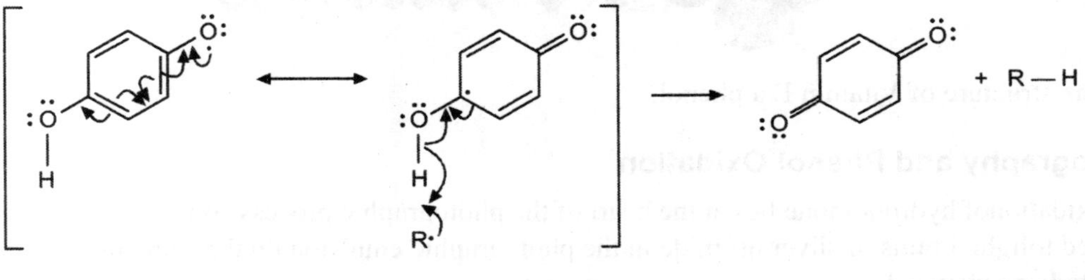

Hydroquinone thus terminates free-radical chain reactions by intercepting free-radical I intennediates R. and reducing them to RH.

The effectiveness of several widely used food preservatives is based on reactions such I as these. Examples of such preservatives are "butylated hydroxy toluene" (BHT) and "buty-lated hydroxy anisole" (BHA).

BHT
2,6-di-*tert*-butyl-4-methylphenol

2-*tert*-butyl-4-methoxyphenol 3-*tert*-butyl-4-methoxyphenol BHA

Oxidation involving free-radical processes is one way that foods discolor and spoil. A preservative such as BHT inhibits these processes by donating its OH hydrogen atom to free radicals in the food. The BHT is thus transformed into a phenoxy radical, which is too stable and unreactive to propagate radical chain reactions. Although the use of BHT and BHA as food additives has generated some controversy because of their potential side effects, without such additives foods could not be stored for any appreciable length of time or transported over long distances.

Recent research indicates that vitamin E, a phenol, is the major compound in the blood responsible for preventing oxidative damage by free radicals.

The structure of Vitamin E a phenol.

Photography and Phenol Oxidation

The oxidationof hydroquinone lies at the heart of the photographic process. When photographic film is exposed tolight, grains of silver bromide in the photographic emulsion on the film absorb light and are activated or sensitized.

$$AgBr + light \longrightarrow [AgBr]^*$$

sensitized
silver bromide

Because silver bromide is trapped in the photographic emulsion, it is immobile. Thus sensitized silver bromide molecules provide a faithful record of the positions on the film that have been struck by light. Now, sensitied silver bromide is a much better oxidazing agent than silver bromide that has not been exposed toligh. When exposed film is treated with a solution of hydroquinone (a common photographic developer), [AgBr]* oxidizes hydroquinone to p-benzoquinone (which is subsequent washed away), and the Ag(I) is reduced to finely divided silver metal [Ag(O)], which remains trapped in the photographic emulsion.

2[AgBr]* + HO–C6H4–OH ⟶ O=C6H4=O + 2Ag(black) + 2HBr

Hydroquinone   *p*-benzoquinone

Because unactivated AgBr oxidizes hydroquinone much more slowly, silver metal forms only where light has impinged on the film. This precipitated silver is the black part of a black and white negative.

Special Reactions

1. Condensation with Diazomium Salts: Phenol in alkaline solution undergoes coupling reaction with benzene diazomium chloride yielding Azo-dyes.

OH

phenol + benzene diazomium chloride (N=N–Cl) ⟶ p-hydroxy azo - benzene (HO, N=N) + HCl

2. Condensation Reaction with Phthalic Anhydride: Phenol reacts with Phthalic anhydride in the presence of sulphuric acid to give phenolphthalein.

phthalic anhydride + phenol (OH, H) + phenol (OH, H) $\xrightarrow[\text{conc } H_2SO_4]{\text{heat}}$ phenolphthalein + H_2O

(this we have read in the chapter 28)

3. Kolbe's Reaction: When dry sodium phenate is heated with dry CO_2 under pressure at 140°C sodium salicylate is produced. This on further treatment with HCl gives salicylic acid. The reaction is used fort the preparation of salicylic acid.

sodium phenolate + CO_2 —140°C→ sodium 2-carboxyphenolate —HCl→ salicylic acid

4. *Reimer Tienmann's Reaction:* **When heated with caustic alkalies and chloroform, phenol gives o - hydroxyl benzaldehyde or salicylaldehyde)**

sodium phenolate + 3KOH + $CHCl_3$ —60°C→ salicylaldehyde

If carbon tetrachloride is used instead of chloroform then salicylic acid is produced instead of the salicylaldehyde.

sodium phenolate —KOH→ potassium phenolate —CCl_4→ potassium 2-(trichloromethyl)phenolate —3KOH→ potassium 2-(trihydroxymethyl)phenolate → potassium 2-carboxyphenolate → salicylic acid

5. *Uffelman's Reaction:* **This reaction is done when ferric chloride and phenol or carbolic acid reacts to form a violet blue precipitate.***

***This reaction is used as the detection of phenols and used as Uffelman's Reagent for detecting lactic acid.**

OH

phenol + $FeCl_3$ ⟶ Fe^+ O Cl Cl Cl

ferroxytrichlorobenzene

7. ***Formaldehyde Resin:*** It produces formaldehyde resin, also known as Bakelite, which is hard substance and can be moulded (when hot) into a desired shape. It is a very important compound in plastic industry.

Test for Phenols

1. Liebermann's Reaction
2. Uffelman's Reaction
3. Azo dyes with diazomium salts
4. Condensation reaction with Phthalic anhydride producing Phenolphthalein

Uses of Phenol

1. The phenol has its typical use and carbolic acid in domestic houses
2. As in Bakelite
3. In insulation plastics
4. Preparation of picric acid
5. preparation of dyes
6. In lab as an analytical reagent.

Derivatives of Monohydric Phenols

Cresols

Here we are going to discuss about the derivatives of monohydric phenols mainly these derivatives are cresols.

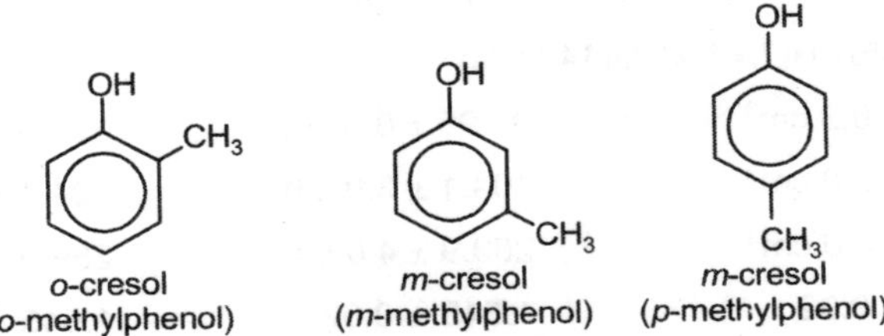

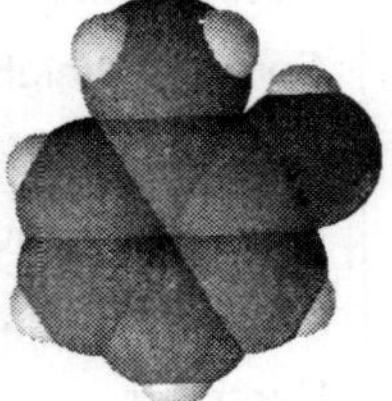

In its chemical structure, a cresol molecule has a methyl group substituted onto the benzene ring of a phenol molecule. There are three forms of cresols that are only slightly different in their chemical structure: ***ortho-cresol*** (o-cresol), ***meta-cresol*** (m-cresol), and ***para-cresol*** (p-cresol)

Manufacture

o, m and p occur in the middle oil with the phenol in the coal tar distillation and thus obtained as a mixture, after separation of phenol.

By efficient fractional distillation, o - cresol may be obtained practically free from m and p. Fractional crystallization of the residual mixture more or less pure m and p-cresols. More effectively by sulphonation of the mixture of m and p isomers with warm 94% sulphuric acid and cooling slowly p - cresolsulphonic acid crystallizes out and is separated by centrifugation. Pure p-cresol is obtained on distillation the sulphonic acid with superheated steam at 160°C and m-cresol from the mother liquor, after crystallization of the isomer by distillation with steam at 125°C.

Formation of p- and m cresol in 1: 1 ratio (but not o) on hydrolysis of p- chlorotoluene with aqueous NaOH at 340°C (followed by acidification) is of academic interest because it involves the benzene intermediate.

CH_3 / Cl — 1-chloro-4-methylbenzene $\xrightarrow[-HCl]{OH^-\ 340C}$ [CH_3] toluene intermediate $\xrightarrow{H_2O}$ CH_3 / OH p-cresol + CH_3 / OH m-cresol

Physical Properties

COMPOUND	o cresol	m cresol	p -cresol
Molecular Formula	C_7H_8O	C_7H_8O	C7H8O
Melting point	29.8°C	11.8°C	35.5°C
Boiling point	191.0°C	202.0°C	201.9°C
Formula Weight	108.13782	108.13782	108.13782
Composition	C(77.75%) H(7.46%) O(14.80%)		
Molar Refractivity	32.95 ± 0.3 cm^3	32.95 ± 0.3 cm^3	32.95 ± 0.3 cm^3
Molar Volume	104.1 ± 3.0 cm^3	104.1 ± 3.0 cm^3	104.1 ± 3.0 cm^3
Parachor	259.9 ± 4.0 cm^3	259.9 ± 4.0 cm^3	259.9 ± 4.0 cm^3
Index of Refraction	1.545 ± 0.02	1.545 ± 0.02	1.545 ± 0.02

Surface Tension	38.8 ± 3.0 dyne/cm	38.8 ± 3.0 dyne/cm	38.8 ± 3.0 dyne/cm
Density	1.038 ± 0.06 g/cm³	1.038 ± 0.06 g/cm³	1.038 ± 0.06 g/cm³
Polarizability	13.06 ± 0.5 10^{-24} cm³	13.06 ± 0.5 10-24cm³	13.06 ± 0.5 10^{-24} cm³
Monoisotopic Mass	108.057515 Da	108.057515 Da	108.057515 Da
Nominal Mass	108 Da	108 Da	108 Da
Average Mass	108.1378 Da	108.1378 Da	108.1378 Da

Chemical Properties

Cresols are practically insoluble in water, they are slightly weaker acids (Ka = 0.63 and 0.98 10-10 for o m and p cresols) than phenol (Ka 1.30×10^{-10}) due to electron repelling CH_3 group. All become blue with ethanoic ferric chloride, cresols are highly toxic (therefore used as bactericides) also corrosive for the skin o-cresol differs form m- and p in forming tri-bromotoluequinone, with bromine water; bromination and oxidation occur simultaneously m- cresol thus yields 2, 4, 6 tribromo-m-cresol.

o-cresol $\xrightarrow[H_2O]{Br_2}$ 2,3,5-tribromo-6-methyl -1,4-benzoquinone

m-cresol $\xrightarrow[H_2O]{Br_2}$ 2,4,6-tribromo-3-methylphenol

m cresol when reacts with phosphorous oxychloride tricresyl phosphate is produced.

o-cresol + phosphoric trichloride ⟶ tri cresyl phosphate

tricresyl phosphate is used as a plasticizer in nitrocellulose and acrylate lacquers and varnishes and in polyvinyl chloride, a flame retardant in plastics and rubbers, as a gasoline additive as a lead scavenger for tetra-ethyl lead, in hydraulic fluids, as a heat exchange medium, for waterproofing of materials, as a solvent for extractions, a solvent for nitrocellulose and other polymers, and an intermediate in organic synthesis. It is also used as an AW additive and EP additive in lubricants, and as a hydraulic fluid. As a gasoline additive, it also helps preventing engine misfires.

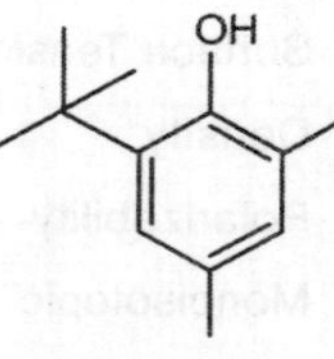

2,4-Dimethyl-6-tert-butylphenol is an alkylated phenol used industrially as an antioxidant, eg. to prevent gumming in fuels, and as an ultraviolet stabilizer. It is used in jet fuels, gasolines, and avgas. It is a clear liquid with melting point 21-23 °C and boiling point 248-249 °C. It is also used in printing industry. It is stable in acidic, neutral, and alkaline solutions, and is considered to not be readily biodegradable. It does not absorb UV light, therefore direct photodegradation is not suspected.

One of its trade names is **Topanol A**.

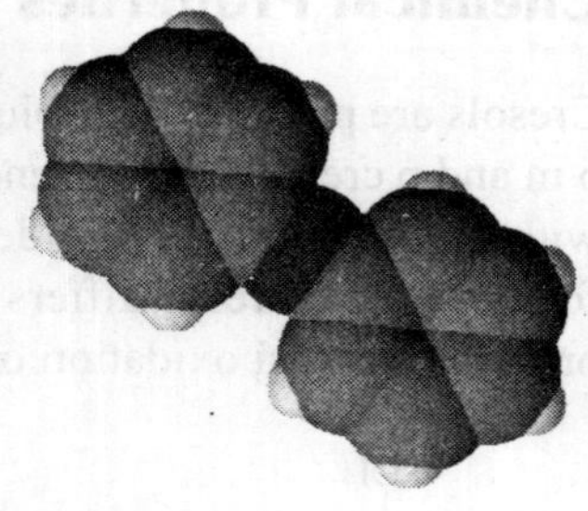

2-Phenylphenol, or o-phenylphenol, is an organic compound that consists of two linked benzene rings and a phenolic hydroxyl group. It is a white or buff-colored, flaky crystalline solid with a melting point of about 57°C. It is a biocide used as a preservative under the trade names Dowicide, Torsite, Preventol, Nipacide and many others.

The primary use of 2-phenylphenol is as an agricultural fungicide. It is generally applied post-harvest. It is a fungicide used for waxing citrus fruits. As a food additive, it has E number E231.

It is also used for disinfection of seed boxes. It is a general surface disinfectant, used in households, hospitals, nursing homes, farms, laundries, barber shops, and food processing plants. It can be used on fibers and other materials. It is used to sterilize hospital and veterinary equipment. Other uses are in rubber industry and as a laboratory reagent. It is also used in the manufacture of other fungicides, dye stuffs, resins and rubber chemicals.

2-Phenylphenol is found in low concentrations in some household products such as spray disinfectants and aerosol or spray underarm deodorants.

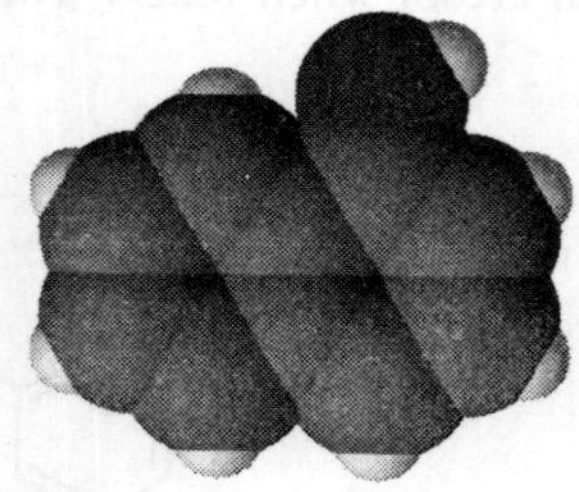

Naphthol, or hydroxynaphthalene or naphthalenol, is either of two colorless crystalline solids with the formula $C_{10}H_7OH$. They are positional isomers differing by the location of the hydroxyl group on naphthalene. β-Naphthol is naphthalen-1-ol, and α-naphthol is naphthalen-2-ol. Naphthol is the naphthalene homologue of phenol, with the hydroxyl group being more reactive than in the phenols. They are soluble in simple alcohols, ethers, and chloroform. They can be used in the production of dyes and further organic synthesis. For example, β-naphthol reacts to form BINOL.

2 OH $\xrightarrow[\text{(+)amphetamine}]{CuCl_2}$ OH OH S-BINOL

Metol (or Elon, in Eastman Kodak parlance) is a developing agent used in photographic developers. In its pure form, it is a solid rather light-sensitive chemical which is the sulfate salt of para-(methylamino) phenol. It is still also commonly known by its old chemical name *monomethyl-p-aminophenol sulfate.*

$$\left(HO-C_6H_4-\overset{+}{N}H_2-CH_3\right)_2 SO_4^{2-}$$

Because it has been in use for this purpose for over 100 years and often by amateur photographers, there is a wide body of evidence about the health problems that contact with Metol can cause. These are principally local dermatitis of the hands and fore-arms as well as some evidence of sensitization dermatitis in which subsequent exposure triggers of a chronic condition that is resistant to medication. The use of Metol in highly caustic solutions and the presence of other materials in dark-rooms that have been implicated in dermatitis such as Cr(VI) salts, may exacerbate some health impacts.

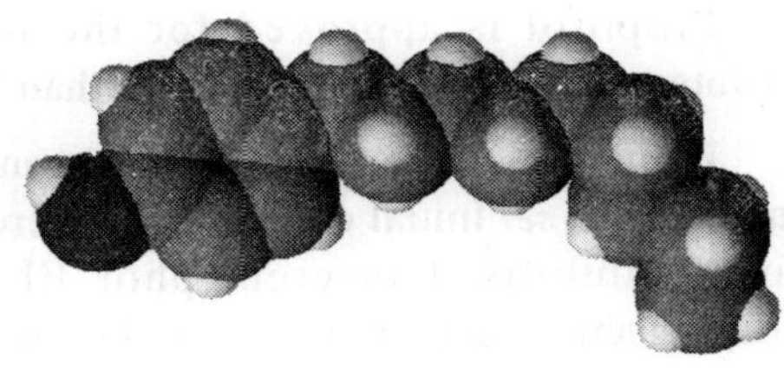

Nonylphenol is an organic compound of the wider family of alkylphenols. It is a product of industrial synthesis formed during the alkylation process of phenols, particularly in the synthesis of polyethoxylate detergents. Because of their man-made origins, nonylphenols are classified as xenobiotics. In nonylphenols, a hydrocarbon chain of nine carbon atoms is attached to the phenol ring in either the *ortho* (2), *meta* (3), or *para* (4) position, with the most common ring isomers being *ortho* or *para* (e.g. figure 1 para-nonylphenol). Moreover, the alkyl chains can exist as either linear n-alkyl chains, or complex branched chains. Nonylphenol is commonly obtained as a mixture of isomers, and is thus usually found as a pale yellow liquid at room temperature with a melting point of -10°C and a boiling point of 295-320°C. However, pure isomers of nonylphenol crystallize readily at room temperatures and for example, para-*n*-nonylphenol, forms white crystals at room temperature.

Ethoxylated alkylphenols, alkylphenol ethoxylates (APE), are used as industrial surfactants in manufacture of wool and metal, as emulsifiers for emulsion polymerization, in laboratory detergents, and pesticides. APEs are a component of some household detergents outside of Europe; within Europe, due to environmental concerns, they are replaced by more expensive but safer alcohol ethoxylates. Nonoxynol-9, one of the APEs, is used as a surfactant in cleaning and cosmetic products, and as a spermicide in contraceptives.Nonylphenol, and a related compound tert-octylphenol, were first detected as an air pollutant in New York City and New Jersey, probably due to its evaporation from the Hudson river and other smaller rivers in the region that routinely receive municipal wastewaters. It is possible that the atmosphere is a destructive sink for nonylphenol as it is probably reactive with atomspheric radicals and/or is photoactive.

Nonylphenol and nonyphenol ethoxylates have been banned in the European Union as a hazard to human and environmental safety.Biochemically, p-nonylphenol and many of its derivatives act as a xenoestrogen.

Propofol is a short-acting intravenous anesthetic agent used for the induction of general anesthesia in adult patients and pediatric patients older than 3 years of age; maintenance of general anesthesia in adult patients and pediatric patients older than 2 months of age; and sedation in medical contexts, such as intensive care unit (ICU) sedation for intubated, mechanically ventilated adults, and in procedures such as colonoscopy.

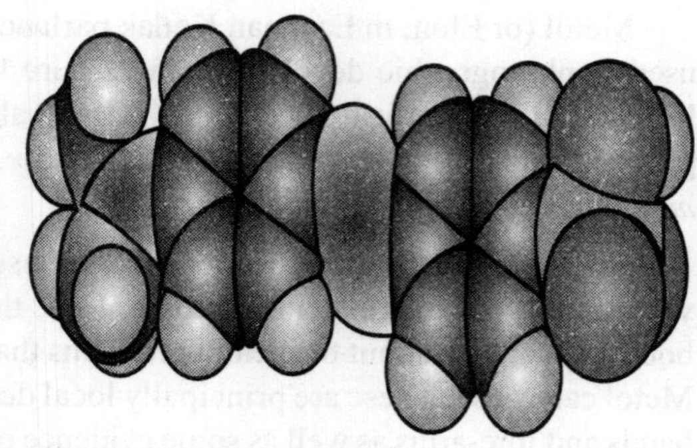

Propofol is approved for the induction and maintenance of anesthesia in more than 50 countries.

Propofol is a water-immiscible oil and so cannot be injected *per se*. Initial clinical trials were in 1977, in a form solubilised in cremophor EL, but due to anaphylactic reactions it was withdrawn from the market. It was subsequently reformulated as an emulsion of a soya oil/propofol mixture in water, and re-launched in 1986 by AstraZeneca with the brand name **Diprivan (Shortened version of DI-isoPRopyl IV ANesthetic)**. The current preparation is 1% propofol, 10% soybean oil and 1.2% purified egg phospholipid (emulsifier), with 2.25% of glycerol as a tonicity adjusting agent, and sodium hydroxide to adjust the pH. Diprivan contains EDTA as an antimicrobial agent. Newer generic formulations contain sodium metabisulfite or benzyl alcohol. Propofol emulsion appears as a highly opaque white fluid due to the scattering of light from the tiny (~150 nm) oil droplets that it contains.

Pharmacology

Propofol is highly protein bound *in vivo* and is metabolised by conjugation in the liver. Its rate of clearance exceeds hepatic blood flow, suggesting an extrahepatic site of elimination as well. Its mechanism of action is uncertain, but it is postulated that its primary effect may be potentiation of the GABA-A receptor, possibly by slowing the closing channel time. Recent research has also suggested the endocannabinoid system may contribute significantly to Propofol's anesthetic action and to its unique properties.[1]

The elimination half-life of propofol has been estimated to be between 2-24 hours. However, its duration of clinical effect is much shorter because propofol is rapidly distributed into peripheral tissues.

Sesamol is a natural organic compound which is a component of sesame oil. It is a white crystalline solid that is a derivative of phenol. It is sparingly soluble in water, but miscible with most oils.Sesamol has been found to be an antioxidant that may prevent the spoilage of oils, and may protect the body from damage from free radicals.[3] It also may prevent the spoilage of oils by acting as an antifungal. Sesamol is used as a chemical intermediate in the industrial synthesis of the pharmaceutical drug paroxetine (Paxil).

Syringol is a dimethyl ether of pyrogallol. It is slightly soluble in water. It is combustible, with flash point of 140 °C.

Together with guaiacol, syringol (and its derivates) is a characteristic product of pyrolysis of lignin. Its presence in smoke is characteristical for wood smoke. In preparation of food by smoking, syringol is the main chemical responsible for the smoky aroma, while guaiacol contributes mainly to taste.

Thymol is a monoterpene phenol derivative of cymene, $C_{10}H_{13}OH$, isomeric with carvacrol, found in oil of thyme, and extracted as a white crystalline substance of a pleasant aromatic odor and strong antiseptic properties. It is also called "hydroxy cymene". It has been found to be useful in controlling varroa mites in bee colonies. A minor use is in bookbinding: before rebinding, books with mould damage can be sealed in bags with thymol crystals to kill fungal spores. It is also used as a preservative in halothane, an anaesthetic.

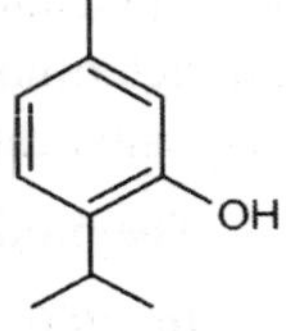

Chemical name	5-methyl-2-(1-methylethyl)phenol
Chemical formula	$C_{10}H_{14}O$
Molecular mass	150.22 g/mol
Density	0.96 g/cm^3
Melting point	48-52 °C
Boiling point	232 °C

Paracetamol

History: In ancient and medieval times, known antipyretic agents were compounds contained in white willow bark (a family of chemicals known as salicins, which led to the development of aspirin), and compounds contained in cinchona bark. Cinchona bark was also used to create the anti-malaria drug quinine. Quinine itself also has antipyretic effects. Efforts to refine and isolate salicin and salicylic acid took place throughout the middle- and late-19th century, and was accomplished by Bayer chemist Felix Hoffmann (this was also done by French chemist Charles Frédéric Gerhardt 40 years earlier, but he abandoned the work after deciding it was too impractical).

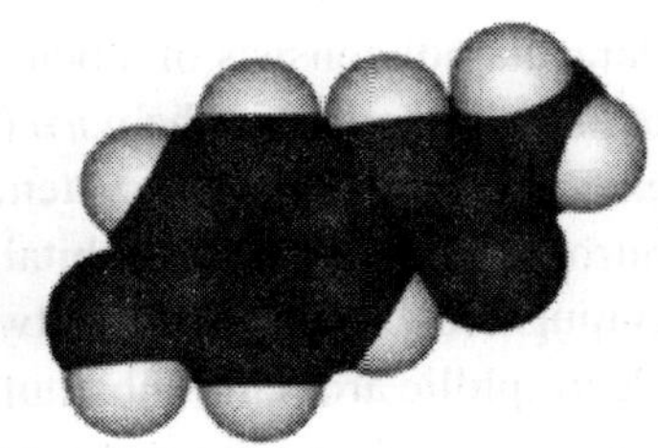

When the cinchona tree became scarce in the 1880s, people began to look for alternatives. Two alternative antipyretic agents were developed in the 1880s: acetanilide in 1886 and phenacetin in 1887. By this time, paracetamol had already been synthesized by Harmon Northrop Morse via the reduction of *p*-nitrophenol with tin in glacial acetic acid. While this was first performed in 1873, paracetamol was not used medically for another two decades. In 1893, paracetamol was discovered in the urine of individuals who had taken phenacetin, and was concentrated into a white, crystalline compound with a bitter taste. In 1899, paracetamol was found to be a metabolite of acetanilide. This discovery was largely ignored at the time.

In 1946, the Institute for the Study of Analgesic and Sedative Drugs awarded a grant to the New York City Department of Health to study the problems associated with analgesic agents. Bernard Brodie and Julius Axelrod were assigned to investigate why non-aspirin agents were associated with the development of methemoglobinemia, a condition that decreases the oxygen-carrying capacity of blood and is potentially lethal. In 1948, Brodie and Axelrod linked the use of acetanilide with methemoglobinemia and determined that the analgesic effect of acetanilide was due to its active metabolite paracetamol. They advocated the use of paracetamol, since it did not have the toxic effects of acetanilide.

The product went on sale in the United States in 1955 under the brand name Tylenol.

In 1956, 500 mg tablets of paracetamol went on sale in the United Kingdom under the trade name **Panadol**, produced by Frederick Stearns & Co, a subsidiary of Sterling Drug Inc. Panadol was originally available only by prescription, for the relief of pain and fever, and was advertised as being "gentle to the stomach," since other analgesic agents of the time contained aspirin, a known stomach irritant. In June 1958 a children's formulation, **Panadol Elixir**, was released.

In 1963, paracetamol was added to the *British Pharmacopoeia*, and has gained popularity since then as an analgesic agent with few side-effects and little interaction with other pharmaceutical agents.

The U.S. patent on paracetamol has expired and generic versions of the drug are widely available under the Drug Price Competition and Patent Term Restoration Act of 1984, although certain Tylenol preparations are protected until 2007. U.S. patent 6,126,967 filed September 3, 1998 was granted for "Extended release acetaminophen particles."

Structure and Reactivity

Paracetamol consists of a benzene ring core, substituted by one hydroxyl group and the nitrogen atom of an amide group in the *para* (1,4) pattern. The amide group is in fact acetamide (ethanamide). It is an extensively conjugated system, as the lone pair on the hydroxyl oxygen, the benzene pi cloud, the nitrogen lone pair, the p orbital on the carbonyl carbon and the lone pair on the carbonyl oxygen are all conjugated. The presence of two activating groups also make the benzene ring highly reactive towards electrophilic aromatic substitution. As the substituents are ortho,para directing and para with respect to each other, all positions on the ring are more or less equally activated. The conjugation also greatly reduces the basicity of the oxygens and the nitrogen, while making the hydroxyl acidic through delocalisation of charge developed on the phenoxide anion.

Triclosan (chemically 5-chloro-2-(2,4-dichlorophenoxy)phenol) is a potent wide spectrum antibacterial and antifungal agent.

Chemistry

This organic compound is a white powdered solid with a slight aromatic/ phenolic odor. It is a chlorinated aromatic compound which has functional groups representative of both ethers and phenols. Phenols often show anti-bacterial properties. Triclosan is slightly soluble in water, but soluble in ethanol, diethyl ether, and stronger basic solutions such as 1 M sodium hydroxide, like many other phenols.

Cl OH O Cl Cl

Uses

It is found in soaps, deodorants, toothpastes, mouthwashes, and cleaning supplies and is infused in an increasing number of consumer products, such as kitchen utensils, toys, bedding, socks, and trash bags, sometimes as the proprietary Microban treatment. It has been shown to be effective in reducing and controlling bacterial contamination on the hands and on treated products. More recently, showering or bathing with 2% triclosan has become a recommended regime for the decolonization of patients whose skin is carrying methicillin resistant *Staphylococcus aureus* (MRSA) (PMID 16581155) following the successful control of MRSA outbreaks in several clinical settings. (PMID 2255283) (PMID 7677266)

Triclosan is regulated by both the U.S. Food and Drug Administration and by the European Union. In the environment, triclosan is removed during normal waste treatment processes as shown by extensive environmental studies, and any of it that remains after waste treatment quickly breaks down into other compounds in the environment. However, one study showed that triclosan was broken down into dioxins in river water, because of the presence of sunlight.

Chavicol, or p-allylphenol, is a natural organic compound. Its chemical structure consists of a benzene ring substituted with a hydroxy group and a propenyl group. It is a colorless liquid found together with terpenes in betel oil. It is miscible with alcohol, ether, and chloroform.

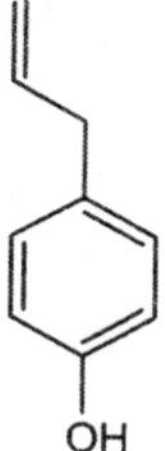

Chavicol is used as an odorant in perfumery.

Physical Properties

Chemical name	4-(2-propenyl)phenol
Chemical formula	$C_9H_{10}O$
Molecular mass	134.18 g/mol
Density	1.020 g/cm^3
Melting point	16°C
Boiling point	238°C 123°C at 16 mmHg

Carvacrol, or **cymophenol**, $C_6H_3CH_3(OH)(C_3H_7)$, is a constituent of the ethereal oil of *Origanum hirtum*, oil of thyme, oil obtained from pepperwort, and wild bergamot. It may be synthetically prepared by the fusion of cymol sulfonic acid with caustic potash; by the action of nitrous acid on 1-methyl-2-amino-4-propyl benzene; by prolonged heating of five parts of camphor with one part of iodine; or by heating carvol with glacial phosphoric acid or by preforming a dehydrogenation of carvone with a Pd/C catalyst. It is extracted from Origanum oil by means of a 50% potash solution. It is a thick oil which sets at 20°C to a mass of crystals of melting point 0°C, and boiling point 236-237°C. Oxidation with ferric chloride converts it into dicarvacrol, whilst phosphorus pentachloride transforms it into chlorcymol.

Physical Properties

Chemical name	2-methyl-5-(1-methylethyl)phenol
Chemical formula	$C_{10}H_{14}O$
Molecular mass 1	50.22 g/mol
Melting point	0 °C
Boiling point	236-237 °C

Butylated hydroxyanisole (BHA) is a mixture of two isomeric organic compounds, 2-*tert*-butyl-4-hydroxyanisole and 3-*tert*-butyl-4-hydroxyanisole. It is prepared from 4-methoxyphenol and isobutylene. It is a waxy solid that exhibits antioxidant properties.

Like butylated hydroxytoluene (BHT), the conjugated aromatic ring of **BHA** is able to stabilize free radicals, sequestering them. By acting as free radical scavengers, further free radical reactions are prevented.

Evidence that BHA is a carcinogen has been obtained from animal trials.

Butylated hydroxytoluene (BHT) is a fat-soluble organic compound primarily used as an antioxidant food additive (E number E321). It is also used as an antioxidant in cosmetics, pharmaceutical drugs, jet fuels, rubber, petroleum products, and embalming fluid.

BHT is produced by the reaction of *p*-cresol with isobutylene. It was patented in 1947 and received approval of the Food and Drug Administration for use as a food additive and preservative in 1954. BHT reacts with free radicals, slowing the rate of autoxidation in food, preventing changes in the food's color, odor, and taste.

In the chemical industry it is added to tetrahydrofuran and diethyl ether in order to inhibit the formation of dangerous organic peroxides.

Controversy

Concerns have been raised about the use of BHT in food products. The compound has been banned for use in food in Japan (1958), Romania, Sweden, and Australia. The US has barred it from being used in infant foods. However, some food industries have voluntarily eliminated it from their products, including

Dihydroxy Phenols

Benzenediols or **dihydroxybenzenes** are aromatic chemical compounds in which two hydroxyl groups are substituted onto a benzene ring. Because they have at least one hydroxyl group covalently bonded directly to a carbon atom in a benzene ring, they are in a class of organic compounds called phenols. There are three isomers of bezenediol, each of which has its particular name as shown in the mini-table below. Various other ways of naming these three chemical compounds are also shown:

ortho isomer	meta isomer	para isomer
Pyrocatechol	Resorcinol	Hydroquinone
1,2-benzenediol	1,3-benzenediol	1,4-benzenediol
o-benzenediol	*m*-benzenediol	*p*-benzenediol
1,2-dihydroxybenzene	1,3-dihydroxybenzene	1,4-dihydroxybenzene
o-dihydroxybenzene	*m*-dihydroxybenzene	*p*-dihydroxybenzene
catechol	resorcin	

All three of these compounds are colorless to white granular solids at room temperature and pressure, but upon exposure to oxygen they may darken. All three isomers have the chemical formula $C_6H_6O_2$. The chemical structures are shown here:

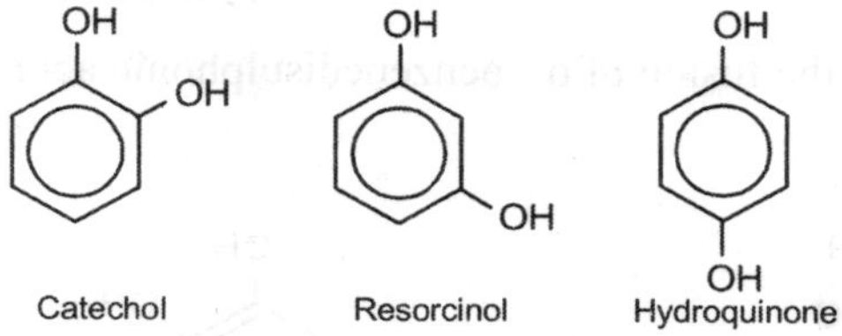

OH OH OH OH OH OH

Catechol Resorcinol Hydroquinone

Similar to other phenols, the hydroxyl groups on the aromatic ring of a benzenediol are weakly acidic. Each benzenediol can lose an H^+ from one of the hydroxyls to form a type of phenolate ion. Hydroquinone can lose an H^+ from both to form a diphenolate ion. The disodium diphenolate salt of hydroquinone is used as an alternating comonomer unit in the production of the polymer PEEK.

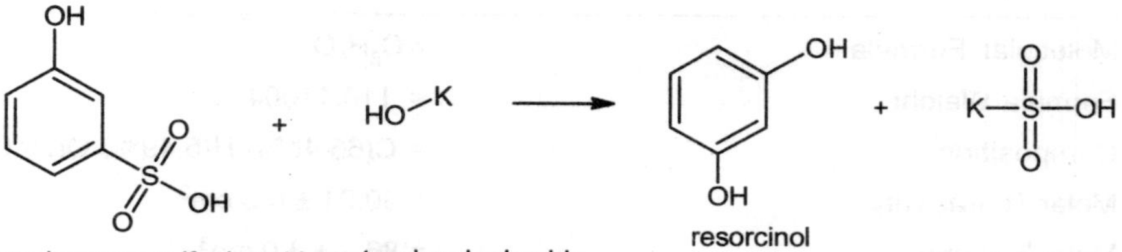

OH + HO—K → OH + K—S(=O)(=O)—OH

3-hydroxybenzenesulfonic acid potassium hydroxide resorcinol

According to the relative position of the two hydroxyl groups the di hydric phenols show characteristic differences in chemical behaviour.

The *ortho* compounds frequently give green colouration to ferric chloride and readily yields heterocyclic derivatives in groups are replaced by a divalent radical.

The meta compounds usually give a deep violet colouration, with ferric chloride, and undergo the flurescein reaction, i.e. when heated with Phthalic anhydride they produce thaleins shows a green flurescein in alkaline solution.

The para compounds on oxidation very easily pass into quinones which are readily recognized. Below some dihydric phenols are being explained with their relevant information

Catechol, It occurs in living and as well as fossil plants, thus it has been obtained from catechin, moringatannic acid, kinotannic acid and all those sources containing tannic acid, it is seen that cathecol is also available in crude wood tar, crude beet sugar and tar waters from bituminous coal.

Preparation

Catechol is generally prepared by the oxidation of phenol with hydrogen peroxide can carbon

3 phenol + hydrogen peroxide $\xrightarrow{\text{Carbon}}$ 2 pyrocatechol + benzaldehyde

Catechol is also prepared by the fusion of o - benzenedisulphonic acid and alkali

benzene-1,2-disulfonic acid + 2NaOH ⟶ catechol + $Na_2S_2O_5$ + H_2O

Physical Properties

Molecular Formula	= $C_6H_6O_2$
Formula Weight	= 110.11064
Composition	= C(65.45%) H(5.49%) O(29.06%)
Molar Refractivity	= 30.01 ± 0.3 cm^3
Molar Volume	= 86.2 ± 3.0 cm^3
Parachor	= 237.3 ± 4.0 cm^3
Index of Refraction	= 1.612 ± 0.02

Surface Tension	= 57.1 ± 3.0 dyne/cm
Density	= 1.275 ± 0.06 g/cm^3
Polarizability	= 11.89 ± 0.5 10^{-24} cm^3
Monoisotopic Mass	= 110.036779 Da
Nominal Mass	= 110 Da
Average Mass	= 110.1106 Da

Chemical Properties

Catechol reacts with ferric chloride producing a green coloured complex compound $[Fe(C_6H_4O_2)_3H_3]$

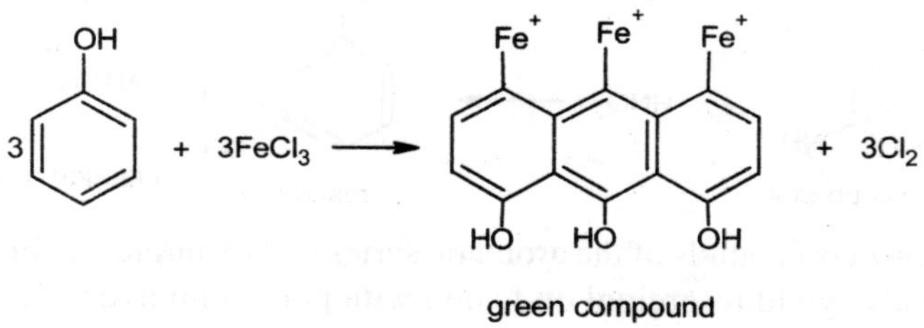

$$3\,C_6H_5OH + 3FeCl_3 \longrightarrow \text{green compound} + 3Cl_2$$

green compound

This compound changes to red colour in alkaline solution

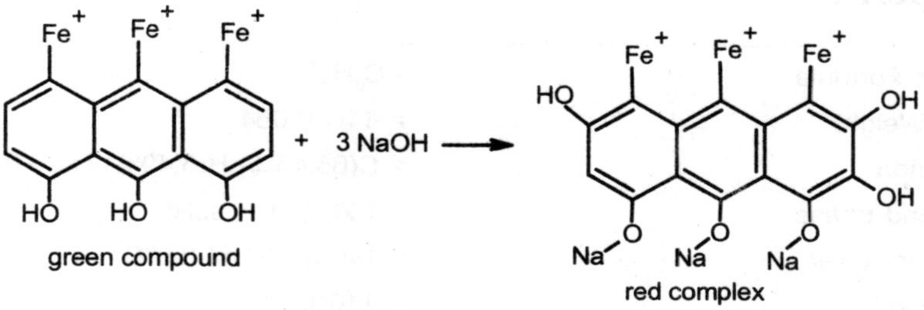

$$\text{green compound} + 3\,NaOH \longrightarrow \text{red complex}$$

green compound

red complex

The other phenols and phenol - carboxylic acids also give deep colouration with ferric chloride in alkaline solutions.

Resorcinol Also known as m - dihydroxybenzene

Resorcinol (or **resorcin**) is a chemical compound from the dihydroxy phenols. it is the 1,3-isomer of benzenediol. It is also known with a variety of other names, including: m-dihydroxybenzene, 1,3-benzenediol, 1,3-dihydroxybenzene, 3-hydroxyphenol, m-hydroquinone, m-benzenediol, and 3-hydroxycyclohexadien-1-one.

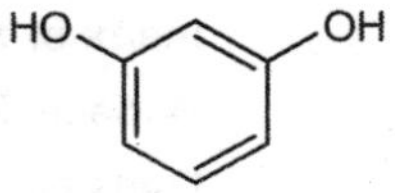

Preparation

It is obtained on fusing many resins (galbanum, asafoetida, etc.) with potassium hydroxide, or by the distillation of Brazilwood extract.(the name resorcinol thus come into knowledge) It may be prepared synthetically by fusing 3-iodophenol, phenol-3-sulfonic acid or benzene-1,3-disulfonic acid with potassium carbonate;

benzene-1,2-disulfonic acid + K_2CO_3 → resorcinol + $K_2S_2O_4$ + CO_2

By the action of nitrous acid on 3-aminophenol;

3-aminophenol + HNO_2 → resorcinol + nitrogen oxide

Many *ortho-* and *para-*compounds of the aromatic series (for example, the bromophenols, benzene-*para*-disulphonic acid) also yield resorcinol on fusion with potassium hydroxide.

Physical Properties

Molecular Formula	= $C_6H_6O_2$
Formula Weight	= 110.11064
Composition	= C(65.45%) H(5.49%) O(29.06%)
Density and phase	= 1.28 g/cm^3, solid
Solubility in water	= 140 g/100 ml (50°C)
Melting point	= 110°C
Boiling point	= 280°C
Molar Refractivity	= 30.01 ± 0.3 cm^3
Molar Volume	= 86.2 ± 3.0 cm^3
Parachor	= 237.3 ± 4.0 cm^3
Index of Refraction	= 1.612 ± 0.02
Surface Tension	= 57.1 ± 3.0 dyne/cm
Density	= 1.275 ± 0.06 g/cm^3
Polarizability	= 11.89 ± 0.5 10-24cm^3
Monoisotopic Mass	= 110.036779 Da
Nominal Mass	= 110 Da
Average Mass	= 110.1106 Da

Chemical Properties

Resorcinol crystallizes from benzene as colorless needles which are readily soluble in water, alcohol and ether, but insoluble in chloroform and carbon disulfide. It reduces Fehling's solution, and ammoniacal silver solutions. It does not form a precipitate with lead acetate solution, as the isomeric pyrocatechol does. Iron(III) chloride colors its aqueous solution a dark violet, and bromine water precipitates tribromoresorcin. Sodium amalgam reduces it to dihydroresorcin, which when heated to 150 to 160 °C with concentrated barium hydroxide solution gives γ-acetylbutyric acid; when fused with potassium hydroxide, resorcinol yields phloroglucin, pyrocatechol and diresorcin. It condenses with acids or acid chlorides, in the presence of dehydrating agents, to oxyketones, e.g. with zinc chloride and glacial acetic acid at 145°C it yields resacetophenone $(HO)_2C_6H_3 = CO.CH_3$. With the anhydrides of dibasic acids it yields fluoresceins. When heated with calcium chloride-ammonia to 200°C it yields meta-dioxydiphenylamine. With sodium nitrite it forms a water-soluble blue dye, which is turned red by acids, and is used as an indicator, under the name of lacmoid. It condenses readily with aldehydes, yielding with formaldehyde, on the addition of catalytic hydrochloric acid, methylene diresorcin $[(HO)C_6H_3(O)]_2oCH_2$, whilst with chloral hydrate, in the presence of potassium bisulfate, it yields the lactone of tetra-oxydiphenyl methane carboxylic acid.

Resorcinol gives green colouration with ferric chloride as Catechol. It yields flurescein on being heated with phthaleic anhydride.

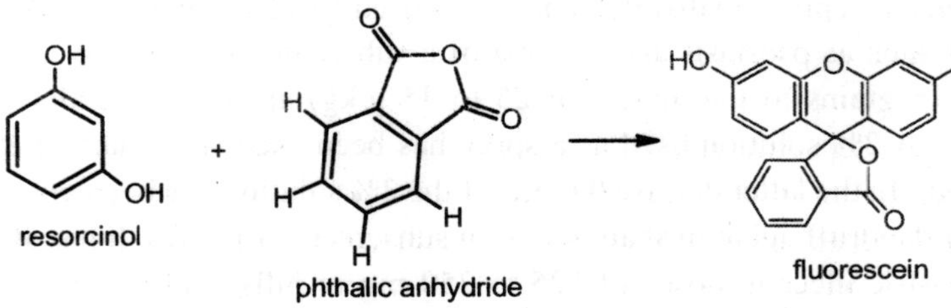

resorcinol + phthalic anhydride → fluorescein

With cold nitric acid yields a trinitro derivative also known as styphnic acid.

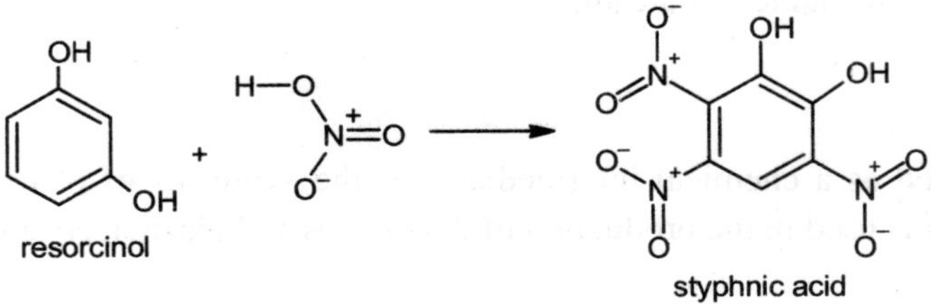

resorcinol + nitric acid → styphnic acid

It has been shown that the action of ethyl iodide and potassium hydroxide on resorcinol leads to the formation of resorcinol diethyl ethyl ether accompanied by not inconsiderable amounts of tri and tetraethyl derivatives of resorcinol. The latter compound possesses the structure 1 thus proving that, during the ethylation, of the resorcinol reacted in the keto - enolic form. In other words, we have here an example of the tautomerism of resorcinol.

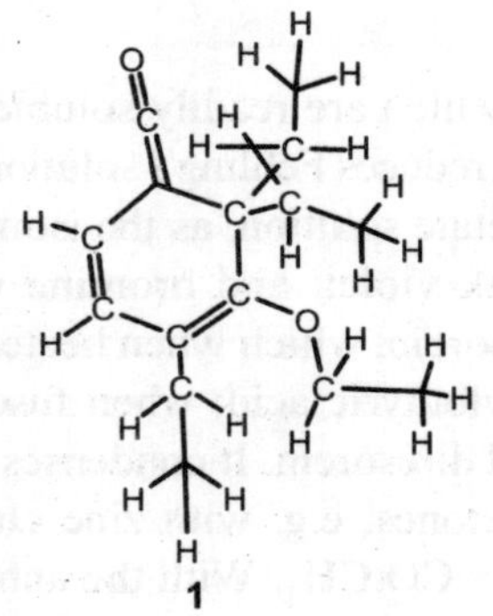

Resorcinol combines with sodium bisulphate to give a product which is in all probability the bisulphate compound of 3: 5 diketohexamethylene-I-sulphonic acid. The formation of this substance may be derived from the tautomeric form of resorcinol 2.

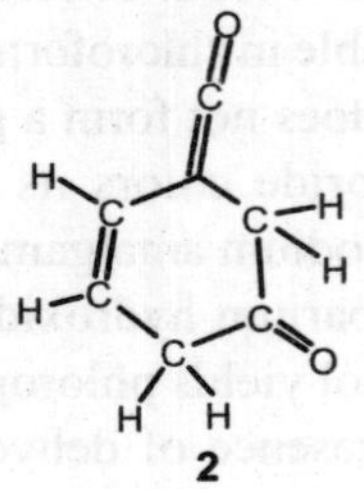

Applications

Medical

Used externally it is an antiseptic and disinfectant, and is used 5 to 10% in ointments in the treatment of chronic skin diseases such as psoriasis and eczema of a sub-acute character. Weak, watery solutions of resorcinol (10 or 15 grains to the ounce, or 25 to 35 g/kg) are useful in allaying the itching in erythematous eczema. A 2% solution used as a spray has been used with marked effect in hay fever and in whooping cough. In the latter disease 0.6 mL of the 2% solution has been given internally. It can be included as an anti-dandruff agent in shampoo or in sunscreen cosmetics. It has also been employed in the treatment of gastric ulcer in doses of 125 to 250 mg in pills, and is said to be analgesic and haemostatic in its action. In large doses it is a poison causing giddiness, deafness, salivation, sweating and convulsions. It is also worked up in certain medicated soaps. Monoacetylresorcinol, $C_6H_4(OH)(O\text{-}COCH_3)$, is used under the name of euresol.

Chemical

Resorcinol is also used as a chemical intermediate for the synthesis of pharmaceuticals and other organic compounds. It is used in the production of diazo dyes and plasticizers and as a UV absorber in resins.

An emerging use of resorcinol is as a template molecule in supramolecular chemistry. The -OH groups on resorcinol form hydrogen bonds to target molecules holding them in the proper orientation for a reaction. Many such reactions are able to be carried out in the solid state thereby reducing or eliminating the use of solvents that may be harmful to the environment.)

Resorcinol is the starting material for resorcinarene molecules.

1,1'-Bi-2-naphthol (BINOL) is an organic compound that is often used as a ligand for transition-metal catalysed asymmetric synthesis. BINOL has axial chirality and the two enantiomers can be readily separated and are stable toward racemisation. The specific rotation of the two enatiomers is +/ – 35.5° (c=1 in THF). It is a precursor for another chiral ligand called BINAP.

OH
OH

Preparation

The organic synthesis of BINOL is not a challenge as such but the preparation of the individual enantiomers is.

(*S*)-BINOL can be prepared directly from an asymmetric oxidative coupling of 2-naphthol with copper(II) chloride. The chiral ligand in this reaction is (+)-amphetamine.

2 (2-naphthol, OH) — $CuCl_2$ / (+)amphetamine → S-BINOL

Racemic BINOL can also be produced using iron(III) chloride as an oxidant. The mechanism involves complexation of iron(III) into the hydroxyl, followed by a radical coupling reaction of the naphthol rings initiated by iron(III) oxidizing into iron(II).

Optically active binol can also be obtained from racemic binol by optical resolution. In one method, the alkaloid N-benzylcinchonidinium chloride form a crystalline inclusion compound. The inclusion compound of the *S*-enantiomer is soluble in acetonitrile but that of the *R* enantiomer is not.

In another method binol is reacted with the acid chloride pentanoyl chloride to obtain the di-ester compound. The enzyme cholesterol esterase is then added in the form of bovine pancreas acetone powder which is able to hydrolize the (*S*)-di-ester but not the (*R*)-di-ester. The (*R*)-dipentanoate is hydrolysed in a second step with sodium methoxide.

Bisphenol A was first synthesized by A.P. Dianin in 1891. Bisphenol A was investigated in the 1930s during the search for synthetic estrogens. At that time, another synthesized compound, diethylstilbestrol, turned out to be more powerful an estrogen, so bisphenol A was not used as a synthetic estrogen.

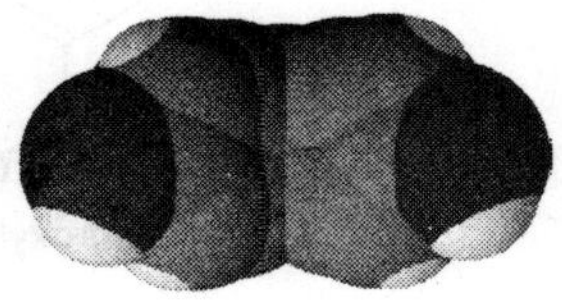

It is prepared by reaction of two equivalents of phenol with one equivalent of acetone.

2 phenols + acetone → 4,4'-propane-2,2-diyldiphenol

Its current uses are as a primary monomer in polycarbonate plastic and epoxy resins. Bisphenol A is also used as an antioxidant in plasticizers and as a polymerization inhibitor in PVC.

Polycarbonates are widely used in many consumer products, from sunglasses and CDs to water and food containers and shatter-resistant baby bottles. Some polymers used in dental fillings also contain bisphenol A, while epoxy resins containing bisphenol A are popular coatings for the inside of cans used for canning food.

Orcin is 3,5-dihydroxytoluene, $C_6H_3(CH_3)(OH)_2$, found in many lichens including *Rocella tinctoria* and *Lecanora*, and formed by fusing extract of aloes with potash.

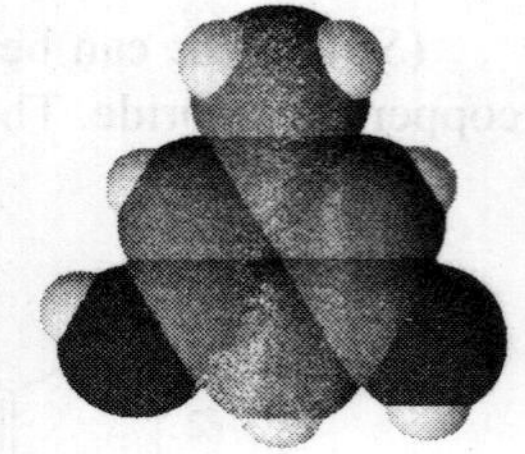

It may be synthesized from toluene; more interesting is its production when acetone dicarboxylic ester is condensed with the aid of sodium. It crystallizes in colorless prisms with one molecule of water, which redden on exposure. Ferric chloride gives a bluish-violet coloration with the aqueous solution. Unlike resorcinol it does not give a fluorescein with phthalic anhydride. Oxidation of the ammoniacal solution gives orcein, $C_{28}H_{24}N_2O_7$, the chief constituent of the natural dye archil. Homo-pyrocatechin is an isomer (CH1: OH: OH= 1 3 :4), found as its methyl ether (creosol) in beech-wood tar.

Tri and Polyhydric Phenols

Pyrogallol or **benzene-1,2,3-triol** is a white crystalline powder and a powerful reducing agent. It was first prepared by Scheele 1786 by heating gallic acid (as the name suggests pyros = heat, gallol = gallic acid). An alternate preparation is heating para-chlorophenoldisulphonic acid with potassium hydroxide.

2,3,4-trihydroxybenzoic acid →(Δ) benzene-1,2,3-triol

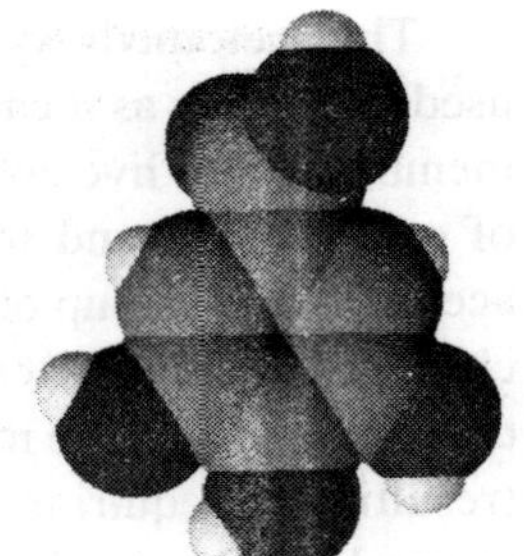

Gallic acid is an organic acid, also known as 3,4,5-trihydroxybenzoic acid, found in gallnuts, sumac, witch hazel, tea leaves, oak bark, and other plants. The chemical formula is $C_6H_2(OH)_3CO_2H$. Gallic acid is found both free and as part of tannins. It is commonly used in the pharmaceutical industry. Gallic acid can also be used to synthesize the hallucinogenic alkaloid, mescaline, also known as 3,4,5-trimethoxyphenethylamine.

Salts and esters of gallic acid are termed gallates.

Gallic acid was one of the substances used by Angelo Mai among other early investigators of palimpsests to clear the top layer of text off and reveal hidden manuscripts underneath. Mai was the first to employ it, but did so "with a heavy hand", often damanging manuscripts for future study.

Physical Properties

Molecular Formula	= $C_7H_6O_5$
Formula Weight	= 170.11954
Composition	= C(49.42%) H(3.55%) O(47.02%)
Molar Refractivity	= 38.82 ± 0.3 cm^3
Molar Volume	= 97.2 ± 3.0 cm^3
Parachor	= 314.4 ± 4.0 cm^3
Index of Refraction	= 1.730 ± 0.02
Surface Tension	= 109.2 ± 3.0 dyne/cm
Density	= 1.749 ± 0.06 g/cm^3
Polarizability	= 15.39 ± 0.5 10-24cm^3
Monoisotopic Mass	= 170.021523 Da
Nominal Mass	= 170 Da
Average Mass	= 170.1195 Da

Pyrogallol is prepared from gallic acid

Calcein, also known as **fluorexon, fluorescein complex**, is a fluorescent dye with an excitation and emission wavelengths of 495/515nm, respectively. Calcein also self-quenches at concentrations above 100mM. It is used as a complexometric indicator for titration of calcium ions with EDTA, and for fluorometric determination of calcium. It has the appearance of orange crystals.

The acetomethoxy derivate of calcein (**calcein AM**) is used in biology as it can be transported through the cellular membrane into live cells, which makes it useful for testing of cell viability and for short-term marking of cells. The acetomethoxy group obscures the part of the molecule that chelates calcium. After transport into the cell the enzymes cut off the group, the molecule binds to calcium within cell (resulting in acquiring strong green fluorescence), and gets trapped inside. As dead cells lack this enzyme, only live cells are marked. Calcein is also used for marking freshly hatched fish and for labeling of bones in live animals.

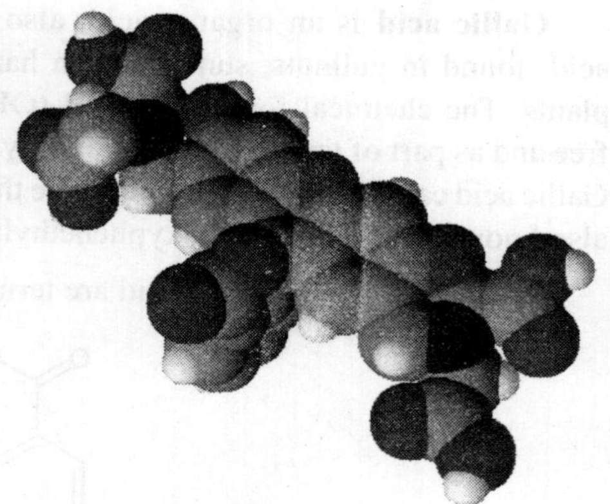

Capsaicin the chemical compound (8-methyl-N-vanillyl-6-nonenamide) is the active component of chili peppers, which are plants belonging to the genus Capsicum. It is an irritant for mammals, including humans, and produces a sensation of burning in any tissue it comes in contact with. Capsaicin and several related compounds are called **capsaicinoids** and are produced as a secondary metabolite by chili peppers, probably as deterrents against herbivores. Pure capsaicin is a hydrophobic, colorless, odorless, crystalline to waxy compound.

Capsaicinoids

Capsaicin is the main capsaicinoid in chili peppers, followed by dihydrocapsaicin. These two compounds are also about twice as potent to the taste and nerves as the minor capsaicinoids nordihydrocapsaicin, homodihydrocapsaicin, and homocapsaicin. Dilute solutions of pure capsaicinoids produced different types of pungency; however, these differences were not noted using more concentrated solutions.

Capsaicinoid name	Abbrev.	Typical relative amount	Scoville heat units	Chemical structure
1	2	3	4	5
Capsaicin	C	69%	15,000,000	
Dihydrocapsaicin	DHC	22%	15,000,000	

1	2	3	4	5
Nordihydrocapsaicin	NDHC	7%	9,100,000	
Homodihydrocapsaicin	HDHC	1%	8,600,000	
Homocapsaicin	HC	1%	8,600,000	

Natural Function

Capsaicin is present in large quantities in the seeds and fleshy fruits of plants in the genus *Capsicum*. Such fruits have typically evolved to aid in seed dispersal by attracting animals, which consume the fruits and swallow the seeds, which pass through the digestive tract and are subsequently deposited elsewhere. Thus, it may seem paradoxical that a plant would expend the considerable amount of resources necessary to produce large, juicy fruits, only to fill them with a compound that acts as a powerful deterrent.

The seeds of *Capsicum* plants, however, are predominantly dispersed by birds, in which capsaicin has analgesic properties rather than acting as an irritant. Chili pepper seeds consumed by birds pass through the digestive tract unharmed, whereas those consumed by mammals do not germinate at all. The presence of capsaicin in the fruits therefore protects them from being consumed by mammals, which wouldn't provide any benefit to the plant, while allowing them to be eaten by its preferred seed carriers.

Uses in Food

Because of the burning sensation caused by capsaicin when it comes in contact with human flesh, it is commonly used in food products to give them added spice or "heat" (piquancy). The degree of heat found within a food is measured on the Scoville scale. Typically the capsaicin is obtained from chili peppers. Hot sauce is an example of a product customarily containing large amounts of capsaicin and may contain chili peppers or pure capsaicin.

Capsaicin is a nonpolar molecule, and is therefore hydrophobic. Consequently, drinking water to reduce the burning caused by the molecule is ineffective, as the nonpolar capsaicin is unable to dissolve in the polar water molecules, and is instead spread across the surface of the mouth. This works by the same principle that causes oil and water to separate.

Instead, consuming foods high in fats and oils, such as milk or bread and butter, will help alleviate the burning. The lipophilic capsaicin is able to mix freely with the fats in the food and is removed from the surface of the mouth. Alcohol and alcoholic beverages also dissolve capsaicin due to the solvent characteristics of ethanol. Over time, the capsaicin will also dissipate on its own.

Uses in Medicine

Capsaicin is currently used in topical ointments to relieve the pain of peripheral neuropathy such as post-herpetic neuralgia caused by shingles. It may be used in concentrations of between 0.025% and 0.075%.

It may also be used as a cream for the temporary relief of minor aches and pains of muscles and joints associated with arthritis, simple backache, strains and sprains. The treatment typically involves the application of a topical anesthetic until the area is numb. Then the capsaicin is applied by a therapist wearing rubber gloves and a face mask. The capsaicin remains on the skin until the patient starts to feel the "heat", at which point it is promptly removed. Capsaicin is also available in large adhesive bandages that can be applied to the back.

The result appears to be that the nerves are overwhelmed from the burning sensation and are unable to report pain for an extended period of time. With chronic exposure to capsaicin, neurons are depleted of neurotransmitters and it leads to reduction in sensation of pain and blockade of neurogenic inflammation. If capsaicin is removed, the neurons recover.

Capsaicin is being explored as a cure for diabetes by researchers in Toronto, Canada. Early work curing diabetic mice looks promising. Capsaicin was injected into pancreatic sensory nerves of mice with Type 1 diabetes because of a suspected link between the nerves and diabetes.

The American Association for Cancer Research reports studies suggesting capsaicin is able to kill prostate cancer cells by causing them to undergo apoptosis. The studies were performed on tumors formed by human prostate cancer cell cultures grown in mouse models, and showed tumors treated with capsaicin were about one-fifth the size of the untreated tumors.

Another study carried out at the University of Nottingham suggests capsaicin is able to trigger apoptosis in human lung cancer cells as well.

Dihydrocapsaicin is a capsaicinoid and analog and congener of capsaicin in chili peppers (Capsicum). Like capsaicin it is an irritant. Dihydrocapsaicin accounts for about 22% of the total capsaicinoids mixture and has about the same pungency as capsaicin. Pure dihydrocapsaicin is a lipophilic colorless odorless crystalline to waxy compound.

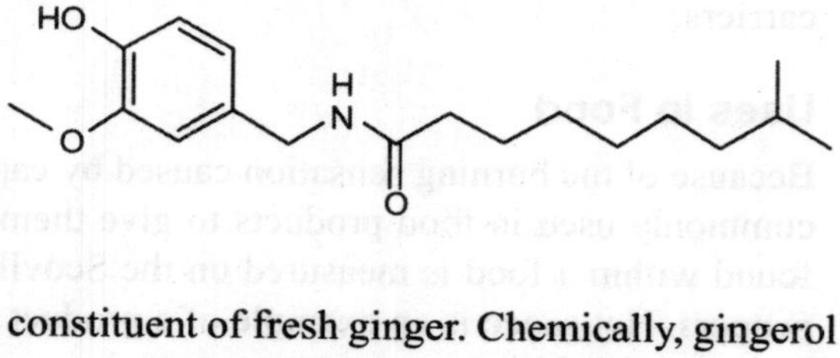

Gingerol, or sometimes [6]-gingerol, is the active constituent of fresh ginger. Chemically, gingerol is a relative of capsaicin, the compound that gives chile peppers their spiciness. It is normally found as a pungent yellow oil, but also can form a low-melting crystalline solid.

Cooking ginger transforms gingerol into zingerone, which is less pungent and has a spicy-sweet aroma.

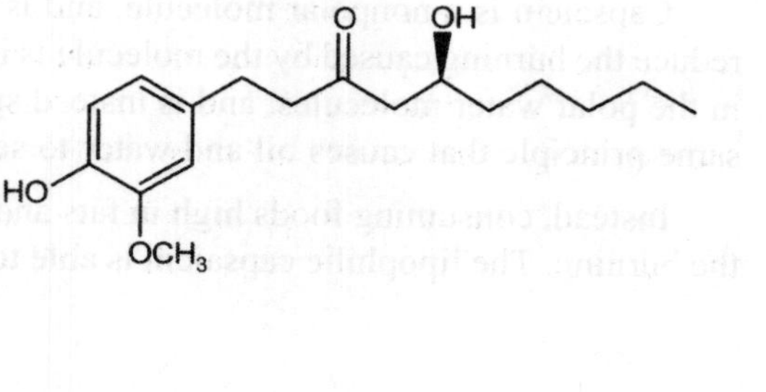

Gingerol may reduce nausea caused by motion sickness or pregnancy[1] and may also relieve migraine.

Gossypol $C_{30}H_{30}O_8$ is a polyphenol derived from the cotton plant (genus Gossypium, family Malvaceae). Gossypol is a polyphenolic aldehyde that permeates cells and acts as an inhibitor for several dehydrogenase enzymes. It is a yellow pigment.

Among other things, it has been tested as a male oral contraceptive in China. In addition to its contraceptive properties, gossypol has also long been known to possess anti-malarial properties.Other researchers are investigating the anti-cancer properties of gossypol.

Heroin (INN) **Diacetylmorphine**, (BAN) **diamorphine**) is a semi-synthetic opioid. It is the 3,6-diacetyl derivative of morphine (hence *diacetylmorphine*) and is synthesised from it by acetylation. The white crystalline form is commonly the hydrochloride salt **diacetylmorphine hydrochloride**. It mimics endorphins and creates a sense of well-being upon entering the bloodstream (usually via intravenous injection). It is thus used both as a pain-killer and a recreational drug. Frequent administration has a high potential for causing addiction and may quickly lead to tolerance, especially as compared to other substances, though occasional use may not lead to symptoms of withdrawal. It should be noted that withdrawal symptoms from heroin can be felt in as little as three days of continual use if stopped abruptly. This is a lot sooner than traditional painkillers such as oxycodone and hydrocodone. Internationally, heroin is controlled under Schedules I and IV of the Single Convention on Narcotic Drugs.[3] It is illegal to manufacture, possess, or sell heroin in the United States; however, under the name **diamorphine**, heroin is a legal prescription drug in the United Kingdom. Popular street names for heroin are *gear, diesel, smack, B, skag, Bobby, black tar, horse, junk, jack, jenny, brown, brown sugar, dark, dope, dragon, bitch, gak, boy,* and *H.*

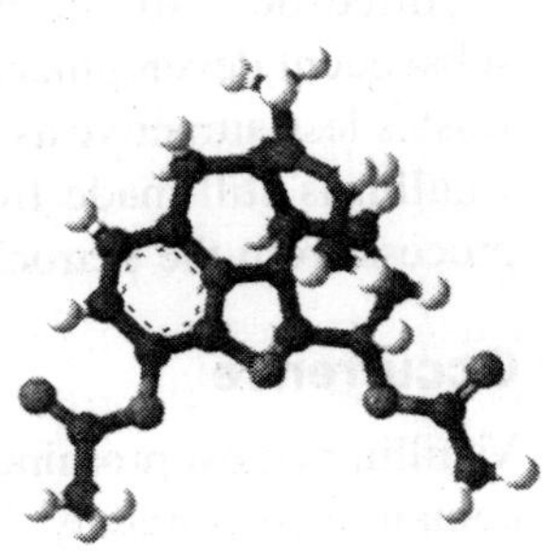

Preparation

Heroin was first synthesized in 1874 by C.R. Alder Wright, an English chemist working at St. Mary's Hospital Medical School in London, England. He had been experimenting with combining morphine with various acids. He boiled anhydrous morphine alkaloid with acetic anhydride over a stove for several hours and produced a more potent, acetylated form of morphine, now called *diacetylmorphine.*

Dianin's compound is 4-p-hydroxyphenyl-2,2,4-trimethylchroman and invented by A.P. Dianin in 1914 [1]. This compound is a condensation isomer of bisphenol A and acetone and of special importance in host-guest chemistry because it can form a large variety of clathrates with suitable guest molecules. One example is the clathrate of Dianin's compound with morpholine. Slow evaporation of a solution containing both organic compounds yields crystals. Each asymmetric unit cell making up the crystal contains 6 chroman molecules of which two are deprotonated and 2 protonated morpholine molecules. The 6 chroman molecules are racemate pairs.

Vanilin or Vanilla was cultivated as a flavoring by pre-Columbian Mesoamerican peoples; at the time of their conquest by Hernándo Cortés, the Aztecs used it as a flavoring for chocolate. Europeans became aware of both chocolate and vanilla around the year 1520.

Vanillin was first isolated as a relatively pure substance in 1858 by Nicolas-Theodore Gobley, who obtained it by evaporating a vanilla extract to dryness, and recrystallizing the resulting solids from hot water. In 1874, the german scientists Ferdinand Tiemann and Wilhelm Haarmann deduced its chemical structure, at the same time finding a synthesis for vanillin from coniferin, a glycoside of isoeugenol found in pine bark, and in 1876, Karl Reimer synthesized vanillin from guaiacol. By the late 19th century, semisynthetic vanillin derived from the eugenol found in clove oil was commercially available.

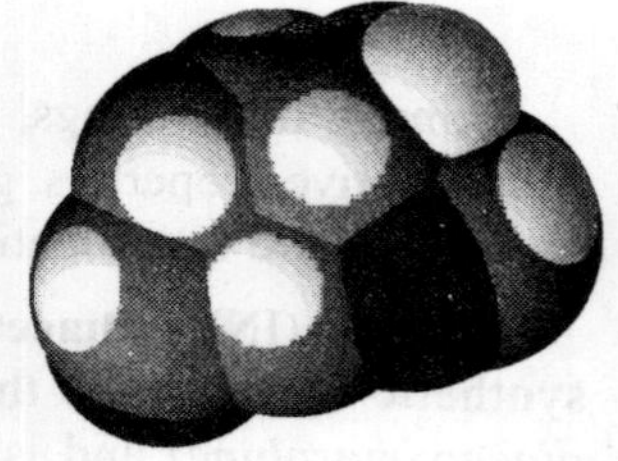

Synthetic vanillin became significantly more available in the 1930s, when production from clove oil was supplanted by production from the lignin-containing waste produced by the Kraft process for preparing wood pulp for the paper industry. By 1981, a single pulp and paper mill in Ontario supplied 60% of the world market for synthetic vanillin. However, subsequent developments in the wood pulp industry have made its lignin wastes less attractive as a raw material for vanillin synthesis. While some vanillin is still made from lignin wastes, most synthetic vanillin is today synthesized in a two-step process from the petrochemical precursors guaiacol and glyoxylic acid.

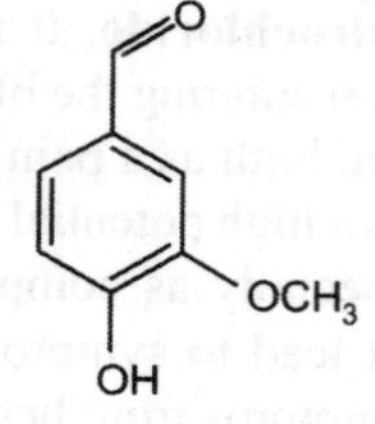

Occurrence

Vanillin is most prominent as the principal flavor and aroma compound in vanilla. Cured vanilla pods contain approximately 2% by dry weight vanillin; on cured pods of high quality, relatively pure vanillin may be visible as a white dust or "frost" on the exterior of the pod.

At smaller concentrations, vanillin contributes to the flavor and aroma profiles of foodstuffs as diverse as olive oil, butter, and raspberry and lychee fruits. Aging in oak barrels imparts vanillin to some wines and spirits. In other foods, heat treatment evolves vanillin from other chemicals. In this way, vanillin contributes to the flavor and aroma of coffee,[15] maple syrup,[16] and whole grain products including corn tortillas and oatmeal.

Production

Natural Production

Natural vanillin is extracted from the seed pods of *Vanilla planifola*, a vining orchid native to Mexico, but now grown in tropical areas around the globe. Madagascar is presently the largest producer of natural vanillin. As harvested, the green seed pods contain vanillin in the form of its β-D-glycoside; the green pods do not have the flavor or odor of vanilla. After being harvested, their flavor is developed by a months-long curing process, the details of which vary among vanilla-producing regions, but in broad terms it proceeds as follows: First, the seed pods are blanched in hot water, to arrest the processes of the living plant tissues. Then, for 1-2 weeks, the pods are alternately sunned and sweated: during the day, they are laid out in the sun, and each night, wrapped in cloth and packed in airtight boxes to sweat. During this process, the pods become a dark brown, and enzymes in the pod release vanillin as the free molecule. Finally, the pods are dried and further aged for several months, during which time their flavors further develop. Several methods have been described for curing vanilla in days rather than months, although they have not been widely developed in the natural vanilla industry,[20] with its focus on producing a premium product by established methods, rather than on innovations that might alter the product's flavor profile. Vanillin accounts for about 2% of the dry weight of cured vanilla beans, and is the chief among about 200 other flavor compounds found in vanilla.

Fig. 29.2. These green seed pods contain vanillin only in its glycoside form, and lack the characteristic odor of vanilla

Physical Properties

Molecular Formula	= $C_8H_8O_3$
Formula Weight	= 152.14732
Composition	= C(63.15%) H(5.30%) O(31.55%)
Molar Refractivity	= 41.56 ± 0.3 cm^3
Molar Volume	= 123.5 ± 3.0 cm^3
Parachor	= 324.0 ± 4.0 cm^3
Index of Refraction	= 1.587 ± 0.02
Surface Tension	= 47.3 ± 3.0 dyne/cm
Density	= 1.231 ± 0.06 g/cm^3
Polarizability	= 16.47 ± 0.5 $10^{-24}cm^3$
Monoisotopic Mass	= 152.047344 Da
Nominal Mass	= 152 Da
Average Mass	= 152.1473 Da

Chemical Synthesis

Vanillin was first synthesized from eugenol (found in oil of clove) in 1874-75, less than 20 years after it was first identified and isolated. Vanillin was commercially produced from eugenol until the 1920s. Later it was synthesized from lignin-containing sulfite liquor, a byproduct of wood pulp processing in paper manufacture. Counter-intuitively, even though it uses waste materials, the lignin process is no longer popular because of environmental concerns, and today most vanillin is produced from the petrochemical raw material guaiacol Several routes exist for synthesizing vanillin from guaiacol. At present, the most significant of these is the two-step process practiced by Rhodia since the 1970s, in which guaiacol reacts with glyoxylic acid by electrophilic aromatic substitution. The resulting vanilmandelic acid is then converted to vanillin by oxidative decarboxylation.

Uses

The largest single use of vanillin is as a flavoring, usually in sweet foods. The ice cream and chocolate industries together comprise 75% of the market for vanillin as a flavoring, with smaller amounts being used in confections and baked goods. Vanillin is also used in the fragrance industry, in perfumes, and to mask unpleasant odors or tastes in medicines, livestock fodder, and cleaning products. Vanillin has been used as a chemical intermediate in the production of pharmaceuticals and other fine chemicals. In 1970, more than half the world's vanillin production was used in the synthesis of other chemicals, but as of 2004 this use accounts for only 13% of the market for vanillin. Vanillin can also be found in bottles of Buckfast Tonic Wine (made by the monks of Buckfast Abbey) sold in the United Kingdom. It is not however found in the bottles sold in the Republic of Ireland.

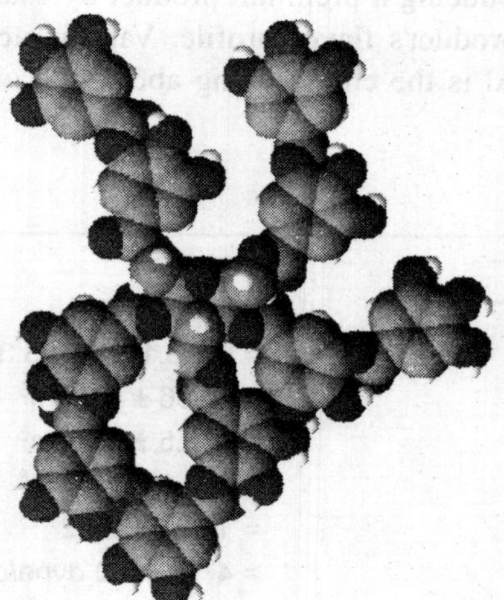

Tannic acid, a commercial form of tannin, is a polyphenol. Its weak acidity (pKa around 10) is due to these phenol groups in the structure. Tannic acid is a basic ingredient in the chemical staining of wood. The tannic acid or tannin is already present in woods like oak, walnut, and mahogany. Tannic acid can be applied to woods low in tannin so chemical stains that require tannin content will react.

Tannic acid is the most common mordant for cellulose fibers such as cotton. Tannin is often combined with alum and/or iron. The tannin mordant should be done first as metal mordants combine well with the fiber-tannin complex.

The presence of tannic acid in the bark of redwood (*sequoia*) is a strong natural defense against wildfire, decomposition and infestation. It is found in the seeds, bark, cones, and heartwood.

The chemical formula for commercial tannic acid is often given as $C_{76}H_{52}O_{46}$, but in fact it contains a mixture of related compounds. Its structure is based mainly on glucose esters of gallic acid. It is a yellow to light brown amorphous powder which is highly soluble in water; one gram dissolves in 0.35 mL of water.

It is said that soaking feet in tannic acid (or strong tea) can help prevent blisters.

But the use of tea for toughening skin appears to be apocryphal, inasmuch as tea is said to be incapable of tanning leather.

Physical Properties

Molecular Formula	$= C_{78}H_{56}O_{46}$
Formula Weight	= 1729.25164
Melting Point	= 218°C
Composition	= C(54.18%) H(3.26%) O(42.56%)
Molar Refractivity	= 388.87 ± 0.4 cm^3
Molar Volume	= 830.7 ± 5.0 cm^3
Parachor	= 3096.8 ± 6.0 cm^3
Index of Refraction	= 1.907 ± 0.03
Surface Tension	= 193.0 ± 5.0 dyne/cm
Density	= 2.08 ± 0.1 g/cm^3
Polarizability	= 154.16 ± 0.5 10^{-24} cm^3
Monoisotopic Mass	= 1728.204274 Da
Nominal Mass	= 1728 Da
Average Mass	= 1729.2516 Da

Chemical Properties

Tannic acid generally reacts with ferric ion producing a black dye which is also used as ink.

Xylenol or **dimethylphenol** is an arene compound with two methyl groups and a hydroxyl group. 6 isomers exist of xylenol of which 2,6-xylenol with both methyl group in an ortho position with respect to the hydroxyl group is the most important. The name xylenol is a contraction of two similar compounds xylene and phenol.

2,6-dimethylphenol 2,5-dimethylphenol 2,4-dimethylphenol

2,3-dimethylphenol 3,4-dimethylphenol 3,5-dimethylphenol

Physical Properties

The physical properties of the 6 xylenol isomers are very similar.

Isomer	2,6	2,5	2,4	2,3	3,4	3,5
CAS	576-26-1	95-87-4	105-67-9	526-75-0	95-65-8	108-68-9
Mp °C	43-45	63 - 65	22-23	70-73	62 - 68	61-64
Bp °C	203	212	211-212	217	227	222
Density g/mL		0.971	1.011			

Uses

Together with cresols and cresylic acid, xylenols are an important class of phenolics with great industrial importance. Xylenols are used as pesticides and used in the manufacture of antioxidants. Xylenol orange is a redox indicator built on a xylenol skeleton.

2,6-xylenol is a monomer for Poly(p-phenylene oxide) engineering resins through carbon-oxygen oxidative coupling. Carbon to carbon dimerization is also possible. In one study 2,6-xylenol is oxidized with iodosobenzene diacetate with an 5 fold excess of the phenol.

In the first step of the proposed reaction mechanism the acetyl groups in the iodine compound are replaced with the phenol. This complex dissociates into an aryl radical anion and a phenoxy residue. The two aryl radicals recombine forming a new carbon carbon covalent bond and subsequently lose two protons in a rearomatization step. The immediate reaction product is a diphenoquinone as result of a one-step 4-electron oxidation. It is nevertheless possible to synthesize the biphenol compound via a comproportionation of the quinone with xylenol already present. In this reaction sequence the hypervalent iodine reagent is eventually reduced to phenyliodine.

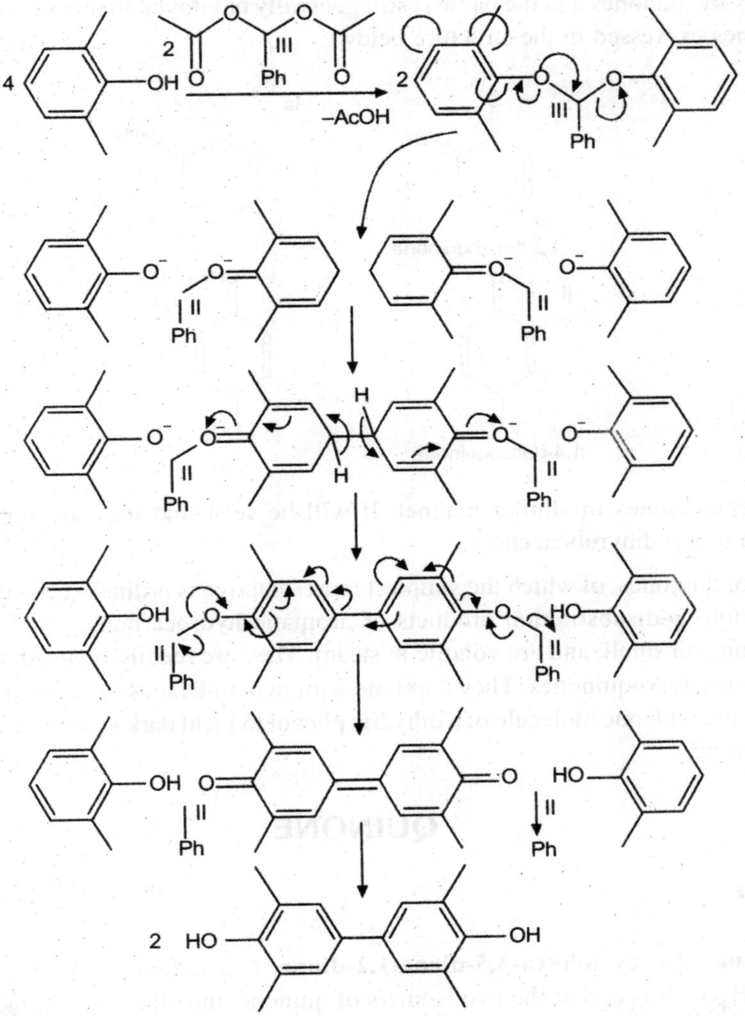

3,5,3',5'-tetramethyl-biphenyl-4,49-diol is used as a reducing agent for silver in photographic applications and as a constituent in epoxy resins.

Quinone and Quinonoid Compounds

By quinones we understand compounds which have two hydrogen atoms of the benzene nucleus are replaced by two oxygen atoms. The products are differentiated as *o* and *p* quinones according to the relative position of the substituent. Up to the present no *m*-quinone has been isolated and *o*-quinones of benzene and its alkyl and halogen derivative have only been prepared in rare cases and for the most part comparatively recent. For this reason the *p*-compounds, which were the first to be discovered, were described shortly by quinones and the name is still generally employed in this sense. The constitution of the benzoquinones expressed in the structure below:

I 1,2-benzoquinone; Ia (o-quinodimethane, CH_2 / CH_2)

II 1,4-benzoquinone; (p-quinodimethane, CH_2 / CH_2)

and that of other quinones in similar manner. It will be seen that they are represented as di keto derivatives of an *o* or *p* dihyrobenzene.

p-quinones or quinones, of which the simplest representative is ordinary quinone are formed by the oxidation of various *p*-disubstitution products of aromatic hydrocarbons. Most of them are yellow compounds of pungent smell, and are volatile in steam. They are readily reduced, taking up two atoms of hydrogen to form hydroquinones. They combine with two molecules of monohydric phenols to give phenol quinones and with one molecule of a dihydric phenol to yield dark - coloured addition compounds such as quinhydrone.

QUINONE

Introduction

1,2-Benzoquinone, also **cyclohexa-3,5-diene-1,2-dione**, is a ketone, with formula $C_6H_4O_2$. It is one of the two isomers of quinone, the other being 1,4-benzoquinone. 1,2-Benzoquinone is produced on oxidation of catechol.

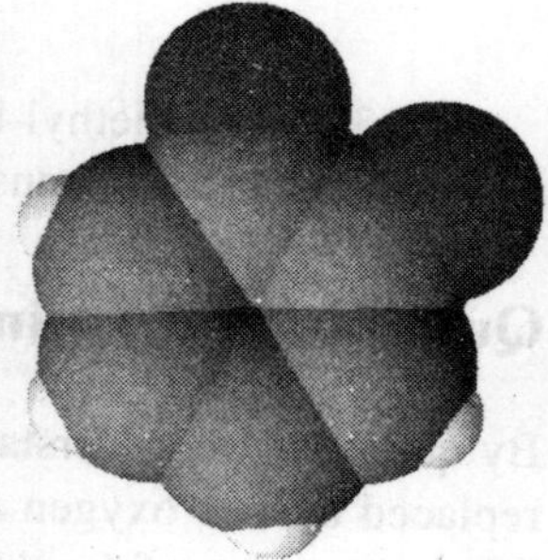

pyrocatechol $\xrightarrow[\text{reflux}]{K_2Cr_2O_7/H_2SO_4}$ 1,2-benzoquinone

Benzoquinone is also prepared by the distillation of quinic acid with manganese dioxide and sulphuric acid.

HO, OH, HO, OH, OH — quinic acid — MnO_2/H_2SO_4, distill → 1,2-benzoquinone

Benzoquinone is also prepared by the oxidation of aminophenol

OH, NH_2 — $K_2Cr_2O_7$, H_2SO_4 → + NH_4OH

Oxidation of aniline with sodium dichromate in presence of sulphuric acid

NH_2 — $Na_2Cr_2O_7$, H_2SO_4 → + NH_3

Benzoquinone forms golden yellow crystals possessing a peculiar pungent smell.

Physical Properties

Molecular Formula	= $C_6H_4O_2$
Formula Weight	= 108.09476
Composition	= C(66.67%) H(3.73%) O(29.60%)
Molar Refractivity	= 27.13 ± 0.3 cm^3
Molar Volume	= 86.0 ± 3.0 cm^3
Parachor	= 226.2 ± 6.0 cm^3
Index of Refraction	= 1.543 ± 0.02
Surface Tension	= 47.8 ± 3.0 dyne/cm
Density	= 1.256 ± 0.06 g/cm^3
Polarizability	= 10.75 ± 0.5 10-24cm^3
Monoisotopic Mass	= 108.021129 Da
Nominal Mass	= 108 Da
Average Mass	= 108.0948 Da

Chemical Properties

Benzoquinone reacts with hydroquinone yielding quinhydrone

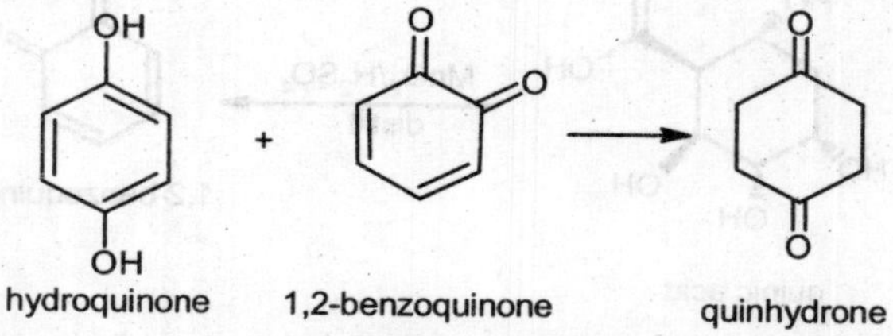

hydroquinone 1,2-benzoquinone quinhydrone

Quinhydrone crystallizes in green crystal prisms of metallic lusture and occur's as an intermediate product in the reduction. Quinone undergoes chlorination forming tetrachloroquinone, or chloranil

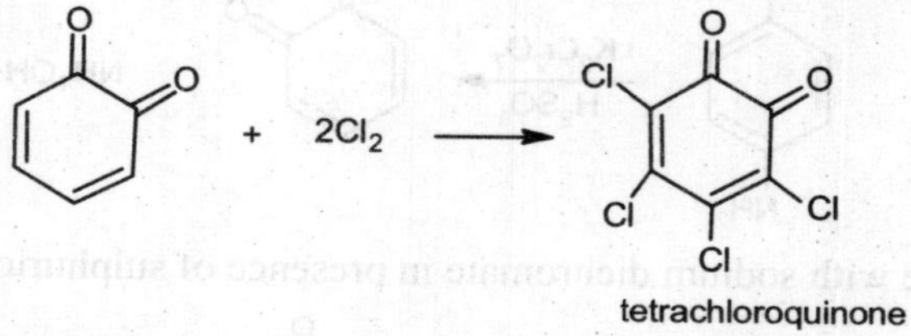

+ $2Cl_2$ →

tetrachloroquinone

It is used as an oxidizing agent in the manufacture of dyes.

Quinhydrones

The simplest member of this group is the above mentioned quinhydrone, invented by Wöhler in 1884, the constitution of which has given rise to much discussion. It is now almost generally agreed that the quinhydrones are treated as molecular compounds. It was confirmed by Pfeiffer, who postulated that in the quinhydrone, the carbonyl oxygen atoms of the quinonoid structural components are united to the unsaturated carbon atoms of the benzoic structural components.

Since partially all quinhydrones, as well as the phenol-ether and the hydrocarbon compounds of the quinones are composed of I mole quinone 1 or 2 mol benzenoid derivative, we are lead to an assumption that each carbonyl oxygen atom can link upto one benzoid molecule as indicated in the following formula.

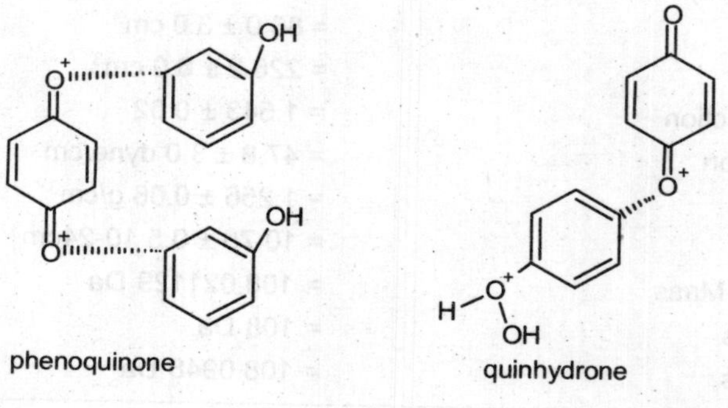

phenoquinone quinhydrone

The union is affected in such a manner that the subsidiary valency of the oxygen atom is saturated by a uniform field of affinity which is produced by all, at all events by the majority, of the unsaturated carbon atoms of the benzene derivative.

This conception enables us to represent graphically the close relationship existing between quinhydrones, the coloured molecular compounds of nitro derivative and the coloured additive compounds formed by ketones with acids and metallic salts.

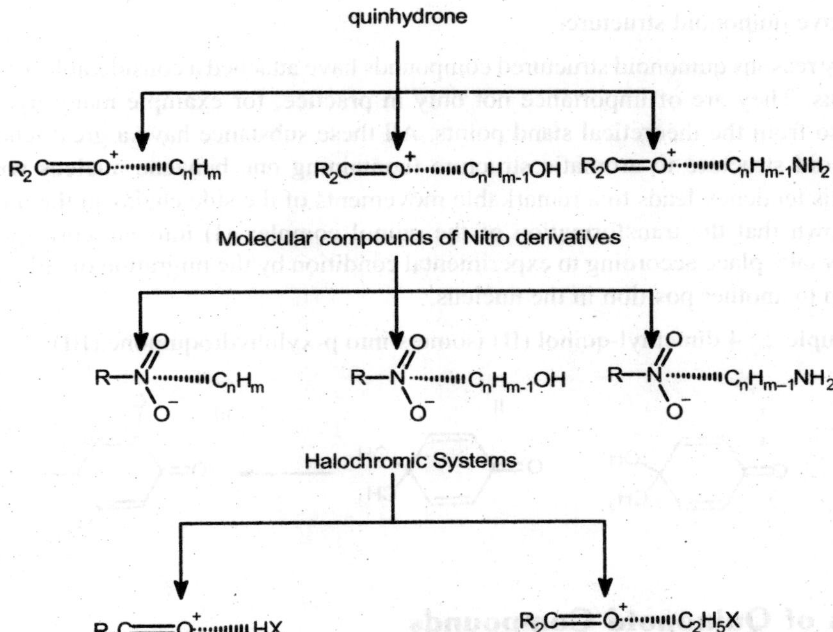

The connecting link between these organic molecular compounds and similar inorganic complexes is provided by the compounds formed by ketones with metallic salts. The latter stands in the closest relationship, two of the best known groups to two of the best known groups of inorganic compounds of e.g

$R_2C{=}O^{+}\cdots SnCl_4 \cdots {}^{+}O{=}CR_2$ $\qquad$ $KCl \cdots SnCl_4 \cdots KCl$

Quinonoid Compounds

Quinonoid compounds are derived from a and p-quinones by the replacement of hydrogen or ketonic oxygen by monovalent or polyvalent atoms or groups. Such compounds, which still contain the arrangement of linking characteristics of the quinones (see formula I and II below)

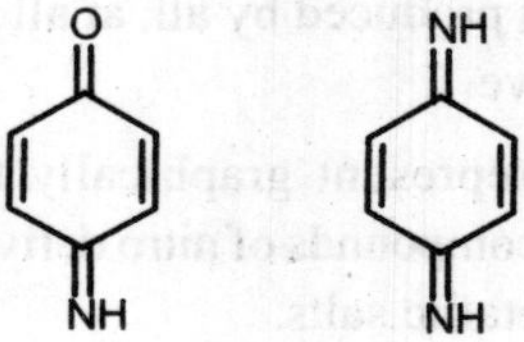

are said to have quinonoid structure.

For many reasons quinonoid structured compounds have attached a considerable amount of attention from chemists. They are of importance not only in practice, for example many dyes belong to this class, but also from the theoretical stand points. All these substance have a great tendency to change from quinonoid structure to aromatic structure (containing one benzene nucleus with three double bonds and this tendency leads to a remarkable movements of the side chains in the molecules. Thus it has been shown that the transformation of the quinol complex (I) into the corresponding aromatic structure may take place according to experimental condition by the migration of either the hydroxyl or methyl group to another position in the nucleus.

For example 2 : 4 dimethyl-quinol (II) isomers into p-xylohydroquinone (III)

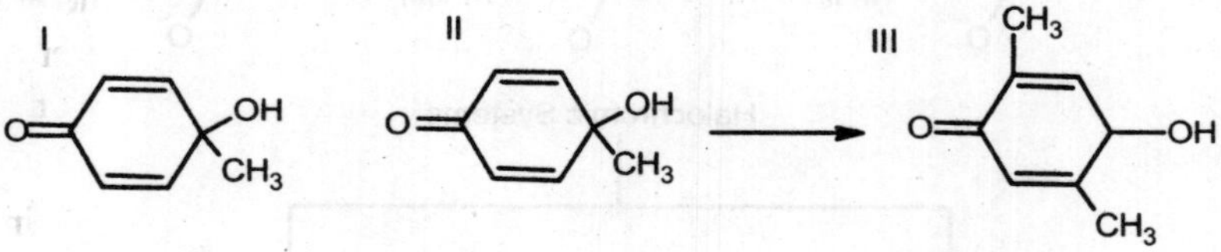

Synthesis of Quinonoid Compounds

The synthesis of quinonoid compounds is attracting considerable attention because of their implications in many areas of chemistry and biology. More specifically, benzoquinonemonoimines occupy a central role in many commercial applications and have considerable potential in colour chemistry, and in supramolecular, coordination and organometallic chemistry. It has been recently reported that new N-substituted benzoquinonemonoimines of type 1 (R = alkyl or aryl), which are rare examples of zwitterions being more stable than their canonical forms.

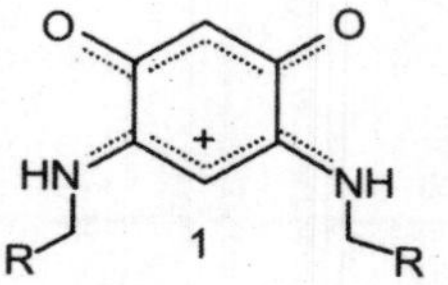

The first member of this novel class of potentially anti-aromatic $6\pi + 6\pi$ electron molecules (R = *t* – Bu) immediately caught the attention of theoreticians owing to its fundamental importance. Furthermore, these zwitterions appear to be reagents of choice: (i) in organic chemistry (ii) in coordination chemistry, as new ligands in the stepwise synthesis of polynuclear complexes, and (iii) in biochemistry as precursors to a bioinhibitors-related OH-substituted aminoquinone.

A multistep procedure consisting of the reaction of diaminoresorcinol 2·2HCl with RC(O)Cl in CH_3CN and excess NEt_3, to afford diamidodiesters 3 in high yield equation below, followed by their reduction with $LiAlH_4$ and an aerobic work-up, led to the zwitterions 1

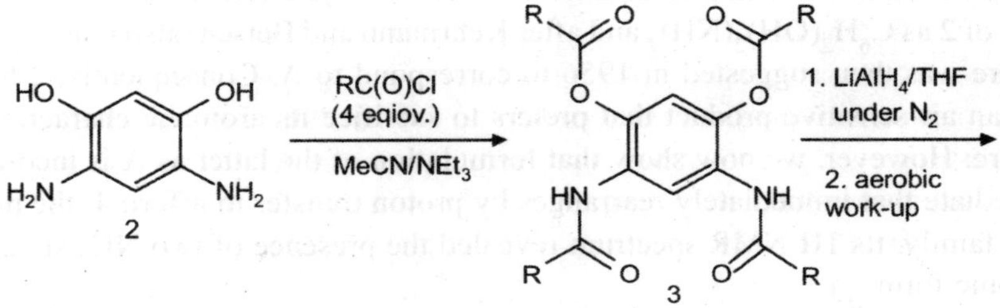

RC(O)Cl (4 eqiov.), MeCN/NEt_3: 2 → 3; 1. $LiAlH_4$/THF under N_2, 2. aerobic work-up: 3 → 1 ...(1)

However, the preparation of a wide range of zwitterions could not be achieved owing to the need for highly reactive acid chlorides, in which the presence of other functional groups is limited, and whose reduction exclusively leads to -NCH_2R substituted compounds. For further applications of this class of colorants, an atom-economic and greener synthesis (i.e. without organic solvent) would be of great interest. Here we report a new and extremely efficient preparation in water of new zwitterions, for which the N-alkyl substituent (i.e. the properties) can be easily varied by using an unprecedented transamination reaction on a quinonoid ring.

2·2HCl reacted smoothly with a large excess of various primary amines RNH_2 in water (or alcohol) at room temperature under air, to afford high yields of the corresponding zwitterions 5-9 which have been fully characterized

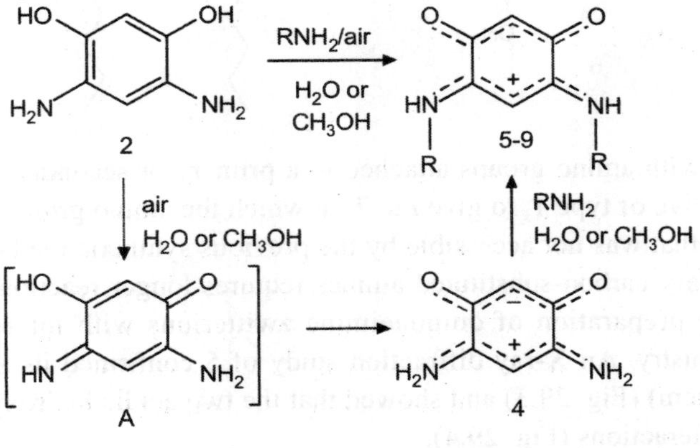

2 → (RNH_2/air, H_2O or CH_3OH) → 5-9; 2 → (air, H_2O or CH_3OH) → [A] → 4 → (RNH_2, H_2O or CH_3OH) → 5-9

Scheme 1 One-pot synthesis of the zwitterions. 5 R = CH_2CH_2OH; 6 R = CH_3; 7 R = $CH(CH_3)CH_2CH_3$; 8 R = $(CH_2)_3CH_3$; 9 R = $CH_2CH_2N(CH_3)_2$.

In contrast to 7 and 8 which are only soluble in organic solvents, 5 is almost insoluble in most organic solvents but soluble in water owing to the presence of hydrophilic groups (-OH). In addition, 6 and 9 are soluble in both organic solvents and water. Therefore, we can now fine-tune the solubility of these molecules, which is a key point for their applications.

The one-pot synthesis of the new zwitterions 5-9 results from deprotonation of 2·2HCl in the presence of amine, air oxidation to afford intermediate A (not isolated) which rearranges to 4, followed by *in situ* reaction of the latter with excess amine. In 1883, Typke (erroneously) formulated the air oxidation product of 2 as $C_6H_2(OH)_2(NH)_2$ and after Kehrmann and Betschl also noted its violet-brown colour, its structure was then suggested in 1956 to correspond to A. Consequently, 2 has since been considered to be an air-sensitive product that prefers to sacrifice its aromatic character in favor of a quinonoid structure. However, we now show that formulation of the latter as A is incorrect since this is only an intermediate that immediately rearranges by proton transfer to afford 4, the parent member of this zwitterion family. Its 1H NMR spectrum revealed the presence of two NH_2 signals, consistent with the zwitterionic form

Zwitterions 5-9 can also be obtained directly from isolated 4 in high yield, supporting its role as a reaction intermediate. Interestingly, treatment of N-substituted zwitterions, such as 6, with primary amines leads to compounds such as 8 in which the N-substituent has been exchanged Although nucleophilic substitution reactions can occur smoothly at quinonoid compounds, and biochemically relevant quinone-dependent transamination reactions have been reported, this reversible reaction represents the first example, to the best of our knowledge, of a transamination occurring on a quinonoid ring.

$$\mathbf{6} \underset{CH_3NH_2}{\overset{n\text{-BuNH}_2}{\rightleftharpoons}} \mathbf{8} \quad \ldots(2)$$

Primary amines with amino groups attached to a primary or secondary carbon reacted smoothly with 4, or any derivative of type 1, to give *i.a.* 7, in which the amino group is attached to a secondary carbon, a compound that was not accessible by the previous synthetic method. For steric reasons, the reaction with secondary carbon-substituted amines requires longer reaction times.This new efficient synthesis allows the preparation of quinoneimine zwitterions with interesting functionalities for supramolecular chemistry. An X-ray diffraction study of 5 confirmed its zwitterionic structure (i.e. fully delocalized system) (Fig. 29.3) and showed that the two acidic hydroxyl protons are involved in hydrogen bonding interactions (Fig. 29.4).

Therefore, in contrast to 1 (R = t-Bu) which formed a head-to-tail but zig-zag arrangement in the solid state, zwitterion 5 forms a head-to-tail and coplanar supramolecular network.

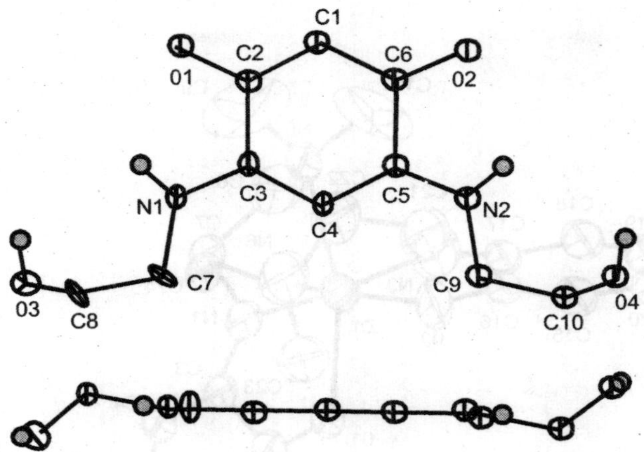

Fig. 29.3. Top and side views of the structure of 5 in the crystal. Thermal ellipsoids are drawn at the 50% probability level.

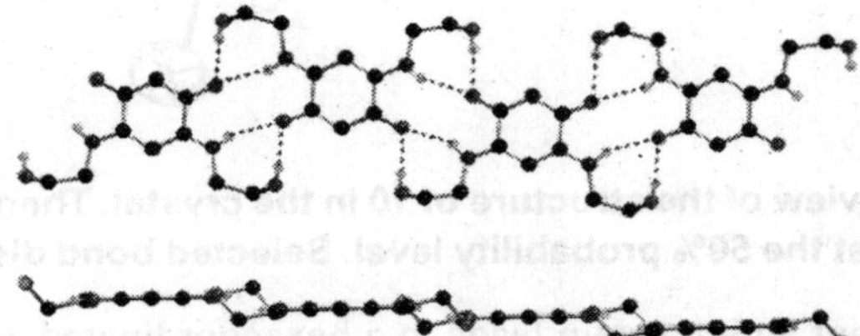

Fig. 29.4. Crystalmaker top and side views of the supramolecular array generated by 5 in the solid stateColour coding: nitrogen, blue; oxygen, red; hydrogen, green.

Furthermore, zwitterionic 9 with a pendant coordinating arm was reacted with $Zn(acac)_2$ in dichloromethane solution at room temperature, to afford complex 10, which was fully characterized, including by X-ray diffraction (Fig. 29.5).

9 —[$Zn(acac)_2$, CH_2Cl_2]→ 10 ...(3)

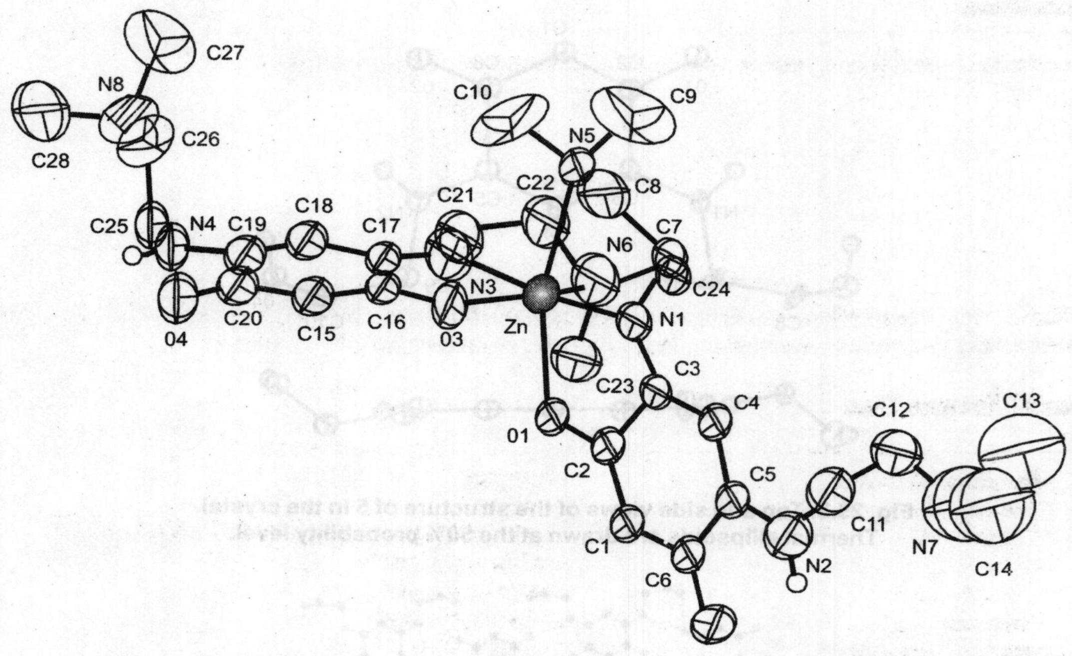

Fig. 29.5. ORTEP view of the structure of 10 in the crystal. Thermal ellipsoids are drawn at the 50% probability level. Selected bond distances

Coordination of the pendant amino group leads to a hexacoordinated zinc centre. Examination of the bond distances within the quinoneimine core reveals an alternation of single and double bonds, which is consistent with localized systems. The Zn-O(1) and Zn-N(1) distances of 2.129(3) and 2.035(4) , respectively, are in the expected range.

The nature and geometry of the coordination sphere of Zn(II) complexes in the presence of multifunctional ligands, is currently attracting considerable attention in bioinorganic chemistry and homogeneous catalysis.

Quinone Imine

These are derived from quinones by replacing one or both of the ketonic oxygen by the introducing imino group : NH or the aryl or the alkyl or aryl amine group : NR. The simplest imines are quinone - imines and quinonediimine

NH

O

4-iminocyclohexa-2,5-dien-1-one
quinone imine

NH

NH

cyclohexa-2,5-diene-1,4-diimine
quinone diimine

Preparation

Quinone imine is prepared by the oxidation of p-aminophenol with silver oxide.

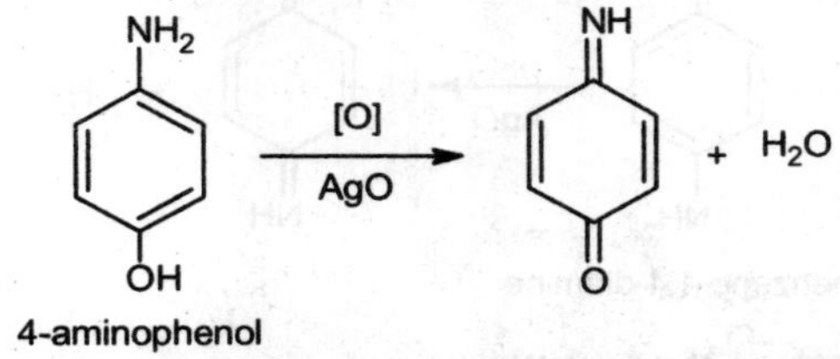

NH$_2$ — [O], AgO → NH (quinone imine) + H_2O

4-aminophenol

It forms colourless crystals smelling like quinone, which rapidly darkens in colour. It is extremely unstable and decomposes in atmosphere. Still physical properties are given below:

Physical Properties

Molecular Formula	= C_6H_5NO
Formula Weight	= 107.11
Composition	= C(67.28%) H(4.71%) N(13.08%) O(14.94%)
Molar Refractivity	= 30.40 ± 0.5 cm^3
Molar Volume	= 94.2 ± 7.0 cm^3
Parachor	= 244.0 ± 8.0 cm^3
Index of Refraction	= 1.558 ± 0.05
Surface Tension	= 44.8 ± 7.0 dyne/cm
Density	= 1.13 ± 0.1 g/cm^3
Polarizability	= 12.05 ± 0.5 10-24 cm^3
Monoisotopic Mass	= 107.037114 Da
Nominal Mass	= 107 Da
Average Mass	= 107.11 Da

Chemical Properties

On warming with sulphuric acid it is hydrolyzed to quinone and ammonia.

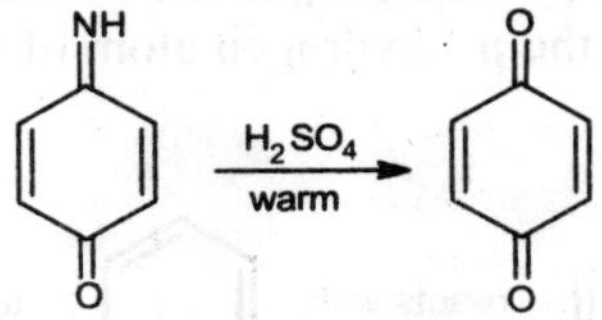

NH — H_2SO_4, warm → O

With stannous chloride and hydrochloric acid it is reduced to p-aminophenol

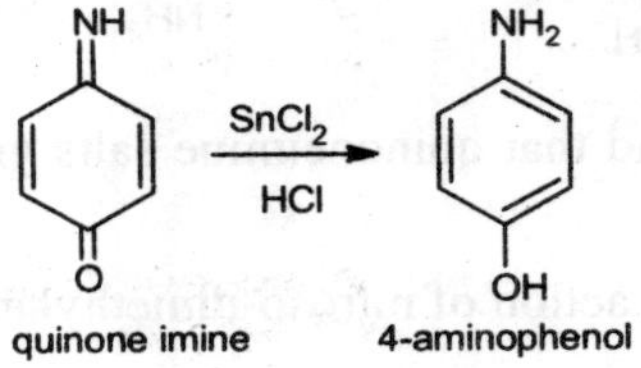

NH — $SnCl_2$, HCl → NH$_2$ / OH

quinone imine 4-aminophenol

Quinone diimine is also formed in similar manner by the oxidation of p-phenylene diamine

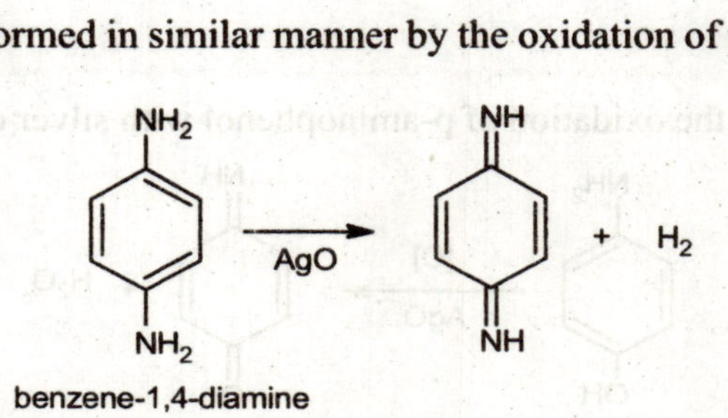

NH_2 ... AgO → NH ... + H_2

NH_2 NH

benzene-1,4-diamine

And by the reduction of quinone dichlorodiimine

Cl Cl N N $SnCl_2$ HCl → NH + HCl NH

quinone dichlorodiimine

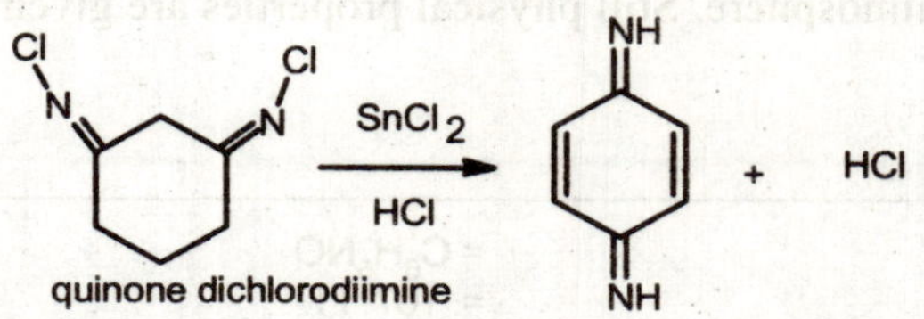

It crystallizes in colourless needles with a melting point of 124°C, which is very unstable. With stannous chloride in hydrochloric acid it is reduced to p-phenylene diamine and with sulphuric acid it produces ammonia and quinone.

As already stated, both the above imines are colourless in the crystalline state, and by comparison with quinone and the fluenes, it would appear that, in contradiction to the view long held the group C = NH is a weaker chromophore than C = O or C = C.

Quinone-diimine is the parent substance of large classes of dye-stuff chief among which are indamines and azines. Recently, certain other types of dyes particularly in the diphenyl and tri-phenyl methane series have also been formulated in the same manner as the quinone-imines.

Indamines are readily prepared by oxidizing mixtures of monoamines and p-diamines, or by the interaction of amines and quinone dichlorodiimine.

In the former reaction it may be supposed that quinone diimine is first produced, when then in the course of further oxidation reacts with the p - hydrogen atom of the amine as following:

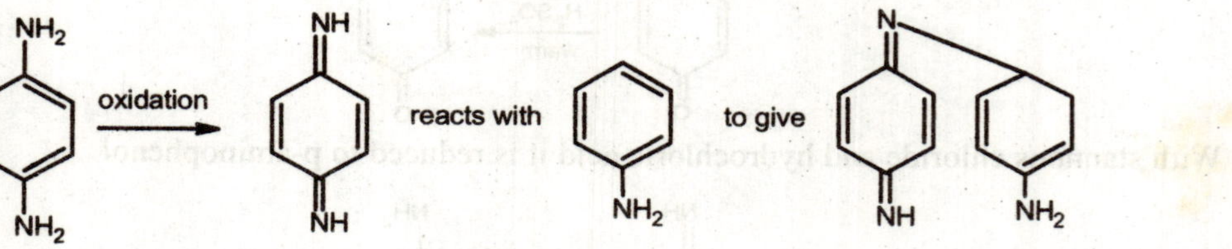

NH_2 oxidation → NH reacts with ... to give N

NH_2 NH NH_2 NH NH_2

In confirmation of this, it is found that quinoneimine salts yield with amines intensely coloured solution of indamines.

Indamines are also formed by the action of nitroso-dimethylaniline on amines.